Progress in Inflammation Research

Series Editor

Prof. Dr. Michael J. Parnham
PLIVA
Research Institute
Prilaz baruna Filipovica 25
10000 Zagreb
Croatia

Forthcoming titles:

(Already published titles see last page.)

Muscarinic Receptors in Airways Diseases

Johan Zaagsma
Herman Meurs
Ad F. Roffel

Editors

Springer Basel AG

Editors

Prof. Dr. Johan Zaagsma
Dept. of Molecular Pharmacology
University of Groningen
A. Deusinglaan 1
9713 AV Groningen
The Netherlands

Prof. Dr. Herman Meurs
Dept. of Molecular Pharmacology
University of Groningen
A. Deusinglaan 1
9713 AV Groningen
The Netherlands

Dr. Ad F. Roffel
Department of Research Management
Pharma Bio-Research Group B.V.
P.O. Box 200
9470 AE Zuidlaren
The Netherlands

A CIP catalogue record for this book is available from the Library of Congress, Washington D.C., USA

Deutsche Bibliothek Cataloging-in-Publication Data
Muscarinic receptors in airways diseases / Johan Zaagsma ... ed. - Basel ; Boston ; Berlin :
Birkhäuser, 2001
(Progress in inflammation research)
ISBN 978-3-0348-9532-3   ISBN 978-3-0348-8358-0 (eBook)
DOI 10.1007/978-3-0348-8358-0

ISBN 978-3-0348-9532-3

© 2001 Springer Basel AG
Originally published by Birkhäuser Verlag in 2001
Softcover reprint of the hardcover 1st edition 2001

Printed on acid-free paper produced from chlorine-free pulp. TCF ∞
Cover design: Markus Etterich, Basel
Cover illustration: Bronchial tree. Michiel Duyvendak, NL-Groningen

ISBN 978-3-0348-9532-3

9 8 7 6 5 4 3 2 1

Contents

List of contributors

Darryl J. Adamko, Johns Hopkins Asthma and Allergy Center, 5501 Hopkins Bayview Circle, Baltimore, MD 21224, USA;
e-mail: dadamko@welchlink.welch.jhn.edu

Peter J. Barnes, Department of Thoracic Medicine, National Heart and Lung Institute, Imperial College, Dovehouse St., London SW3 6LY, UK;
e-mail: p.j.barnes@ic.ac.uk

Kenneth R. Chapman, Asthma Centre of The Toronto Western Hospital, University Health Network, Suite 4-011 ECW, 399 Bathurst Street, Toronto, Ontario, Canada M5T 2S8; e-mail: kchapman@inforamp.net

Bernd Disse, Boehringer Ingelheim Clinical Research Institute, Respiratory Medicine, Boehringer Ingelheim GmbH, 55216 Ingelheim/Rhein, Germany;
e-mail: disse@ing.boehringer-ingelheim.com

Richard M. Eglen, DiscoveRx Corporation, 42501 Albrae Street, Suite 100, Fremont, CA 94538, USA; e-mail: reglen@discoverx.com

Carolina R.S. Elzinga, Department of Molecular Pharmacology, University of Groningen, A. Deusinglaan 1, 9713 AV Groningen, The Netherlands;
e-mail: c.r.s.elzinga@farm.rug.nl

Allison D. Fryer, Johns Hopkins Asthma and Allergy Center, 5501 Hopkins Bayview Circle, Baltimore, MD 21224, USA

David B. Jacoby, Johns Hopkins Asthma and Allergy Center, 5501 Hopkins Bayview Circle, Baltimore, MD 21224, USA;
e-mail: djacoby@welchlink.welch.jhu.edu

Charles J. Kirkpatrick, Institute of Pathology, University of Mainz, Obere Zahlbacher Strasse 67, 55101 Mainz, Germany;
e-mail: kirkpatrick@pathologie.klinik.uni-mainz.de

Herman Meurs, Department of Molecular Pharmacology, University of Groningen, A. Deusinglaan 1, 9713 AV Groningen, The Netherlands;
e-mail: h.meurs@farm.rug.nl

Allen C. Myers, The John Hopkins University School of Medicine, Johns Hopkins Asthma & Allergy Center, 5501 Hopkins Bayview Circle, Baltimore, MD 21224, USA; e-mail: almy@jhmi.edu

Ad F. Roffel, Department of Research Management, Pharma Bio-Research Group B.V., P.O. Box 200, 9470 AE Zuidlaren, The Netherlands; e-mail: ARoffel@pbr.nl

Duncan F. Rogers, Thoracic Medicine, National Heart & Lung Institute (Imperial College), Dovehouse Street, London SW3 6LY, UK; e-mail: duncan.rogers@ic.ac.uk

Bradley J. Undem, The John Hopkins University School of Medicine, Johns Hopkins Asthma & Allergy Center, 5501 Hopkins Bayview Circle, Baltimore, MD 21224, USA; e-mail: bundem@mail.jhmi.edu

Nikki Watson, Department of Pharmacology, University of Virginia, 5316 Jordan Hall, 1300 Jefferson Park Ave., Charlottesville, VA 22908, USA;
e-mail: hjw6n@unix.mail.virginia.edu

Ignaz K. Wessler, Institute of Pharmacology, University of Mainz, Obere Zahlbacher Strasse 67, 55101 Mainz, Germany; e-mail: wessler@mail.uni-mainz.de

Johan Zaagsma, Department of Molecular Pharmacology, University of Groningen, A. Deusinglaan 1, 9713 AV Groningen, The Netherlands;
e-mail: j.zaagsma@farm.rug.nl

Preface

In humans, as well as in most other mammalian species, the muscarinic cholinergic system comprises the major neural pathway involved in airway constriction and mucus secretion, and increased cholinergic tone has been implicated in the pathophysiology of asthma and chronic obstructive pulmonary disease (COPD).

Despite the long-standing use of antimuscarinic drugs as bronchodilators in asthma and COPD, the cholinergic mechanisms that control airway caliber in health and disease are only partially understood. Over the last decade it has become increasingly clear that the cholinergic system in the airways is a highly complex neural (and non-neural) system due to its complex anatomy, heterogeneity of muscarinic receptor subtypes and the rapidly expanding number of regulatory mechanisms of cholinergic activity by integrated neural systems and non-neural mediators that modulate the ultimate functional responses. In particular, recent findings that both neurogenic and non-neurogenic inflammatory mechanisms may have a major impact on muscarinic receptor function and cholinergic reflex activity in the airways are of crucial importance for unravelling the role of the cholinergic system in chronic inflammatory diseases such as asthma and COPD and the development of new therapeutic approaches.

The purpose of this volume in the series *Progress in Inflammation Research* is to provide a comprehensive and integrated overview of the current knowledge on the distribution, signalling, function and regulation of muscarinic receptor subtypes in the airways and their importance for the pathophysiology and pharmacotherapeutic treatment of asthma and COPD.

In Chapter 1, Undem and Myers give a general overview of the vagal control of the airways and present recent findings on the complex modulation of cholinergic outflow to the airways by cholinergic and noncholinergic neural pathways and inflammatory mediators at the level of ganglionic synaptic transmission and at the neuro-effector junction.

It has been well established that in addition to being a classical neurotransmitter, acetylcholine is widely expressed in various pro- and eucaryotic non-neuronal cells. In Chapter 2, Wessler and Kirkpatrick describe the presence and distribution of non-neural choline acetyltransferase and/or acetylcholine in the airways and present evidence that non-neural acetylcholine may play an important role in the local interplay between the epithelium and immune cells, modulating the release of proinflammatory cytokines and mediators from these cells by autocrine- and paracrine mechanisms.

The identification, localization and function of the muscarinic receptor subtypes mediating acetylcholine action in the airways, as well as possible dysfunction of these receptor subtypes in obstructive lung diseases, are reviewed by Roffel, Meurs and Zaagsma (Chapter 3). Of the five cloned muscarinic receptor subtypes, four (M_1, M_2, M_3 and M_4) have been identified and localized in the lung, in a species-dependent manner. M_3 receptors play a prominent role in airway smooth muscle contraction and mucus secretion and prejunctional M_2 receptors are importantly involved in the autoinhibition of neurotransmitter release in virtually all mammalian species, including man; the functional roles and/or relevance of airway and alveolar M_1 and (in some species) M_4 receptors still largely remain to be established.

This may also apply to the postjunctional M_2 receptor, which is the preponderant muscarinic receptor in airway smooth muscle. The presumed role of M_2 receptors in airway smooth muscle contraction and in functional antagonism of β-adrenoceptor-induced relaxation has thus far been difficult to substantiate and is still a matter of debate. However, in Chapter 4, Eglen and Watson discuss exciting new evidence for a coordinate role of postsynaptic M_2 and M_3 receptors in promoting contraction and reversing relaxation of airway smooth muscle, respectively.

As already indicated in Chapter 3, there is substantial evidence for a prejunctional M_2 receptor dysfunction in asthma, which may contribute to the enhanced cholinergic reflex activity in this disease. In Chapter 5, Adamko, Fryer and Jacoby describe the loss of prejunctional M_2 receptor function that may be induced by various environmental factors known to cause airway inflammation and hyperresponsiveness in asthmatics, including allergen challenge, viral infection and exposure to ozone. Multiple inflammatory mechanisms that may participate in these effects are discussed.

A dysfunction of postjunctional M_2 and M_3 receptors in asthma or COPD has thus far not been demonstrated convincingly. However, in Chapter 6, Meurs, Roffel, Elzinga and Zaagsma present various mechanisms of cross-talk between the signalling pathways of postjunctional M_2 and M_3 receptors and $β_2$-adrenoceptors in the airways that may be involved in (the modulation of) disease- and β-agonist-induced diminished $β_2$-adrenoceptor function in patients with asthma. In addition, the possible involvement of M_2 and M_3 receptors and of muscarinic receptor-β-adrenoceptor cross-talk in airway smooth muscle proliferation is discussed.

Barnes (Chapter 7) reviews the recent advances made in the elucidation of the mechanisms involved in muscarinic receptor gene regulation. Particular attention is paid to effects of inflammatory mediators such as cytokines and growth factors on the expression of the M_2 receptor, which may obviously be important for the inflammation-induced dysfunction of this receptor.

The cholinergic control of airways mucus secretion is extensively reviewed by Rogers (Chapter 8). This chapter deals with the neural pathway involved in mucus secretion by the surface epithelium and submucosal glands, the muscarinic receptor subtypes involved, and the identification of novel inhibitory mechanisms that may serve to develop new drugs for hypersecretion in airways diseases, including asthma and COPD.

A historical overview of anticholinergic drug therapy in asthma and COPD is described by Chapman in Chapter 9. With respect to current treatment, arguments for combining anticholinergic and β_2-adrenergic bronchodilators in acute and severe asthma as well as in the chronic therapy of COPD are discussed in depth.

Although the inhaled anticholinergics currently available for clinical use, of which ipratropium bromide is the main representative, have proven to be effective and safe bronchodilators, these drugs are non-selective towards the various muscarinic receptor subtypes and have a relatively short duration of action. In the final chapter (Chapter 10), Disse describes a number of concepts in the development of novel compounds to overcome these disadvantages. The most promising development in this respect is tiotropium bromide, a long-acting M_3 and M_1 receptor selective antimuscarinic drug, which is presently undergoing extensive clinical evaluation in COPD.

It is our hope that this book will provide the reader with a useful survey of recent developments in the physiology, pathophysiology, pharmacology and therapeutics of the cholinergic system in the airways and that it may serve as a scientific basis for further research in this field. Finally, we would like to thank the authors of the chapters for their expert contributions.

Johan Zaagsma
Herman Meurs
Ad F. Roffel

Cholinergic and noncholinergic parasympathetic control of airway smooth muscle

Bradley J. Undem and Allen C. Myers

The Johns Hopkins University School of Medicine, Johns Hopkins Asthma & Allergy Center, 5501 Hopkins Bayview Circle, Baltimore, MD 21224, USA

Introduction

In mammals, including humans, drugs that block cholinergic muscarinic receptors cause significant and, in some cases, near maximal bronchodilation [1–3]. Moreover, in animal studies, the bronchodilating effect of the anticholinergic drugs can be mimicked by sectioning the right and left vagus nerves [4, 5]. Considered together, these observations support the hypothesis that the parasympathetic branch of the autonomic nervous system is the major regulator of mammalian airway smooth muscle tone.

The neural control of airway smooth muscle must be considered in the context of a complex integrated nervous system that functions as a "reflex" process. In simple terms, the neuronal reflex can be segregated into three major components: (1) the primary afferent nerve fibers, (2) the integrating centers in the brain and (3) the parasympathetic efferent nerves. This chapter will provide an overview of the literature pertaining principally to the final branch of this reflex arc that controls airway caliber, namely, the parasympathetic nerves which regulate airway smooth muscle tone. In accordance with the nomenclature put forward by Langley, individual parasympathetic neurons can be categorized as preganglionic neurons and postganglionic neurons. This chapter provides a general overview of the properties and potential regulation of airway parasympathetic preganglionic and postganglionic neurons. Reviews that deal more extensively with central regulation [2, 6, 7], airway ganglia [8–10] and postganglionic innervation of airway smooth muscle [11, 12] can be found elsewhere.

Preganglionic parasympathetic airway neurons

Anatomical characterization

The preganglionic nerve fibers travel down the right and left vagi to innervate the parasympathetic ganglia that are located near or within the airway wall. The nerve

Muscarinic Receptors in Airways Diseases, edited by Johan Zaagsma, Herman Meurs and Ad F. Roffel
© 2001 Birkhäuser Verlag Basel/Switzerland

fibers innervating the tracheal ganglia leave the vagus nerves and enter the recurrent and superior laryngeal nerves. The preganglionic nerve fibers innervating bronchial ganglia enter the bronchi *via* peribronchial nerves off the right and left vagus nerves.

The somata of preganglionic neurons innervating the airway ganglia are located in various nuclei within the brain stem. These brainstem nuclei include the dorsal motor nucleus of the vagus nerve, the nucleus ambiguus, the nucleus retroambigualis and the nucleus dorsalmedialis [7, 13–18]. There is also heterogeneity among airway preganglionic fibers with respect to their size and degree of myelination [19–21]. The data would seem to support the contention that subsets of preganglionic fibers regulate discrete parasympathetic functions in the airway, although these have been scarcely studied. For example, reflex mucus secretion has been observed independently of reflex bronchoconstriction [22].

As discussed in more detail below, airway postganglionic parasympathetic neurons can be subdivided into those that use acetylcholine as their transmitter, and those that do not. Some postganglionic nonadrenergic, noncholinergic (NANC) neurons are effective in causing smooth muscle relaxations in airways of numerous species, including humans [23]. With this in mind, it is intriguing that in the guinea pig [24] and cat [25], the evidence indicates that the population of preganglionic fibers that innervate the cholinergic postganglionic neurons is different from the preganglionic fibers innervating the NANC neurons.

Regulation of preganglionic activity

The preganglionic parasympathetic nerves innervating airway ganglia are active during eupnic breathing [26–28]. This results in the cholinergic tone observed in the airways of mammals when studied *in vivo*. Electrophysiological studies demonstrate that during normal breathing, preganglionic neurons innervating the airway discharge action potentials at relatively high frequencies, ranging from 1 to about 20 Hz [26, 27]. To put this in perspective, continuous synchronous stimulation of postganglionic cholinergic nerves in the airways at frequencies as low as 1 Hz can lead to near maximal smooth muscle contraction [29]. The "spontaneous" tone of airway smooth muscle can be quite extensive. This has been quantified in the guinea pig, where the parasympathetic nerve activity results in airways that are constricted by 20–50% of the maximum obtainable smooth muscle contraction (see [30]).

The pattern of action potential discharge traversing the preganglionic fibers coincides with respiration, increasing either during inspiration or expiration, depending on the particular preganglionic neuron (Fig. 1) [26]. This led naturally to the hypothesis that respiratory centers in the central nervous system (CNS) govern the rate of preganglionic activity. Alternatively, one could argue that respiration *per se*,

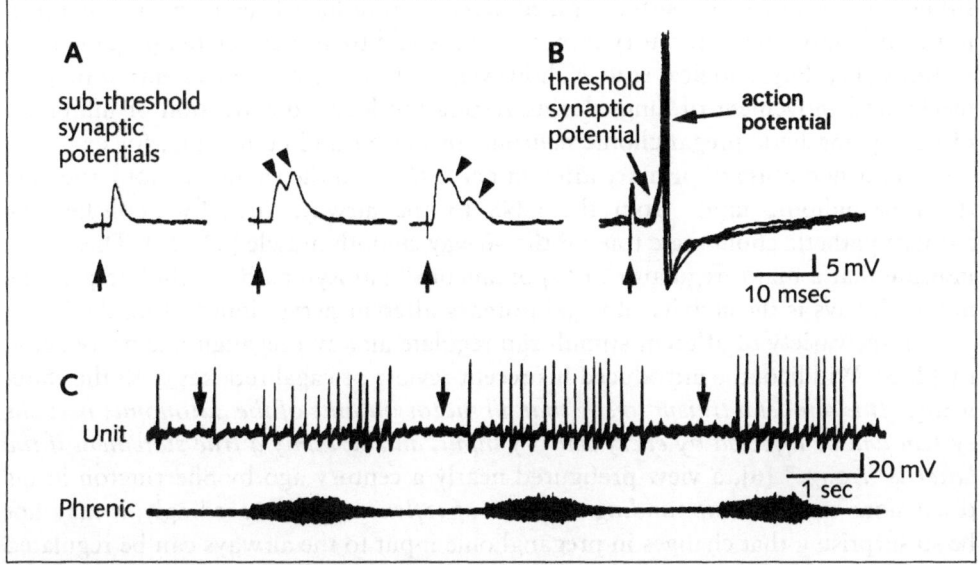

Figure 1

Evidence for integration and filtering by lower airway parasympathetic ganglia neurons. In A, traces show sub-threshold nicotinic post-synaptic potentials recorded from three guinea pig bronchial ganglia neurons in vitro. Despite synchronous, maximal stimulation of the vagus nerve (at vertical arrows), the corresponding synaptic potentials do not generate action potentials in the majority of neurons studied. The traces also demonstrate that more than one preganglionic axon can "converge" on a single neuron generating more than one temporally distinct post-synaptic potential (arrow heads at middle and right traces in A). In B, five over-laid traces show threshold nicotinic post-synaptic potentials recorded from a guinea pig bronchial ganglia neuron in vitro. Here, synchronous, maximal stimulation of the vagus nerve (at vertical arrow) causes synaptic potentials that do generate action potentials (A and B from Myers et al., 1990). In C, traces show simultaneous intracellular recording of the spontaneous activity of a cat trachea parasympathetic ganglion neuron in vivo (upper trace) and the corresponding phrenic nerve activity (lower trace). Note the numerous (5–10 Hz) sub-threshold excitatory post-synaptic potentials (epsps, downward arrows) throughout the respiratory cycle and most action potential generation (vertical spikes) during inspiration. Neurons were also recorded in the same preparation that generated action potentials corresponding to expiration (modified from Mitchell et al., 1987).

via activation of mechanically sensitive afferent nerve endings in the lungs, reflexively regulates the activity of preganglionic parasympathetic nerve fibers. This "chicken-or-egg" argument for control of parasympathetic tone in the airways is yet to be fully resolved. In favor of a "central drive" hypothesis are anatomical tracing studies in rats, in which a subset of medullary neurons have been identified that project to both phrenic respiratory motor neurons and to airway-related preganglionic neurons [17]. Physiological studies, however, seems to favor the contention that the mechanical movement of lungs during respiration leads to activation or inhibition of parasympathetic preganglionic neurons. In the cat and guinea pig, for example, selective denervation of primary afferent nerve fibers in the vagus, without affecting the preganglionic input from the CNS to the airway, virtually abolishes the parasympathetic cholinergic tone of the airway smooth muscle [31, 32]. These data indicate that a major regulator of "spontaneous" parasympathetic cholinergic tone in the airways is the activity of vagal primary afferent nerves innervating the lungs.

A large variety of afferent stimuli can regulate airway preganglionic nerve activity [2, 6]. Widdicombe introduced his recent review of vagal reflexes with the statement, "*It is almost a truism to say that all motor outputs of the autonomic nervous system will be affected by every sensory input, and certainly a true statement if the latter is strong.*" [6], a view prefigured nearly a century ago by Sherrington in his discussions on the *compounding together of reflexes* [33]. Accordingly, it may not be so surprising that changes in preganglionic input to the airways can be regulated by reflexes initiated outside the airways in skeletal muscle contraction, arterial baroreceptor activation, and by afferent input from the esophagus, heart, even the ear [6]. Thorough and scholarly reviews of reflex control of airway tone can be found elsewhere [2, 6].

Notwithstanding the protean nature of sensory inputs modulating airway-related preganglionic nerve activity, perhaps the most clear examples of changes in vagal reflexes to the airways are initiated by reflexes arising from the airways themselves. Stimulation of slowly adapting stretch receptors during an inspiration leads to a bronchodilation that is secondary to a virtual abolition of action potential discharge in airway-related preganglionic neurons [1]. By contrast, mechanical or chemical stimulation of rapidly adapting receptors and C-fibers in the airways lead to coughing and increases in cholinergic tone in the airways [34]. The activity of rapidly adapting Aδ fibers and C-fibers is increased by the process of inflammation [2, 34], which theoretically may lead to increases in parasympathetic cholinergic tone of airway smooth muscle in subjects with inflammatory airway disease. As discussed below, changes in synaptic transmission and changes in neurotransmitter release at the nerve-effector cell junction are also sites at which inflammation can modulate the activity of parasympathetic nerves innervating the airways. Not only can afferent input from the lower airway affect airway-related preganglionic parasympathetic nerve activity, but stimulation of afferent nerves in the larynx and nasopharynx can increase cholinergic tone in the bronchi [35, 36].

Evidence from animal models supports the hypothesis that the preganglionic control of the NANC relaxant innervation of the airways may be separate from the preganglionic control of the cholinergic contractile pathways [24, 25]. There is relatively little information, however, on the precise mechanisms regulating airway

NANC-related preganglionic neurons [37–39]. Ideally, two conditions need to be met to rigorously investigate reflex regulation of NANC relaxant innervation. First, the strong inputs from the cholinergic pathways need to be inhibited and second, there needs to be tone in the smooth muscle in order to observe and quantify the NANC-mediated bronchodilation. The difficulty lies in the fact that most of the tone in mammalian smooth muscle is cholinergic, so that inhibiting the cholinergic input with a muscarinic receptor antagonist causes a near abolition of airway smooth muscle tone. Therefore one would need to cause some degree of contraction with a non-muscarinic contractile agonist after first blocking the cholinergic inputs with an antimuscarinic drug [38, 39]. These difficulties may explain why relatively little is known about the role of central reflexes in the control of relaxant parasympathetic innervation to human airway smooth muscle.

The neurotransmitters involved in the regulation of airway-related preganglionic nerve activity are not completely understood. Transneuronal labeling studies indicate that airway preganglionic parasympathetic neurons are innervated by a network of brainstem neurons that lie in the same region as those involved with regulation of respiration [40]. Several neurotransmitter mechanisms are involved in the stimulation or inhibition of the airway-related preganglionic neurons. Primary afferent neurons typically stimulate second-order neurons *via* excitatory amino acid (i.e., glutamate) neurotransmission. Excitatory neurotransmission is also likely to be involved in the regulation of airway preganglionic nerve fibers. Inhibition of glutaminergic receptors in the brainstem, especially the non-N-methyl-D-aspartate (NMDA) subtype of receptors, decreases cholinergic tone in canine airways [41]. Stimulation of midline medullary neurons with glutamate increases airway cholinergic tone by a mechanism that can be inhibited by local application of methysergide, implicating serotinergic neurons in the regulation of airway preganglionic nerves [42]. Tachykinins [43] and imidazoline receptor agonists [44] may also increase the activity of airway-related preganglionic nerve activity.

Postganglionic parasympathetic neurons

The postganglionic parasympathetic neurons innervating airway smooth muscle are of at least two major phenotypes. One type of neuron is cholinergic and, when stimulated, contracts airway smooth muscle *via* the release of acetylcholine and subsequent stimulation of muscarinic cholinergic receptors on airway smooth muscle. The other type of postganglionic neuron uses NANC transmitters such as nitric oxide, and/or neuropeptides, such as vasoactive intestinal peptide (VIP), that when released lead to relaxation of airway smooth muscle. A third, less frequently observed phenotype may co-transmit acetylcholine with nitric oxide or VIP. The NANC relaxant innervation is seen in the vast majority of mammals studied, including humans, where it appears to provide the only relaxant innervation to bronchial smooth muscle [23, 45]. It should be pointed out that other neuropeptides have been localized to postganglionic parasympathetic neurons in the airways, including substance P [46], but functional roles for these neurons at the level of the smooth muscle have not been defined.

Airway parasympathetic ganglia

The somata of postganglionic parasympathetic nerve fibers are located in small clusters near or within the airways [10, 47, 48]. These clusters of parasympathetic neurons are termed parasympathetic or intrinsic ganglia. The airway parasympathetic ganglia, which typically contain fewer than 40 neurons, integrate information arising from the CNS *via* the preganglionic fibers, and then transmit that information to a localized region of the airway *via* action potential discharge in the postganglionic nerve fibers. Whereas modulation of preganglionic outflow may cause a more or less generalized change in parasympathetic tone in the airways, modulation of this outflow by ganglionic synaptic activity serves as a regional regulator of airway parasympathetic tone.

Physiological studies on airway parasympathetic ganglia have been limited to a rather small number of investigations on the trachea of guinea pig [49, 50], ferret [51–53], rat [54–56], cat [26] and in the bronchi of guinea pig [57] and humans [58]. These ganglia can be found throughout the airway tree, but are much more often associated with the large airways [48].

In the guinea pig airway, the cholinergic neurons are found in ganglia within the tracheal wall, whereas the noncholinergic (nitric oxide-, VIP-producing) neuronal cell bodies involved in the relaxant innervation of the smooth muscle are located in ganglia within the myenteric plexus of the neighboring esophagus [59–62]. The NANC and cholinergic neurons are also localized in separate ganglia in the ferret trachea. The cholinergic neurons cluster as ganglia in the superficial plexus of the ferret trachea, whereas VIP-NO parasympathetic neurons cluster as small ganglia in the submucosal plexus [63]. The specific location of NANC *vis-à-vis* cholinergic neurons has not been extensively studied in other species.

Synaptic transmission

Preganglionic nerves can form synaptic junctions at the postganglionic dendrites (axo-dendritic synapse) or the soma (axo-somatic synapses). In guinea pig bronchial ganglia, the vast majority (> 90%) of synapses at cholinergic neurons are axo-dendritic [64]. The nature of the synapses at NANC neurons has not been studied.

In all cases studied thus far, the chemical nature of the fast excitatory neurotransmission in airway parasympathetic ganglia is cholinergic *via* nicotinic receptors. Action potentials invading the preganglionic axon terminals cause the quantal release of acetylcholine into the synaptic cleft. The acetylcholine binds to nicotinic receptors, resulting in an increase in the conductance of cations across the postsynaptic membrane. The ions flowing through the nicotinic receptor (ion channel) cause a postsynaptic membrane depolarization that is referred to as an excitatory postsynaptic potential (epsp). The epsp is then conducted toward the cell body, all the while decaying in amplitude in an electrotonic fashion, until it reaches the axon hillock of the postganglionic neuron. If the epsp is still of sufficient amplitude after it has traversed the distance between the synapse and the hillock, the action potential threshold is reached, thereby generating an action potential that is conducted in an all-or-none-fashion to the postganglionic varicosities innervating the smooth muscle (or other effector cells).

The term given for the fidelity by which an action potential in the preganglionic fiber is transmitted to an action potential in the postganglionic fiber is the "synaptic safety factor". Autonomic ganglia with a large safety factor are those in which all or nearly all preganglionic action potentials lead to corresponding action potentials in the postganglionic neurons. These "relay-type" ganglia, as one might expect, are typified as having very simple, or no, dendritic processes; thus, there is little space between the synapse and the action potential generation zone of the postganglionic neuron. Cardiac parasympathetic ganglia are examples of these relay-type ganglia [65]. By contrast, the synapses in bronchial parasympathetic ganglia have low safety factors; i.e., many if not most of the individual epsps are subthreshold for action potential generation (Fig. 1). This is likely to be due to the large and complex dendritic arborizations seen in bronchial ganglion neurons (Fig. 2). On average, neurons in the guinea pig bronchial ganglia have 7 multi-branching dendritic processes [64]. There is electrophysiological evidence for convergence at the ganglion neuron; such a single neuron receives input from more than one preganglionic nerve fiber (Fig. 1). When an action potential in the preganglionic nerve fiber causes an epsp in the dendrite, the time- and space-constant of the postganglionic neuron is such that, by the time the epsp reaches the axon, it is of insufficient amplitude to cause an action potential. In this way the bronchial ganglia act as a filter of input from the CNS [8, 57, 66]. This concept has been directly addressed *in vivo* in the cat, where indeed many of the epsps caused by preganglionic discharge did not reach the action potential threshold (Fig. 1) [26].

The filtering function of the airway ganglia is important. As mentioned above, the preganglionic nerves to the airway may discharge action potentials at frequencies well in excess of 10 per second. If all the action potentials in the preganglionic fibers were faithfully transmitted as action potentials in the postganglionic nerve, it would be likely to lead to excessive bronchoconstriction. It follows that increasing synaptic efficacy in ganglia with low safety factors, such as airway ganglia, could profoundly increase parasympathetic tone to the effector cells. Various transmitters and inflammatory mediators can increase synaptic neurotransmission in airway ganglia, thus potentially decreasing the filtering function of the ganglia (see below).

Peripheral reflex and synaptic transmission

Although the fast excitatory transmission at airway ganglia is nicotinic in nature, there is evidence of additional types of synaptic transmission. In guinea pig [49, 63, 67] and human [68] airways, tachykinin-containing afferent nerve fibers are found around the principal neurons within the parasympathetic ganglion. The functional significance of the peripheral reflex is supported by the observation that cholinergic and NANC parasympathetic neurons can be stimulated by tachykinins released from the afferent nerve fibers [49, 67, 69, 70]. The major neurokinin (NK) receptor type involved in this tachykinergic synaptic transmission is NK3, although NK1 receptors appear to be involved as well [67, 70]. We have noted in preliminary studies that human bronchial parasympathetic ganglion neurons are stimulated by NK3 receptor activation (personal observations). These studies provide the anatomical and physiological basis for a peripheral parasympathetic reflex arc in the airways (Fig. 2). A peripheral reflex arc is where activation of an afferent nerve fiber results in an axon reflex and release of neuropeptides that directly stimulate postganglionic autonomic neurons independently of the CNS. The peripheral reflex arc in the airways allows for regional modulation of parasympathetic transmission, thus only those ganglia in the vicinity of the sensory C-fiber activation are affected. This is in contrast to the central reflex arc which is thought to provide a more generalized modulation of parasympathetic transmission.

Modulation of ganglionic synaptic transmission

In addition to central and peripheral synaptic neurotransmission, the output from the airway parasympathetic ganglia can be modulated by various inflammatory mediators (Fig. 2). As mentioned above, airway ganglia are sites of considerable filtering of central preganglionic input. Inflammatory mediators can increase the "synaptic efficacy" at the ganglia by either increasing the excitability of the postsynaptic membrane, or by increasing the acetylcholine release from presynaptic terminals [10, 71, 72]. When bronchi are isolated from actively sensitized guinea pigs, antigen challenge leads to activation of tissue mast cells and the release of a variety of mediators associated with allergic inflammation. This process is associated with an increase in synaptic efficacy such that amplitude of subthreshold epsps is increased to the extent that they reach the threshold for action potential generation. In other words, antigen challenge decreases the filtering function of the ganglia neurons [71, 72]. In addition, antigen challenge significantly decreases the accommodative properties of the neurons [71]; this allows for a more sustained activation of the postganglionic neurons. The ionic mechanism of these events is not completely understood [73]; however, histamine and prostaglandin D2 are at least two mediators that are involved with allergen-induced neuromodulation of guinea pig airway ganglionic transmission [71, 74]. Bradykinin is another mediator that can increase the excitability of ganglia neurons in guinea pig airways. Bradykinin, *via*

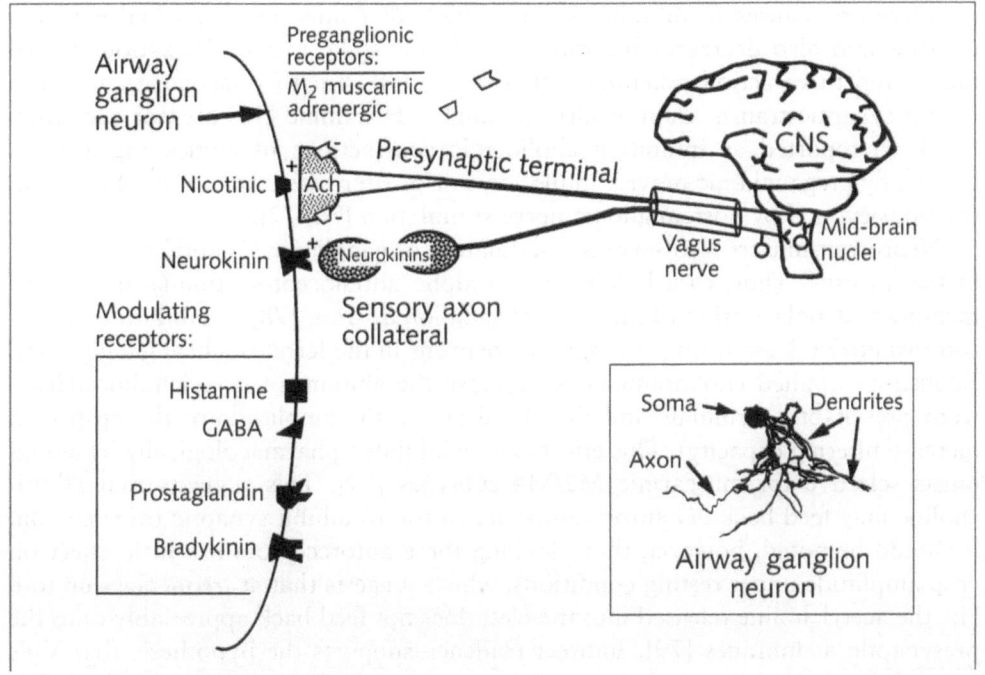

Figure 2
Schematic representation of lower airway ganglion synapse and sensory nerve fiber. The air-ways are innervated by the vagus nerves which carry afferent fibers and preganglionic parasympathetic fibers. The cell bodies of postganglionic airway parasympathetic nerves are located in clusters (ganglia), typically located within or near the lower airway wall.

Preganglionic parasympathetic fibers arise from one of several central nervous system (CNS) brainstem nuclei and release acetylcholine from presynaptic terminals. Acetylcholine binds to nicotinic receptors on the postsynaptic membrane to depolarize the parasympathetic gan-glion neuron, causing a non-regenerative post-synaptic potential. It should be noted that most of these nicotinic synapses are located on dendrites, which can be revealed by dye-injection (a camera-lucida drawing of a guinea pig bronchial ganglia neuron is shown in the inset at lower right; from Myers et al., 1990). This post-synaptic potential decays as it tra-verses the length of the dendrite on its way to the action potential initiation site (typically near the cell body). Many of the post-synaptic potentials do not initiate an action potential (i.e., do not reach threshold), effectively aborting output from the CNS (see text and Fig. 1). Anatomical and electrophysiological evidence indicates that sensory axon collaterals (unmyeli-nated "C-fibers"), containing neurokinins such as substance P, are located near ganglion neu-rons creating a peripheral reflex (see text for details). As shown, autacoid and neurotransmit-ter modulatory receptors have been identified on airway ganglion neurons and on pre-synap-tic terminals in different species. It is uncertain whether the receptors indicated on the airway ganglion neuron are located on dendrites, the cell body (soma), or on both locations.

B2 receptors, causes membrane depolarization of guinea pig bronchial ganglion neurons and also decreases the duration of after-spike hyperpolarization evoked during repetitive action potential activity [75]. Inflammatory mediators may also inhibit synaptic transmission at airway ganglia. Histamine H3 receptor activation has been reported as inhibiting cholinergic contractions of guinea pig airways evoked by preganglionic nerve stimulation to a greater extent than it inhibited contractions evoked by postganglionic nerve stimulation [76, 77].

Neurotransmitters also serve as modulators of ganglionic synaptic transmission in the airways. Thus, catecholamines *via* alpha adrenoceptor stimulation lead to membrane depolarization of rat tracheal ganglion neurons [78] or inhibition of neurotransmitter release from preganglionic neurons in the ferret tracheal ganglia [51]. Exogenous applied cholinomimetics decrease the amount of acetylcholine release from presynaptic terminals and thereby decrease the amplitude of the epsp (i.e., increase filtering capacity). This effect can be inhibited pharmacologically by antagonists selective for muscarinic M2/M4 subtypes [79]. This suggests that acetylcholine may feed back *via* autoreceptor activation to inhibit synaptic transmission. It should be noted, however, that blocking these autoreceptors has little effect on epsp amplitude under resting conditions, which suggests that at frequencies up to 8 Hz, the acetylcholine released into the cleft does not feed back appreciably onto the presynaptic membranes [79]. Indirect evidence supports the hypothesis that VIPs may inhibit synaptic transmission at the airway ganglia [80]. Activation of γ-amino butyric acid (GABA) receptors on rat tracheal parasympathetic ganglia can modulate synaptic transmission *via* an increase in chloride conductance across the membrane [81, 82].

Neuro-effector junction

Unlike the ganglionic synapse, postganglionic autonomic nerve fibers do not typically form specialized fixed junctions at effector cells [83]. For this reason, the point at which the postganglionic nerve fiber communicates with the airway smooth muscle is not considered a true synapse, but rather a "neuro-effector junction". As with other autonomic postganglionic nerves, the postganglionic nerve fibers innervating the airways can be seen as highly branched axons with transmitter-laden varicosities at variable distances from the effector cells they influence [84]. Activation of cholinergic and NANC nerves lead to changes in smooth muscle tone throughout the airway tree from the trachea to peripheral bronchi and bronchioles [85–87].

Cholinergic
Virtually all mammalian airway smooth muscle is contracted upon stimulation of the cholinergic parasympathetic pathway. To date, acetylcholine is the only known

transmitter that participates in vagal parasympathetic contraction of airway smooth muscle. In guinea pigs, vagus nerve stimulation can lead to profound noncholinergic contraction of airway smooth muscle [88]; this is not due to parasympathetic nerve stimulation but to antidromic stimulation of tachykinin release from primary afferent nerve terminals [89].

The extent to which stimulation of postganglionic cholinergic nerves can decrease the caliber of the airway smooth muscle is impressive. Indeed, the airway diameter can be decreased by 70% upon cholinergic nerve stimulation. This can lead to 10-fold increase in airway resistance [5].

Noncholinergic nonadrenergic (NANC)

Electrical stimulation of intrinsic nerves causes nonadrenergic relaxations of precontracted airway smooth muscle. This has been most extensively studied in the guinea pig, but a NANC relaxant input to airway smooth muscle has also been noted in humans, non-human primates, pig, rabbit, and cat. The transmitter for the NANC relaxation has not yet been completely resolved. For a more detailed discussion of the NANC transmitter, see [23]. In human, pig, cat and guinea pig airways, inhibitors of nitric oxide synthase causes a significant inhibition of the NANC relaxations caused by electrical field stimulation [85, 90–94]. Vasoactive intestinal peptide (or a related peptide) also appears to play a role in airway NANC relaxations, especially in guinea pigs [95, 96].

In cats [97] and guinea pigs [98–100] there is direct evidence that the NANC relaxant innervation is vagal and parasympathetic in nature. Stimulation of vagus nerves causes relaxation of the airway smooth muscle, and this effect can be inhibited by blocking synaptic neurotransmission with nicotinic receptor antagonists; similar studies have not yet been carried out in the human airway. However, there is histological evidence of nitric oxide and VIP neurons in human airway parasympathetic ganglia [101, 102].

In addition to the nature of the neurotransmitter, there are other differences in NANC and cholinergic airway innervation. It appears that a major component of the parasympathetic NANC innervation to the airways constitutes a nerve pathway entirely separate from the parasympathetic cholinergic pathways. In the guinea pig, for example, the cell bodies of the NANC neurons innervating the trachea are localized in the myenteric plexus of the esophagus [61, 62, 69, 103]. The preganglionic nerve fibers innervating the cat and guinea pig airway NANC neurons are stimulated at different intensities than are required for stimulation of preganglionic fibers innervating cholinergic neurons [69]. Experiments in vitro indicate that higher frequencies of preganglionic stimulation are required to cause relaxation than contraction of the airway smooth muscle. Thus, whereas synchronously stimulating the vagal preganglionic input with a single impulse to cholinergic ganglion neurons can lead to contractions of the airway smooth muscle, stimulation with synchronous impulses at frequencies in excess of 1 or 2 per second is required to produce significant relaxations of airway smooth muscle [69].

Modulation of cholinergic and NANC airway neuromuscular transmission

The extent to which airway resistance is regulated by the parasympathetic cholinergic nerves is, in large part, regulated by the degree of preganglionic input and synaptic transmission through the ganglia. For example, over a 10-second period, an increase in the frequency of impulse discharge in the preganglionic fibers from 1 per second to 4 per second can cause an increase in smooth muscle contraction from 20% to 40% of the maximum response [66]. The amount of acetylcholine released at the neuromuscular junction, however, may also be modulated by local regulatory mechanisms at the level of the neuro-effector junction. Altering the amount of transmitter released at the neuro-effector junction is termed prejunctional neuromodulation.

The modulation of acetylcholine secretion per electrical impulse has been extensively studied using the isolated trachea/bronchus preparation. The acetylcholine release can be quantified as the amount of acetylcholine or tritiated-choline overflow. The former is a more direct measurement, but the latter is much more sensitive [104]. Using these techniques, with only few exceptions, the numerous mediators and neurotransmitters that directly affect airway smooth muscle tone have also been found to have prejunctional neuromodulatory effects on acetylcholine release. The extent to which a given mediator regulates acetylcholine release is subject to considerable species variation. The literature on this topic is vast and beyond the scope of the present overview; the reader may seek out more thorough reviews on this topic elsewhere [11, 104, 105].

Various neurotransmitters and neuropeptides are effective in facilitating or inhibiting acetylcholine release (see Tab. 1). The most commonly studied neurotransmitter that inhibits acetylcholine release at the airway neuromuscular junction is acetylcholine itself [105]. Cholinergic muscarinic receptor agonists inhibit electrically evoked acetylcholine release in the airways both *in vitro* [106–113] and *in vivo* [108, 114, 115]. The mechanism is secondary to activation of muscarinic receptors with pharmacological characteristics consistent with the M_2 subtype, but with a profile distinct from cardiac M_2 receptors [113]. Electrophysiological studies indicate that activation of these receptors leads to an inhibition of voltage-gated N- and R-type calcium channels [56]. An interesting feature of the muscarinic M_2 receptor is that its function is susceptible to inhibition by the inflammatory process (see [105]). Thus, dysregulation of the M_2 receptor may contribute to increased cholinergic tone in inflammatory airway disease.

Catecholamines promote acetylcholine release from postganglionic varicosities *via* β_2 adrenoceptor activation [116-119]. Paradoxically, β adrenoceptor activation may also lead to an inhibition of acetylcholine release in the airways [120–122]. At

Table 1 - Postganglionic prejunctional modulation by neurotransmitters and autacoids of parasympathetic transmission at airway smooth muscle

	Effect	Receptor	References
Cholinergic responses			
(Neurotransmitters)			
Acetylcholine	Inhibit	M2/ M4?	[106–115]
Catecholamines	Inhibit	α and β	[116, 120–126]
	Enhance	β	[116–119]
Neurokinins	Enhance	NK1, NK2	[66, 127–129]
Opioids	Inhibit	Mu	[130]
Nociceptin	Inhibit	?	[131]
Neuropeptide Y	Inhibit	?	[132, 133]
(Autacoids)			
Prostaglandins	Enhance	?	[135, 138–141]
	Inhibit	EP3	[136, 137]
Histamine	Enhance	H-1	[143–145]
Serotonin	Enhance	5HT-3	[142, 143]
Adenosine	Enhance	A2	[146]
Endothelin	Enhance	ETB	[147]
NANC responses			
Acetylcholine	No Effect		[150]
Catecholamines	No Effect		[125]
Neuropeptide Y	No Effect		[151]

least a component of the inhibitory effect of β adrenoceptor activation may occur *via* the release of prostanoids from the airway epithelium [123]. In the guinea pig and equine airways, activation of α adrenoceptors has been consistently found to prejunctionally inhibit cholinergic contractions [116, 124–126].

Neurokinins are effective neuromodulators of cholinergic contractions in the airways. In contrast to ganglionic transmission, where tachykinins act primarily *via* NK3 receptors [67, 70], the neurokinin receptor involved in the prejunctional potentiation of cholinergic responses appears to involve NK1 or NK2 receptors, depending on the species [66, 127–129].

Opioid agonists are inhibitors of prejunctional release of acetylcholine in the airways. The mechanism of opioid-induced prejunctional inhibition of cholinergic contraction of airways involves the activation of mu-opioid receptors [130]. Nocicpeptin, an opioid-related peptide, also causes prejunctional inhibition of cholinergic contractions of isolated airways, *via* naloxone-insensitive receptors [131]. Neuropeptide Y is another neuropeptide, most often localized in sympathetic nerves, that effectively inhibits the prejunctional release of acetylcholine in the airways [132, 133].

Inflammatory mediators can increase parasympathetic nerve activity indirectly *via* the activation of primary afferent nerve endings [2]. Many autacoids associated with airway inflammation are also direct modulators of acetylcholine release at the level of the airway neuro-effector junction [134].

Prostanoids, such as PGD_2, $PGF_{2\alpha}$ and thromboxane, enhance cholinergic transmission at the airway neuro-effector junction, whereas PGE_2 and prostacyclin inhibit acetylcholine release in the airways [135–141]. The inhibitory effect of prostaglandins appears to be mediated through the EP_3 subtype [137].

Histamine H-1 receptor, serotonin 5 HT-3 receptor, adenosine A_2 receptor and endothelin B receptor activation facilitate cholinergic transmission at the neuro-effector junction in isolated airway preparations [142–147]. It is likely, however, that *in vivo* a major manner in which these mediators augment cholinergic transmission is *via* central reflex actions [2, 148].

In contrast to prejunctional modulation of the cholinergic contractile system in the airways, there is a striking dearth of information regarding the prejunctional modulation of the NANC relaxant innervation. Other than the expected fact that enzymes involved in the production or metabolism of the NANC transmitter can significantly influence the magnitude of the parasympathetic NANC relaxations, little is known about neuromodulation of this system [149]. It is interesting that a number of receptor systems known to modulate acetylcholine release, including muscarinic M_2 [150], α adrenoceptors [125] and NPY [151] do not affect NANC relaxations. This suggests that the repertoire of receptors on the NANC varicosities is different from that on cholinergic varicosities. Alternatively, the process of NANC transmitter secretion may be more resistant to modulation than that for acetylcholine secretion.

Summary

The parasympathetic branch of the autonomic nervous system provides the major control over airway smooth muscle tone in mammals. The dominant regulation of parasympathetic input to the airway occurs in the midbrain where the preganglionic neurons are reflexively stimulated by various peripheral afferent and central inputs. Human airway smooth muscle receives both cholinergic contractile and NANC relaxant innervation from the parasympathetic system. Although there may be nerves that co-release acetylcholine and the NANC transmitters, the evidence to date indicates that the cholinergic and NANC nerves more typically represent distinct parasympathetic neural pathways, each under separate regulatory control.

Impulse generation in the preganglionic nerves of the cholinergic and NANC pathways leads to synaptic neurotransmission at airway-related parasympathetic ganglia. In the guinea pig, the ganglia associated with NANC relaxations are found in the myenteric plexus of the esophagus, whereas the somata of cholinergic postganglionic nerve fibers are found along the serosal surface of the trachea and large bronchi. Regardless of the location, the airway-related parasympathetic ganglia are sites of integration and considerable filtering of preganglionic input.

Airway inflammation can profoundly affect parasympathetic outflow to the airways by several mechanisms. First, airway inflammation is likely to lead to changes in impulse traffic along preganglionic nerve fibers. Second, inflammation can stimulate neuropeptide-containing afferent fibers that synapse directly with bronchial parasympathetic ganglia (so-called peripheral reflex regulation). Third, inflammatory mediators can modulate synaptic efficacy at the ganglia. Fourth, inflammatory mediators may modulate the amount of neurotransmitter release per impulse from the postganglionic nerve varicosities *via* prejunctional neuromodulatory mechanisms.

Acknowledgement

We would like to acknowledge Dr. Michael Carr for his helpful suggestions.

References

1   Widdicombe JG, Nadel JA (1963) Airway volume, airway resistance, and work and force of breathing: Theory. *J Appl Physiol* 18: 863–868

2   Coleridge HM, Coleridge JCG, Schultz HD (1989) Afferent pathways involved in reflex regulation of airway smooth muscle. *Pharmacol Ther* 42: 1–63

3   Severinghaus JW, Stupfel M (1955) Respiratory dead space increase following atropine in man, and atropine, vagal or ganglionic blockade and hypothermia in dogs. *J Appl Physiol* 8: 81–87

4   Colebatch H, Halmagyi F (1963) Effect of vagotomy and vagal stimulation on lung mechanics and circulation. *J Appl Physiol* 18: 881–887

5   Olsen CR, Colebatch HJH, Mebel PE, Nadel JA, Staub NC (1965) Motor control of pulmonary airways studied by nerve stimulation. *J Appl Physiol* 20: 202–208

6   Widdicombe JG, Wells UM (1994) Vagal reflexes. In: D Raeburn, M Gymbiecz (eds): *Airways smooth muscle: structure innervation and neurotransmission.* Birkhäuser Verlag, Basel, 279–307

7   Kalia M (1987) Organization of central control of airways. *Ann Rev Physiol* 49: 595–609

8   Skoogh B-E (1988) Airway parasympathetic ganglia. In: MA Kaliner, PJ Barnes: *The airways neural control in health and disease*. Marcel Dekker, New York, Basel, 217–240

9   Dey RD (1995) Airways ganglia. In: D Raeburn, M. Gymbiecz (eds): *Airways smooth muscle: structure innervation and neurotransmission*. Birkhäuser Verlag, Basel, 79–101

10   Undem BJ, Myers AC (1997) Autonomic ganglia. In: PJ Barnes: *Autonomic control of the respiratory system*. Harwood Academic Publishers, Amsterdam, 87–117

11   Barnes PJ (1992) Modulation of neurotransmission in airways. *Physiol Rev* 72: 699–729

12   Black JL (1997) Innervation of airway smooth muscle. In: PJ Barnes: *Autonomic control of the respiratory system*. Harwood Academic Publishers, Amsterdam, 185–200

13   Kalia M (1981) Brain stem localization of vagal preganglionic neurons. *J Auton Nerv Syst* 3: 451–481

14   Kerr FWL (1969) Preserved vagal visceromotor function following destruction of the dorsal motor nucleus. *J Physiol* 202: 755–769

15   Haselton JR, Solomon KC, Motekaitis AM, Kaufman MP (1992) Bronchomotor vagal preganglionic cell bodies in the dog: an anatomic and functional study. *J Appl Physiol* 73: 1122–1129

16   Kalia M, Mesulam M-M (1980) Brain stem projections of sensory and motor components of the vagus complex in the cat: II Laryngeal, tracheobronchial pulmonary, cardiac, and gastrointestinal branches. *J Comp Neurol* 193: 467–508, 1980

17   Haxhiu MA, Loewy AD (1996) Central connections of the motor and sensory vagal systems innervating the trachea. *J Auton Nerv Syst* 57: 49–56

18   Perez-Fontan JJ, Velloff CR (1997) Neuroanatomic organization of the parasympathetic bronchomotor system in developing sheep. *Am J Physiol* 273: 121–133

19   Jammes Y, Fornaris E, Mei N, Barrat E (1982) Afferent and efferent components of the bronchial vagal branches in cats. *J Auton Nerv Syst* 5: 165–176

20   Evans DHL, Murray JG (1954) Histological and functional studies on the fibre composition of the vagus nerve of the rabbit. *J Anat* 88: 320–337

21   Agostoni E, Chinnock JE, DeBurgh-Daly M, Murray JG (1957) Functional and histological studies of the vagus nerve and its branches to the heart, lungs and abdominal viscera in the cat. *J Physiol* 135: 182–205

22   Widdicombe JG (1988) Vagal reflexes in the airways. In: MA Kaliner, PJ Barnes: *The airways neural control in health and disease*. Marcel Dekker, New York, Basel, 187–202

23   Ellis JL, Undem BJ (1994) Pharmacology of non-adrenergic, non-cholinergic nerves in airway smooth muscle. *Pulm Pharmacol* 7: 205–223

24   Canning BJ, Undem BJ (1993) Capsaicin-sensitive afferents activate neurons mediating NANC relaxations of the guinea pig trachealis: Possible involvement of tachykinins. *FASEB J* 7: A6863

25   Lama A, Delpierre S, Jammes Y (1988) The effects of electrical stimulation of myelinated and non-myelinated vagal motor fibres on airway tone in the rabbit and the cat. *Respir Physiol* 74: 265–274

26   Mitchell RA, Herbert DA, Baker DG, Basbaum CB (1987) *In vivo* activity of tracheal

parasympathetic ganglion cells innervating tracheal smooth muscle. *Brain Res* 437: 157–160

27 Widdicombe JG (1966) Action potentials in parasympathetic and sympathetic efferent fibres to the trachea and lungs of dogs and cats. *J Physiol* 186: 56–88

28 McAllen RM, Spyer KM (1978) Two types of vagal preganglionic motoneurones projecting to the heart and lungs. *J Physiol* 282: 353–364

29 Leblanc PH, Buckner CK, Brunson DB, Laravuso RB, Will JA (1987) Differential effect of ketamine on cholinergic- and noncholinergic-induced contractions of isolated guinea-pig bronchi. *Arch Int Pharmacodyn Ther* 287: 120–132

30 Canning BJ, Undem BJ (1994) Parasympathetic innervation of airways smooth muscle. In: D Raeburn, M Gymbiecz (eds): *Airways smooth muscle: structure innervation and neurotransmission*. Birkhäuser Verlag, Basel, 43–78

31 Jammes Y, Mei N (1979) Assessment of the pulmonary origin of bronchoconstrictor vagal tone. *J Physiol* 291: 305–316

32 Kesler BS, Canning BJ (1999) Regulation of baseline cholinergic tone in guinea-pig airway smooth muscle. *J Physiol* 518.3: 843–855

33 Sherrington SC (1906) *The integrative action of the nervous system*. Yale University Press, New Haven, 49–102

34 Karlsson J-A, Sant'Ambrogio G, Widdicombe JG (1988) Afferent neural pathways in cough and reflex bronchoconstriction. *J Appl Physiol* 65: 1007–1023

35 Kaufman J, Wright GW (1969) The effect of nasal and nasopharyngeal irritation on airway resistance in man. *Am Rev Respir Dis* 100: 626–630

36 Nolte D, Berger D (1983) On vagal bronchoconstriction in asthmatic patients by nasal irritation. *Eur J Respir Dis* 64: 110–114

37 Szarek JL, Gillespie MN, Altiere RJ, Diamond L (1986) Reflex activation of the nonadrenergic noncholinergic inhibitory nervous system in feline airways. *Am Rev Respir Dis* 133: 1159–1162

38 Michoud M-C, Jeanneret-Grosjean A, Cohen A, Amyot R (1988) Reflex decrease of histamine-induced bronchoconstriction after laryngeal stimulation in asthmatic patients. *Am Rev Respir Dis* 138: 1548–1552

39 Lammers J-W, Minette P, McCusker MT, Chung KF, Barnes PJ (1988) Nonadrenergic bronchodilator mechanisms in normal human subjects *in vivo*. *J Appl Physiol* 64: 1817–1822

40 Haxhiu MA, Jansen AS, Nerniack NS, Loewy AD (1993) CNS innervation of airway-related parasympathetic prrganglionic neurons: a transneuronal labeling study using pseudorabies virus. *Brain Res* 618: 115–134

41 Haxhiu MA, Erokwu B, Dreshaj IA (1997) The role of excitatory amino acids in airway reflex responses in anesthetized dogs. *J Auton Nerv Syst* 67: 192–199

42 Haxhiu MA, Erokwu B, Bhardwaj V, Dreshaj IA (1998) The role of the medullary raphe nuclei in regulation of cholinergic outflow to the airways. *J Auton Nerv Syst* 69: 64–71

43 Haxhiu MA, Deal EC, van Lunteren E, Cherniack NS (1989) Central modulatory effects of tachykinin peptides on airway tone. *J Auton Nerv Syst* 28: 105–115

44   Haxhiu MA, Dreshaj IA, McFadden CB, Erokwu BO, Ernsberger P (1998) I1-imidazo-line receptors and cholinergic outflow to the airways. *J Auton Nerv Syst* 71: 167–174

45   Richardson J, Beland J (1976) Nonadrenergic inhibitory nervous system in human air-ways. *J Appl Physiol* 41: 764–771

46   Dey RD, Altemus JB, Michalkiewicz M (1991) Distribution of vasoactive intestinal pep-tide- and substance P-containing nerves originating from neurons of airway ganglia in cat bronchi. *J Comp Neurol* 304: 330–340

47   Fisher WF (1964) The intrinsic innervation of the trachea. *J Anat Lond* 98: 117–124

48   Larsell O (1923) The ganglia, plexuses, and nerve-terminations of the mammalian lung and pleura pulmonalis. *J Comp Neurol* 35: 97–132

49   Myers AC, Undem BJ, Kummer W (1996) Anatomical and electrophysiological com-parison of the sensory innervation of bronchial and tracheal parasympathetic ganglion neurons. *J Auton Nerv Syst* 61: 162–168

50   Lees GM, Pacitti EG, Mackenzie GM (1997) Morphology and electrophysiology of guinea-pig paratracheal neurones. *Anat Rec* 247: 261–270

51   Baker DG, Basbaum CB, Herbert DA, Mitchell RA (1983) Transmission in airway gan-glia of ferrets: inhibition by norepinephrine. *Neurosci Lett* 41: 139–143

52   Cameron AR, Coburn RF (1984) Electrical and anatomic characteristics of cells of fer-ret paratracheal ganglion. *Am J Physiol* 246: C450–C458

53   Coburn RF, Kalia MP (1986) Morphological features of spiking and nonspiking cells in the paratracheal ganglion of the ferret. *J Comp Neurol* 254: 341–351

54   Aibara K, Akaike N (1991) Acetylcholine-activated ionic currents in isolated paratra-cheal ganglion cells of the rat. *Brain Res* 558: 20–26

55   Allen TGJ, Burnstock G (1990) A voltage-clamp study of the electrophysiological char-acteristics of the intramural neurones of the rat trachea. *J Physiol* 423: 593–614

56   Murai Y, Ishibashi H, Akaike N, Ito Y (1998) Acetylcholine modulation of high-volt-age-activated calcium channels in the neurones acutely dissociated from rat paratracheal ganglia. *Br J Pharmacol* 123: 1441–1449

57   Myers AC, Undem BJ, Weinreich D (1990) Electrophysiological properties of neurons in guinea pig bronchial parasympathetic ganglia. *Am J Physiol* 259: L403–L409

58   Myers AC (1997) Evidence for neural integration by human bronchial parasympathetic ganglia neurons. *Am Rev Respir Crit Care Med* 155: A575

59   Canning BJ, Undem BJ (1993) Evidence that distinct neural pathways mediate parasym-pathetic contractions and relaxations of guinea-pig trachealis. *J Physiol* 471: 25–40

60   Canning BJ, Undem BJ, Karakousis PC, Dey RD (1996) Effects of organotypic culture on cholinergic and noncholinergic parasympathetic nerves innervating airway smooth muscle. *Am J Physiol* 271: L698–L706

61   Fischer A, Canning BC, Undem BJ, Kummer W (1998) Evidence for an esophageal ori-gin of VIP and NO synthase-ir nerves innervating the guinea pig trachealis: A retrograde neuronal tracing immunohistochemical analysis. *J Compar Neurol* 394: 326–334

62   Moffatt JD, Dumsday B, McLean JR (1998) Non-adrenergic, non-cholinergic neurons

innervating the guinea-pig trachea are located in the oesophagus: evidence from retrograde neuronal tracing. *Neurosci Lett* 248: 37–40

63   Dey RD, Altemus JB, Rodd A, Mayer B, Said SI, Coburn RF (1996) Neurochemical characterization of intrinsic neurons in ferret tracheal plexus. *Am J Respir Cell Mol Biol* 14: 207–216

64   Myers AC (2000) Anatomical characteristics of tonic and phasic neurons in guinea pig bronchial parasympathetic ganglia. *J Compar Neurol* 419: 439–450

65   McMahan UJ, Kuffler SW (1971) Visual identification of synaptic boutons on living ganglion cells and of varicosities in postganglionic axons in the heart of the frog. *Proc Roy Soc Lond* 177: 485–508

66   Myers AC, Undem BJ (1991) Analysis of preganglionic nerve evoked cholinergic contractions of the guinea pig bronchus. *J Auton Nerv Sys* 35: 175–184

67   Canning BJ, Fischer A, Undem BJ (1998) Pharmacological analysis of the tachykinin receptors that mediate activation of nonadrenergic, noncholinergic relaxant nerves that innervate guinea pig trachealis. *J Pharmacol Exp Ther* 284: 370–377, 1998

68   Lundberg JM, Hokfelt T, Martling C-R, Saria A, Cuello C (1984) Substance P-immunoreactive sensory nerves in the lower respiratory tract of various mammals including man. *Cell Tissue Res* 235: 251–261

69   Canning BJ, Undem BJ (1993) Evidence that the parasympathetic pathway mediating relaxation of guinea-pig trachealis is distinct from that mediating contraction. *J Physiol* 471: 25–40

70   Myers AC, Undem BJ (1993) Electrophysiological effects of tachykinins and capsaicin on guinea-pig bronchial parasympathetic ganglion neurones. *J Physiol* 470: 665–679

71   Myers AC, Undem BJ, Weinreich D (1991) Influence of antigen on membrane properties of guinea pig bronchial ganglion neurons. *J Appl Physiol* 71: 970–976

72   Undem BJ, Myers AC, Weinreich D (1991) Antigen-induced modulation of autonomic and sensory neurons *in vitro*. *Int Arch Allergy appl Immunol* 94: 319–324

73   Myers AC (1998) Ca^{++} and K^+ currents underlying action potential accommodation by neurons in guinea pig bronchial parasympathetic ganglia. *Am J Physiol Lung Cell Mol Phys* L357–L364

74   Myers AC, Undem BJ (1995) Antigen depolarizes guinea pig bronchial parasympathetic ganglion neurons by activation of histamine H1 receptors. *Am J Phys (Lung Cell Mol Physiol)* 12: L879–L884

75   Kajekar R, Myers AC (2000) Bradykinin stimulates guinea pig bronchial parasympathetic ganglia neurons independent of sensory nerve stimulation. *Am J Physiol Lung Cell Mol Physiol* 278: L485–L491

76   McCaig DJ (1986) Electrophysiology of neuroeffector transmission in the isolated, innervated trachea of the guinea-pig. *Br J Pharmacol* 89: 793–801

77   Ichinose M, Barnes PJ (1989) Histamine H_3-receptors modulate nonadrenergic noncholinergic neural bronchoconstriction in guinea-pig *in vivo*. *Eur J Pharmacol* 174: 49–55

78   Reekie FM, Burnstock G (1992) Effects of noradrenaline on rat paratracheal neurones

and localization of an endogenous source of noradrenaline. *Br J Pharmacol* 107: 471–475

79 Myers AC, Undem BJ (1996) Muscarinic receptor regulation of synaptic transmission in airway parasympathetic ganglia. *Am J Physiol* 270: L630–L636

80 Martin JG, Wang A, Zacour M, Biggs DF (1990) The effects of vasoactive intestinal polypeptide on cholinergic neurotransmission in isolated innervated guinea pig tracheal preparations. *Respir Physiol* 79: 111–122

81 Allen TGJ, Burnstock G (1990) GABA$_A$ receptor-mediated increase in membrane chloride conductance in rat paratracheal neurones. *Br J Pharmacol* 100: 261–268

82 Itabashi S, Aibara K, Sasaki H, Akaike N (1992) γ-aminobutyric acid-induced response in rat dissociated paratracheal ganglion cells. *J Neurophysiol* 67: 1367–1374

83 Gabella G (1992) Fine structure of post-ganglionic nerve fibres and autonomic neuroeffector junctions. In: G Burnstock, CHV Hoyle (eds): *Autonomic neuroeffector mechanisms*. Harwood Academic Publishers, Amsterdam, 1–31

84 Gabella G (1987) Innervation of airway smooth muscle: Fine structure. *Ann Rev Physiol* 49: 583–594

85 Ellis JL, Undem BJ (1992) Inhibition by L-N^G-nitro-L-arginine of nonadrenergic-noncholinergic-mediated relaxations of human isolated central and peripheral airways. *Am Rev Respir Dis* 146: 1543–1547

86 Andersson RG, Grundstrom N (1987) Innervation of airway smooth muscle. Efferent mechanisms. *Pharmacol Ther* 32: 107–130

87 Ward JK, Barnes PJ, Springall DR, Abelli L, Tadjkarimi S, Yacoub MH, Polak JM, Belvisi MG (1995) Distribution of human i-NANC bronchodilator and nitric oxide-immunoreactive nerves. *Am J Respir Cell Mol Biol* 13: 175–184

88 Grundstrom N, Andersson RGG, Wikberg JES (1981) Pharmacological characterization of the autonomic innervation of the guinea pig tracheobronchial smooth muscle. *Acta Pharmacol Toxicol* 49: 150–157

89 Undem BJ, Myers AC, Barthlow H, Weinreich D (1990) Vagal innervation of the guinea pig bronchus. *J Appl Physiol* 69: 1336–1346

90 Tucker JF, Brave SR, Charalambous L, Hobbs AJ, Gibson A (1990) L-N^G-nitro arginine inhibits non-adrenergic, non-cholinergic relaxations of guinea-pig isolated tracheal smooth muscle. *Br J Pharmacol* 100: 663–664

91 Kannan MS, Johnson DE (1992) Nitric oxide mediates the neural nonadrenergic, noncholinergic relaxation of pig tracheal smooth muscle. *Am J Physiol* 262: L511–L514

92 Fisher JT, Anderson JW, Waldron MA (1993) Nonadrenergic noncholinergic neurotransmitter of feline trachealis: VIP or nitric oxide? *J Appl Physiol* 74: 31–39

93 Bai TR, Bramley AM (1993) Effect of an inhibitor of nitric oxide synthase on neural relaxation of human bronchia. *Am J Physiol* 264: L425–L430

94 Belvisi MG, Stretton CD, Miura M, Verleden GM, Tadjkarimi S, Yacoub MH, Barnes PJ (1992) Inhibitory NANC nerves in human tracheal smooth muscle: a quest for the neurotransmitter. *J Appl Physiol* 73: 2505–2510

95 Venugopalan CS, Said SI, Drazen JM (1984) Effect of vasoactive intestinal peptide on vagally mediated tracheal pouch relaxation. *Respir Physiol* 56: 205–216

96 Ellis JL, Farmer SG (1989) Effects of peptidases on non-adrenergic, non-cholinergic inhibitory responses of tracheal smooth muscle: a comparison with effects on VIP- and PHI-induced relaxation. *Br J Pharmacol* 96: 521–526

97 Diamond L, O'Donnell M (1980) A nonadrenergic vagal inhibitory pathway to feline airways. *Science* 208: 185–188

98 Chesrown SE, Venugopalan CS, Gold WM, Drazen JM (1980) *In vivo* demonstration of nonadrenergic inhibitory innervation of the guinea pig trachea. *J Clin Invest* 65: 314–320

99 Yip P, Palombini B, Coburn RF (1981) Inhibitory innervation to the guinea pig trachealis muscle. *J Appl Physiol* 50: 374–382

100 Canning BJ, Undem BJ (1993) Relaxant innervation of the guinea-pig trachealis: Demonstration of capsaicin-sensitive and insensitive vagal pathways. *J Physiol* 460: 719–739

101 Fischer A, Hoffman B (1996) Nitric oxide synthase in neurons and nerve fibers of lower airways and in vagal sensory ganglia of man. *Am J Respir Crit Care Med* 154: 209–216

102 Fischer A, Canning BJ, Kummer W (1996) Correlation of vasoactive intestinal peptide and NO synthase with choline acetyltransferase in the airway innervation. *Ann NY Acad Sci* 805: 717–722

103 Canning BJ, Undem BJ, Karakousis PC, Dey RD (1996) Influence of organotypic culture on cholinergic and noncholinergic parasympathetic nerves innervating airway smooth muscle. *Am J Phys (Lung Cell Mol Phys)* 271: L698–L706

104 Barnes PJ (1994) Modulation of neurotransmitter release from airways nerves. In: D Raeburn, M Gymbiecz (eds): *Airways smooth muscle: structure innervation and neurotransmission.* Birkhäuser Verlag, Basel, 209–259

105 Fryer A (1997) The cholinergic control of the airways. In: PJ Barnes: *Autonomic control of the respiratory system.* Harwood Academic Publishers, Amsterdam, 59–86

106 Aas P, Fonnum F (1986) Presynaptic inhibition of acetylcholine release. *Acta Physiol Scand* 127: 335–342

107 Yang Z-J, Biggs DF (1991) Muscarinic receptors and parasympathetic neurotransmission in guinea-pig trachea. *Eur J Pharmacol* 193: 301–308

108 Watson N, Barnes PJ, Maclagan J (1992) Action of methoctramine, a muscarinic M2-receptor antagonist on muscarinic and nicotine cholinoceptors in guinea pig airways *in vivo* and *in vitro*. *Br J Pharmacol* 105: 107–112

109 Doelman CJA, Sprong RC, Nagtegaal JE, Rodrigues-deMiranda JF, Bast A (1991) Prejunctional muscarinic receptors on cholinergic nerves in guinea pig airways are of the M2 subtype. *Eur J Pharmacol* 193: 117–120

110 Minette PA, Barnes PJ (1988) Prejunctional inhibitory muscarinic recptors on cholinergic nerves in human and guinea-pig airways. *J Appl Physiol* 64: 2532–2537

111 Patel HJ, Barnes PJ, Takahashi T, Tadjkarimi S, Yacoub MH, Belvisi MG (1995) Evi-

dence for prejunctional muscarinic autoreceptors in human and guinea pig trachea. *Am J Respir Crit Care Med* 152: 872–878

112 ten Berge RE, Zaagsma J, Roffel AF (1996) Muscarinic inhibitory autoreceptors in different generations of human airways. *Am J Respir Crit Care Med* 154: 43–49

113 Roffel AF, Davids JH, Elzinga CR, Wolf D, Zaagsma J, Kilbinger H (1997) Characterization of the muscarinic recptor subtype(s) mediating contraction of the guinea-pig lung strip and inhibition of the guinea-pig lung strip and inhibition of acetylcholine release in the guinea-pig trachea with the selective muscarinic receptor antagonist tripitramine. *Br J Pharmacol* 122: 133–141

114 Fryer AD, Maclagan J (1984) Muscarinic inhibitory receptors in pulmonary parasympathetic nerves in the guinea-pig. *Br J Pharmacol* 83: 973–978

115 Blaber LC, Fryer AD, Maclagan J (1985) Neuronal muscarinic receptors attenuate vagally-induced contraction of feline bronchial smooth muscle. *Br J Pharmacol* 86: 723–728

116 Zhang XY, Robinson NE, Wang ZW, Lu MC (1995) Catecholamine affects acetylcholine release in trachea: alpha 2-mediated inhibition and beta-mediated augmentation. *Am J Physiol* 268: L368–L373

117 Zhang XY, Zhu FX, Robinson NE (1996) Excitatory prejunctional beta 2-adrenoceptor distribution within equine airway cholinergic nerves. *Respir Physiol* 106: 81–90

118 Belvisi MG, Patel HJ, Takahashi T, Barnes PJ, Giembycz MA (1996) Paradoxical facilitation of acetylcholine release from parasympathetic nerves innervating guinea-pig trachea by isoprenaline. *Br J Pharmacol* 117: 1413–1420

119 Haas JR, Terpstra JS, van der Zwaag J, Kockelbergh PG, Roffel AF, Zaagsma J (1999) Facilitatory beta2-adrenoceptors on cholinergic and adrenergic nerve endings of the guinea pig trachea. *Am J Physiol* 276: L420–L425

120 Vermeire PA, Vanhoutte PM (1979) Inhibitory effects of catecholamines in isolated canine bronchial smooth muscle. *J Appl Physiol: Respirat Environ Exercise Physiol* 46: 787–791

121 Danser AHJ, Ende RVD, Lorenz RR, Flavahan NA, Vanhoutte PM (1987) Prejunctional β_1-adrenoceptors inhibit cholinergic transmission in canine bronchi. *J Appl Physiol* 62: 785–790

122 Rhoden KJ, Meldrum LA, Barnes PJ (1988) Inhibition of cholinergic neurotransmission in human airways by β_2-adrenoceptors. *J Appl Physiol* 65: 700–705

123 Wessler I, Reinheimer T, Brunn G, Anderson GP, Maclagan J, Racke K (1994) Beta-adrenoceptors mediate inhibition of [^{3}H]-acetylcholine release from the isolated rat and guinea-pig trachea: role of the airway mucosa and prostaglandins. *Br J Pharmacol* 113: 1221–1230

124 Grundstrom N, Andersson RGG, Wikberg JES (1981) Prejunctional α_2-adrenoceptors inhibit contraction of tracheal smooth muscle by inhibiting cholinergic neurotransmission. *Life Sci* 28: 2981–2986

125 Thompson DC, Diamond L, Altiere RJ (1990) Presynaptic α-adrenoceptor modulation

of neurally mediated cholinergic excitatory and nonadrenergic noncholinergic inhibitory responses in guinea pig trachea. *J Pharmac Exp Ther* 254: 306–311

126 Thompson DC, Diamond L, Altiere RJ (1992) Atypical presynaptic alpha-adrenoceptor modulation of neurally-mediated cholinergic responses in guinea-pig tracheal smooth muscle. *Pulm Pharmacol* 5: 251–255

127 Tanaka DT, Grunstein MM (1984) Mechanisms of substance P-induced contraction of rabbit airway smooth muscle. *J Appl Physiol: Respirat Environ Exercise Physiol* 57: 1551–1557

128 Belvisi MG, Patacchini R, Barnes PJ, Maggi CA (1994) Facilitatory effects of selective agonists for tachykinin receptors on cholinergic neurotransmission: evidence for species differences. *Br J Pharmacol* 111: 103–110

129 Hey JA, Danko G, del Prado M, Chapman RW (1996) Augmentation of neurally evoked cholinergic bronchoconstrictor responses by prejunctional NK2 receptors in the guinea-pig. *J Auton Pharmacol* 16: 41–48

130 Belvisi MG, Stretton CD, Verleden GM, Ledingham SJ, Yacoub MH, Barnes PJ (1992) Inhibition of cholinergic neurotransmission in human airways by opioids. *J Appl Physiol* 72: 1096–1100

131 Patel HJ, Giembycz MA, Spicuzza L, Barnes PJ, Belvisi MG (1997) Naloxone-insensitive inhibition of acetylcholine release from parasympathetic nerves innervating guinea-pig trachea by the novel opioid, nociceptin. *Br J Pharmacol* 120: 735–736

132 Stretton CD, Barnes PJ (1988) Modulation of cholinergic neurotransmission in guinea-pig trachea by neuropeptide Y. *Br J Pharmacol* 93: 672–678

133 Grundemar WE, Waldeck B, Hakanson R (1988) Neuropeptide Y: prejunctional inhibition of vagally-induced contraction in the guinea pig trachea. *Regul Peptides* 23: 309–314

134 Barnes PJ (1997) Neuromodulation in airways. In: G Burnstock (ed): *Autonomic control of the respiratory system.* Harwood Academic Publishers, Amsterdam, 139–184

135 Shore S, Collier B, Martin JG (1987) Effect of endogenous prostaglandins on acetylcholine release from dog trachealis muscle. *J Appl Physiol* 62: 1837–1844

136 Black JL, Johnson PRA, Alfredson M, Armour CL (1989) Inhibition by prostaglandin E2 of neurotransmission in rabbit but not human bronchus – a calcium related mechanism? *Prostaglandins* 37: 317–330

137 Spicuzza L, Giembycz MA, Barnes PJ, Belvisi MG (1998) Prostaglandin E2 suppression of acetylcholine release from parasympathetic nerves innervating guinea-pig trachea by interacting with prostanoid receptors of the EP3-subtype. *Br J Pharmacol* 123: 1246–1252

138 Leff AR, Munz NM, Tallet J, Cavigelli M, David AC (1985) Augmentation of parasympathetic contraction in tracheal and bronchial airways by prostaglandins F2α *in vitro. J Appl Physiol* 58: 1558–1664

139 Tamaoki J, Sekizawa K, Graf PD, Nadel JA (1987) Cholinergic neuromodulation by prostaglandin D2 in canine airway smooth muscle. *J Appl Physiol* 63: 1396–1400

140 Chung KF, Evans TW, Graf PD, Nadel JA (1985) Modulation of cholinergic neuro-

transmission in canine airways by thromboxane mimetic U46619. *Eur J Pharmacol* 117: 373–375

141 Inoue T, Ito Y (1985) Pre- and post-junctional actions of prostaglandin I_2, carbocyclic thromboxane A_2 and leukotriene C_4 in dog tracheal tissue. *Br J Pharmacol* 84: 289–298

142 Takahashi T, Ward JK, Tadjkarimi S, Yacoub MH, Barnes PB, Belvisi MG (1995) 5-hydroxytryptamine facilitates cholinergic bronchoconstriction in human and guinea pig airways. *Am J Respir Crit Care Med* 152: 377–380

143 Olszewski MA, Robinson NE, Zhu FX, Zhang XY, Tithof PK (1999) Mediators of anaphylaxis but not activated neutrophils augment cholinergic responses of equine small airways. *Am J Physiol* 276: L522–L529

144 Inoue H, Aizawa H, Miyazaki N, Ikeda T, Shigematsu N (1991) Possible roles of the peripheral vagal nerve in histamine-induced bronchoconstriction in guinea-pigs. *Eur Respir J* 4: 860–866

145 Kikuchi Y, Okayama H, Ikayama M, Sasaki H, Takishima T (1984) Interaction between histamine and vagal stimulation on tracheal smooth muscle in dogs. *J Appl Physiol: Respirat Environ Exercise Physiol* 56: 590–595

146 Gustafsson LE, Wiklund NP, Cederqvist B (1986) Apparent enhancement of cholinergic transmission in rabbit bronchi via adenosine A2-receptors. *Eur J Pharmacol* 120: 179–185

147 Fernandes LB, Henry PJ, Rigby PJ, Goldie RG (1996) Endothelin(B) (ET(B)) receptor-activated potentiation of cholinergic nerve-mediated contraction in human bronchus. *Br J Pharmacol* 118: 1873–1874

148 Riccio MM, Reynolds CJ, Hay DWP, Proud D (1995) Effects of intranasal administration of endothelin-1 to allergic and nonallergic indivduals. *Am J Respir Crit Care Med* 152: 1757–1764

149 Barnes PJ (1997) *Autonomic control of the respiratory system*. Harwood Academic Publishers, Amsterdam, The Netherlands, 139–184

150 Altiere RJ, Szarek JL, Diamond L (1985) Neurally mediated non-adrenergic relaxation in cat airways occurs independent of cholinergic mechanisms. *J Pharmacol Exp Ther* 254: 590–597

151 Stretton CD, Belvisi MG, Barnes PJ (1990) Neuropeptide Y modulates non-adrenergic, non-cholinergic neural bronchoconstriction *in vivo* and *in vitro*. *Neuropeptides* 17: 163–170

Role of non-neuronal and neuronal acetylcholine in the airways

Ignaz K. Wessler[1] and Charles J. Kirkpatrick[2]

[1]Institutes of Pharmacology and [2]Pathology, University of Mainz, Obere Zahlbacher Str. 67, D-55101 Mainz, Germany

Introduction

It is well known that acetylcholine represents a dominant neurotransmitter within mammalian airways and that airway functions, like smooth muscle activity and secretion, are under a continuous cholinergic tone. However, the teleology of this basal cholinergic tone, assumed to originate from neuronal activity, appears difficult to understand, whereas neuronal cholinergic reflex activity can be regarded as a rational regulatory pathway to protect the airways from injury [1–3]. Based on recent experimental observations, both phenomena may reflect two different biological roles of acetylcholine, acting first as a universal cytomolecule (non-neuronal) and second as a classical neurotransmitter (neuronal).

The ubiquitous expression indicates that acetylcholine could be a global player in biological systems. The first experimental evidence for chemical neurotransmission was provided by the work of Otto Loewi in 1921, and five years later the vagus-substance "parasympathin" was identified as acetylcholine [4, 5]. However, one should consider that already more than ten years previously Dale and Ewins had provided the first evidence for the presence of acetylcholine in plants [6, 7]. While Dale was investigating the biological effects of ergot extracts, he attributed the depressor effect of this extract to contaminating acetylcholine. In the following seven decades, scientific research was mainly focussed on acetylcholine acting as a neurotransmitter, although increasing experimental evidence showed its expression in non-neuronal organisms and non-neuronal mammalian cells, i.e., in plants, sponges, unicellular organisms, *Planaria torva* and the human placenta, an organ not innervated by cholinergic neurons [8–16]. In 1963, Whittaker stated that "acetylcholine occurs in non-nervous tissues and is so widely distributed in nature to suggest a non-nervous function of it" [17]. Koelle, in the same year, speculated that acetylcholine represents a phylogenetically very old molecule which, in the most primitive structures like plants and unicellular organisms, might be involved in the regulation of transport processes across the cell membrane [18]. About 15 years

Muscarinic Receptors in Airways Diseases, edited by Johan Zaagsma, Herman Meurs and Ad F. Roffel
© 2001 Birkhäuser Verlag Basel/Switzerland

later the first systematic presentation of the expression of the non-neuronal cholinergic system in non-mammals and mammals was published by Sastry and Sadavongvivad [19].

Today it is increasingly recognized that acetylcholine represents an extremely old molecule on the evolutionary time scale (about 3 billion years [20]), is widely expressed in biological systems (bacteria, algae, yeast, plants, fungi, unicellular organisms, nematodes), can be demonstrated in more or less every human cell and, finally, is involved in the regulation of vital cell functions [20–24]. To focus on the early evolutionary event (absence of neuronal tissue) and to distinguish the "universal cell molecule" acetylcholine from the neuronal system the authors have introduced the terms "non-neuronal acetylcholine" and "non-neuronal cholinergic system" [21, 23, 24]. The present article is mainly focussed on the expression and function of non-neuronal acetylcholine within the airways. In the first section the main components of the cholinergic system are discussed with reference to their expression in non-neuronal human cells. The role of neuronal acetylcholine in the airways has been summarized repeatedly in previous articles [3, 25–29].

The cholinergic system

The synthesizing enzyme choline acetyltransferase (ChAT)

Acetylcholine is synthesized from acetylcoenzyme A (acetyl-CoA) and choline by choline acetyltransferase (ChAT), as first described by Nachmansohn and Machado more than 50 years ago [30].The K_m values for choline and acetylCoA are about 1 mM and 10 μM, respectively. The intracellular concentrations of free choline and acetyl-CoA are in the range of about 50 and 5 μM, respectively; i.e., they are considerably lower than the respective K_m values [31]. Thus, ChAT enzyme activity is far away from its V_{max}. The reaction catalysed by ChAT is reversible but its equilibrium is shifted to the right (K_{eq} 12.3). The expression of this enzyme had been used as the most specific marker to identify cholinergic neurons. However, ChAT immunoreactivity and ChAT enzyme activity are now demonstrated in many, if not all, human non-neuronal cells, as well as in non-neuronal animal cells (epithelial cells of human skin, airways, intestine, cornea, placenta, amnionic membrane of chick embryo, endothelial cells of rat brain capillaries, human circulating blood cells and immune cells). In view of these very convincing data, it is no longer justified to regard ChAT as a specific marker for cholinergic neurons.

It is generally accepted that the specificity of ChAT for choline is high, whereas the enzyme can use different acyl group donors. In the rat brain ChAT showed the same affinity for acetyl-CoA, propionyl-CoA and butyryl-CoA [32]. Thus, ChAT enzyme activity may synthesize not only acetylcholine but also other related molecules, for example, propionylcholine or butyrylcholine. Acetylcholine can also be synthesized by carnitine acetyltransferase which is expressed, for example, in mammalian heart and skeletal muscle [33, 34].

Several sources deliver choline for acetylcholine synthesis: dietary choline, synthesis in the liver (methylation of phosphatidylaminoethanol), breakdown of choline containing phospholipids (deacylation of phosphatidylcholine) and hydrolysis of acetylcholine [31, 35]. The choline sources are sufficient to maintain a continuous synthesis of acetylcholine in neuronal and non-neuronal cells. For example, the amount of choline "stored" in phospholipids (phosphatidylcholine) could balance an intracellular level of choline above 1 mM, when the membrane containing phosphatidylcholine would have become utilized.

Choline can be taken up by more or less all cells [36, 37]. The high affinity sodium-dependent uptake system showing a K_T between 0.5 and 2 µM appears specific for neurons [37–39], whereas non-neuronal cells (for example, muscle fibres, heart, erythrocytes, kidney) can take up choline by the low affinity sodium-dependent choline uptake (K_T above 20 µM). The low affinity choline uptake is less dependent on the sodium gradient than the high affinity uptake and exhibits a diminished affinity for the blocking compound hemicholinium-3 [37].

Acetyl groups for the synthesis of acetylcholine originate from pyruvate (glucose, lactate) and acetate by means of the intramitochondrial pyruvate dehydrogenase complex, forming acetyl-CoA. Acetyl-CoA represents an essential intermediate in the metabolism of carbohydrates, fatty acids and some amino acids and is expressed in every type of animal cells. Several mechanisms have been proposed to explain how acetyl-CoA synthesized in the inner mitochondrial matrix can pass the inner mitochondrial membrane [31]. Pyruvate-generated acetyl groups are translocated *via* citrate synthesized in the inner matrix and then *via* the tricarboxylate carrier to the cytosol, where acetyl-CoA can be formed from citrate and ATP *via* the ATP citrate lyase pathway [31]. In addition, the acetyl group carrier can be mediated *via* the transfer of acetylcarnitine and, possibly, *via* calcium-induced hydrophilic pores or channels in the inner mitochondrial membrane [31].

Every type of animal cell contains the two elementary precursors for acetylcholine synthesis in sufficient quantities. This fact implies that acetylcholine can be synthesized in all non-neuronal cells when the synthesizing enzyme ChAT is also present in an active form. As mentioned above, convincing experimental evidence has been amassed in the last five years that ChAT is expressed in the vast majority of human cells. In conclusion, more or less all cells fulfill the three crucial conditions for acetylcholine synthesis: presence of choline, presence of acetyl-CoA and presence of ChAT.

The human ChAT gene has been mapped to chromosome 10 [40–44]. A detailed analysis of the organization of the ChAT gene, particularly regarding the exon/intron boundaries, is available for three species (*Drosophila*, *Caenorhabditis elegans*, rat [43]). The rat ChAT gene consists of at least three 5' noncoding exons (R, N and M types) and 14 coding exons. Analysis of the first intron of the ChAT gene has revealed a 1590 frame encoding the vesicular acetylcholine transporter

(VAChT [43, 45–48]). This "cholinergic gene locus" encodes for the expression of the VAChT- and ChAT-protein and contains the regulatory sequences controlling their expression. In neuronal cells VAChT- and ChAT-expression may be co-regulated [43, 45–48], but less is known about the regulation of the "cholinergic gene locus" in non-neuronal cells. The complex organization of the "cholinergic gene locus" allows the expression of VAChT and ChAT in a cell-specific manner and leads to multiple mRNA transcripts. In particular, various ChAT variants have been identified so far. For example, in the rat at least seven different ChAT mRNA transcripts have been found [43, 44]. The exact number of ChAT isoenzymes is unknown. Western blot analysis of human airway epithelial cell extracts has visualized ChAT-like proteins with a molecular mass of 41 and 54 kDa [23], whereas the neuronal ChAT is a 69 kDa protein [43, 45]. A 51 kDa subunit of the neuronal ChAT protein has also been found and a 84 kDa ChAT protein is composed of 6 identical 14 kDa subunits [49].

Storage and release of acetylcholine

It is unknown whether plants, procaryotes and non-neuronal mammalian cells can store acetylcholine and if so, what subcellular organelles/compartments are involved. One possibility might be that acetylcholine is continuously synthesized within the cell cytosol. An equilibrium may exist between synthesis, liberation into the extracellular space and hydrolysis which occurs intra- and extracellularly, because cholinesterase activity is present in both compartments. Non-neuronal cells may store acetylcholine in subcellular structures unknown so far (endosomes, endoplasmic reticulum, secretory granules) from which acetylcholine may be translocated either into other subcellular compartments or into the extracellular space. In a recent publication, Schäfer and co-workers have mentioned an artefactual staining of granulocytes in the rat small intestine when exposed to an antibody directed against the VAChT protein [50]. It might be possible that granulocytes are endowed with VAChT protein. Effector cells such as human granulocytes, mononuclear cells, macrophages and also human leukemic T-cell lines, contain acetylcholine [51–53]. Human granulocytes express very high ChAT activity (about 0.4 µmol/mg protein/h) in the particulate, i.e., the membrane-bound fraction [53]. In addition, granulocytes may express VAChT protein involved in translocating and storing acetylcholine. In contrast, human circulating mononuclear leukocytes do not express VChT protein, as has been demonstrated by reverse transcriptase polymerase chain reaction (RT-PCR) technique and by Southern blot using specific VChT probes [54]. Very recently, the expression of ChAT and VAChT protein has been reported in human granulosa-luteal cells [55].

Neurons have developed highly improved structures and mechanisms of accumulating, storing and releasing acetylcholine. These structures are the VAChT protein [56], the cholinergic vesicles and the structures involved in vesicular exocytosis. Neuronal acetylcholine is taken up by VAChT, which utilizes a H^+/antiport mechanism for substrate accumulation within the cholinergic vesicles [46, 48, 56, 57]. Membrane-bound (vesicular membrane: synaptophysin, synaptotagmin, synaptobrevin; plasma membrane: neurexins, syntaxin, SNAP-25) as well as soluble proteins (N-methylmaleimide-sensitive factor [NSF], soluble NSF attachment protein (SNAPs)) participate in the process of vesicular exocytosis when a transient increase in cytosolic free calcium occurs [58]. This cascade-like activation of membrane-bound and soluble factors mediates the vesicular release of neuronal acetylcholine in a highly synchronized fashion [58–62]. Thus, threshold levels of neuronal acetylcholine can be established within milliseconds in the synaptic cleft or at neuroeffector junctions. Upregulated expression of acetylcholine-sensitive receptors (for example, at ganglion cell bodies or the postsynaptic motor endplate) and of cholinesterase have additionally optimized the conditions for cholinergic neurotransmission. This organization allows cross-talk between cells to be encoded in a frequency-dependent manner, forming a highly sophisticated communicating pathway.

In contrast, the cellular events involved in the release of non-neuronal acetylcholine from human epithelial cells (skin, intestine), for example, are not known in detail. Nevertheless, it has been repeatedly demonstrated that acetylcholine can leave non-neuronal cells. For example, fibroblasts transfected with the ChAT gene can release acetylcholine in response to a transient increase in intracellular calcium [63, 64]. Epithelial cells of the human placenta release acetylcholine into the extracellular space [65], and the basal release of acetylcholine depends on extracellular calcium [66]. This observation indicates that the basal release of non-neuronal acetylcholine does not simply represent leakage. Nicotine can profoundly stimulate the release of acetylcholine from the human skin [67]. Moreover, stimulation of β-adrenoceptors can increase the release of acetylcholine from human epithelial cells (Wessler et al., unpublished observation). In equine airways, β-adrenoceptor agonists facilitate the electrically stimulated release of acetylcholine [68].

Non-neuronal acetylcholine probably leaves the cells by membrane-bound "gates" or so-called "mediatophores" [69, 70]. In contrast to cholinergic neurons, the non-neuronal cells are not specialized to trigger a rapidly synchronized and high quantity release, but appear to release the cytomolecule acetylcholine continuously in minute quantities.

Receptors at target cells (non-neuronal and neuronal cells)

Functionally active nicotine receptors consisting of the α3-, α5-, β2-and β4-subunits have recently been demonstrated on human airway epithelial cells [71]. These cells also express muscarinic receptors ([72]; see also Fig. 1). Nicotinic and muscarinic receptors are also widely expressed on other non-neuronal cells, for example intestinal and skin epithelial cells, endothelial cells, placenta and circulating blood cells [51, 73–85]. A chimeric nicotinic receptor responding to both nicotinic and muscarinic agonists has been discovered in insect and bovine neurosecretory cells [86, 87] as well as in human keratinocytes [21]. Human skin epithelial cells (keratinocytes) also express the α3-, α5-, and α7-subunits of the nicotinic receptor complex and all subtypes of muscarinic receptors (M1–M5) [75, 76]. Thus, the classical cholinergic receptors represent the target for non-neuronal acetylcholine to mediate auto-/paracrine effects.

In addition to the location on airway epithelial cells, muscarinic receptors have been directly or indirectly detected on airway smooth muscle fibres, glandular tissue and neurons [88–90]. Nicotine receptors are located on intramural ganglion cells and involved in ganglionic transmission. The chapter by Roffel et al. in this volume addresses the expression of muscarinic receptors in the airways; this topic will therefore not be discussed here in more detail. It is evident that these receptors on ganglion cells, axons and effector cells are stimulated by neuronally released acetylcholine.

Signal transduction *via* nicotinic and muscarinic receptors exerts an enormous impact on a cell, because ion channels as well as the key enzymes forming second and third effectors are activated. Thus, nicotinic and muscarinic receptor activation can trigger multiple signalling events: the activation of transmembrane ion flux, the mobilization of intracellular calcium, the increase in cyclic guanine monophosphate (cGMP), the release of nitric oxide (NO), prostanoids and other modulators and the activation of tyrosine kinases, small G-proteins and mitogen-activated protein kinases. Consequently, basic cell functions, such as gene expression, mitogenesis, cell differentiation, cytoskeletal organization, secretion, activity of channels and transporters, are within the scope of biological reactions which could be modulated *via* acetylcholine-sensitive pathways. Nicotinic receptors expressed on airway epithelial cells mediate depolarization and their ion-gating properties resemble ganglionic nicotinic receptors [71]. In non-excitable cells, such as epithelial cells and fibroblasts, muscarinic receptors interact with the three known voltage-independent calcium channels [91]. Muscarinic receptors of the m1-subtype appear to directly activate voltage-independent calcium channels on T lymphocytes [92].

Inactivating enzymes

In vertebrates, acetylcholine is destroyed by acetylcholinesterase (cholinesterase; AChE) and butyrylcholinesterase (BChE). AChE is encoded by a single gene, but alternative splicing generates multiple transcripts. AChE is ubiquitously expressed in, for example, fibroblasts, the non-innervated part of muscle fibres, circulating cells (erythrocytes) and the liver. Erythrocytes are heavily packed with AChE [93]. BChE is found in the liver, lung, smooth muscle and the circulation [94–96]. Erythrocytes, together with plasma cholinesterase (BChE), are very effective humoral mechanisms to eliminate non-neuronal acetylcholine that has escaped into the circulation. This widespread hydrolyzing activity prevents non-neuronal acetylcholine from acting as a hormone; i.e., the effect of non-neuronal acetylcholine is closely limited to the area of its synthesis and release. Finally, it should be considered that specific cholinesterase shows the highest substrate turnover rate of all enzymes characterized so far in biological systems. This outstanding property of cholinesterase limits the action of non-neuronal acetylcholine within the microenvironment of a cell.

Non-neuronal ChAT and acetylcholine in the human lung

Epithelial cells

Within the last 5 years, increasing experimental evidence has demonstrated a wide distribution of ChAT protein in non-neuronal cells of human airways (for example, bronchi with an inner diameter between 3–10 mm, i.e., the fourth to seventh order). ChAT expression was demonstrated by anti-ChAT immunohistochemistry including an immunofluorescence technique, by measuring ChAT enzyme activity in isolated cells and by visualization of ChAT proteins by means of Western blot [20, 23, 24, 97, 98]. Prominent examples of these findings are given in Figure 2. Positive anti-ChAT immunoreactivity was found in airway surface epithelial cells and in epithelial cells of submucosal glands (Fig. 2A, B, C). Freshly isolated as well as cultured (24 h) human airway epithelial cells showed ChAT enzyme activity and contained acetylcholine [23, 24]. Exposure of intact bronchi or of the freshly isolated airway epithelial cells to mono- or polyclonal anti-ChAT antibodies resulted in the staining of the vast majority of the surface cells (Fig. 2A, B, C). Thus, not only ciliated cells but also secretory, basal and brush border cells appear to express ChAT protein. In cytospin experiments it was found that human bronchial ciliated cells expressed ChAT immunoreactivity most prominently at the basal body and rootlet, where the cilia are embedded (Fig. 2C). All observations together demonstrate that human airway epithelial cells synthesize and contain acetylcholine [20, 23, 24].

A comparison of the ChAT enzyme activity and acetylcholine content detected in the surface epithelium and the whole bronchial wall is given in Table 1. ChAT enzyme activity was obviously identical in both compartments but acetylcholine content differed substantially; i.e., epithelial acetylcholine represents less than 2% of transmural acetylcholine. Most likely this discrepancy reflects the different storing potency of non-neuronal *versus* neuronal acetylcholine. Cholinergic neurons are located within the lamina propria and the submucosa and these neurons can concentrate acetylcholine in their vesicles by a factor of more than 1000.

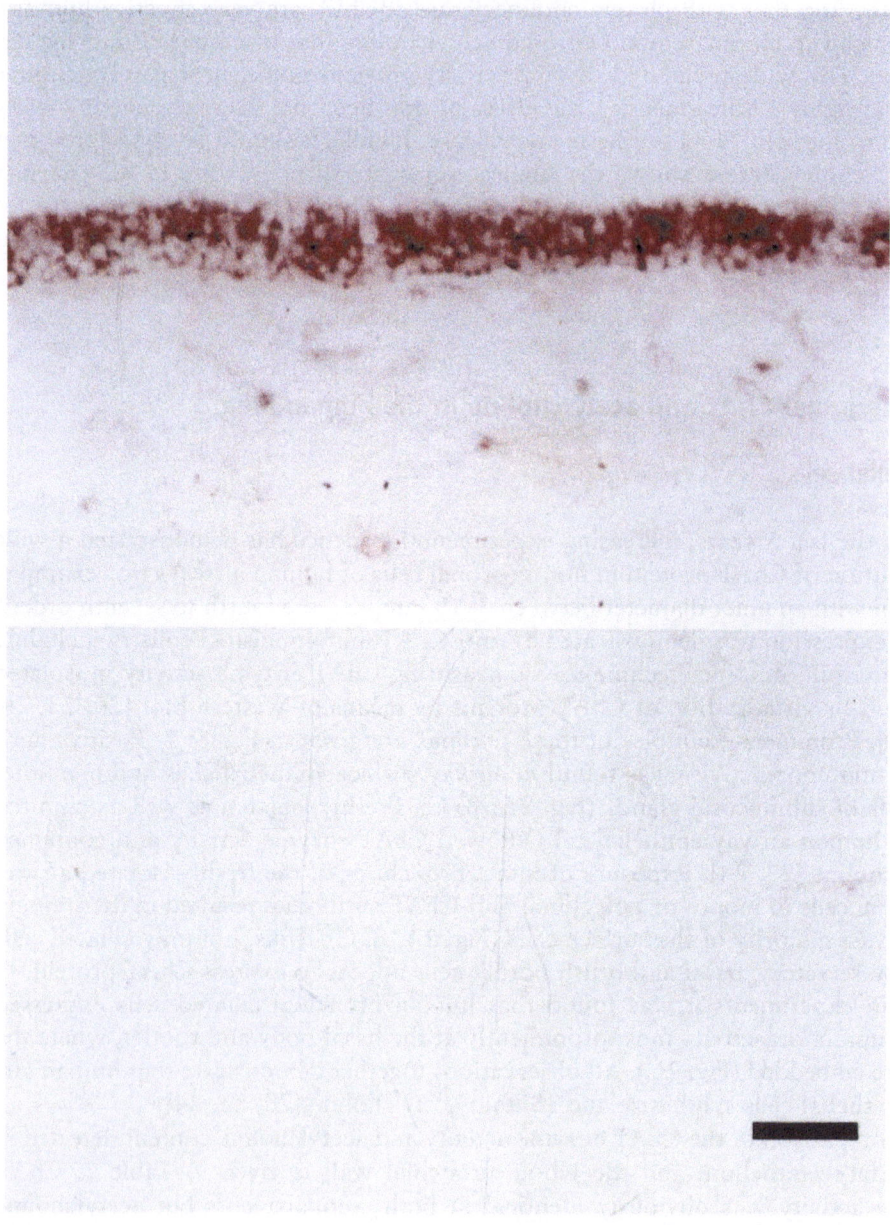

Figure 1
Expression of M1-muscarinic receptor mRNA in human bronchial epithelial cells
Isolated human bronchi were snap frozen and in situ hybridization was applied by using cDNA probes specific for M1 muscarinic receptor. A: M1-anti-sense probe. B: m1-sense probe (magnification × 200; bar represents 100 μm).

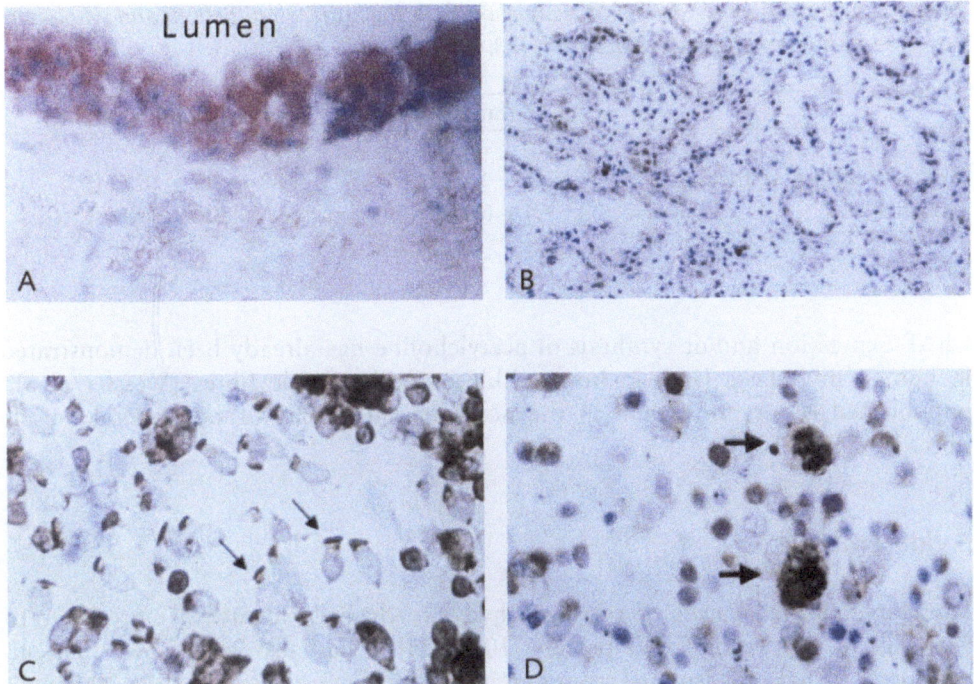

Figure 2
Positive ChAT-immunoreactivity in human airways
A: human bronchus (primary antibody (ab): polyclonal rabbit anti-ChAT; secondary ab: anti-rabbit alkaline phosphatase anti-alkaline phosphate conjugate; positive pink staining; magnification x200); B: submucosal glandular epithelial cells in human bronchi (primary ab: monoclonal mouse anti-ChAT; secondary ab: anti-mouse peroxidase conjugate; positive brown staining; magnification × 200); C: human isolated bronchial ciliated cells (arrows; primary ab: polyclonal rabbit anti-ChAT; secondary ab: anti-rabbit peroxidase conjugate; positive brown staining, magnification × 1000); D: isolated human bronchial cells with alveolar macrophages (arrows; primary ab: polyclonal rabbit anti-ChAT; secondary ab: anti-rabbit peroxidase conjugate; positive brown staining; magnification × 1000).

Smooth muscle cells

In addition to ChAT expression in surface epithelial cells, positive ChAT immunoreactivity was found in secretory (Fig. 2B) and in airway smooth muscle cells (Fig. 3). Anti-ChAT immunofluorescence technique demonstrated that the muscle fibres themselves expressed ChAT protein (Fig. 3). This is not a surprising finding since

Table 1 - Comparison of ChAT and acetylcholine in the surface epithelium and whole airway wall of human bronchi (values from [23, 188])

	ChAT (nmol/mg/h)	Acetylcholine (pmol/g)
bronchial surface epithelium	3.5 ± 1.3 (5)	33 ± 10 (14)
bronchial wall	3.1 ± 0.4 (8)	2500 ± 300 (8)

ChAT expression and/or synthesis of acetylcholine has already been demonstrated in human myoblasts [99], in human skin smooth muscle fibres (Wessler et al., unpublished observation) and in the non-innervated part of rat skeletal muscle fibres [34, 100, 101].

Endothelial cells

Vascular endothelial cells of rat brain had already been identified more than 15 years ago as expressing ChAT protein [102, 103]. It had also been shown that cultured endothelial cells isolated from porcine cerebral microvessels synthesize acetylcholine [104], and recently positive ChAT-immunoreactivity has been detected in endothelial cells of human skin vessels [105]. So far, direct experimental evidence has not been found that endothelial cells in lung tissue expresses the non-neuronal cholinergic system, but it appears to be most likely.

Mesothelial cells

In addition to epithelial surface cells, mesothelial surface cells also express ChAT and synthesize acetylcholine. For example, acetylcholine was detected in wisp samples taken from the human pulmonary pleural surface [23]. Both sheets of the rat pleura contained acetylcholine (Wessler et al., unpublished observation) and strong ChAT-immunoreactivity was found in mesothelial cells of the human pericardium [20, 24].

Immune cells

In addition to the expression of ChAT in surface epithelial, secretory and smooth muscle cells, ChAT expression was also demonstrated in immune cells which invade the airway muscosa/submucosa. Thus, for example, lymphocytes, alveolar macrophages (Fig. 2D) and eosinophils can synthesize acetylcholine [20, 51, 52, 54, 106] and may release inappropriate amounts of acetylcholine upon chronic activation. Acetylcholine can stimulate the activity of immune cells like lymphocytes [107–109] and thereby may modulate not only host defence mechanisms, but also inflammatory cytotoxicity.

Neuronal ChAT and acetylcholine in the mammalian lung

Vagus nerve and the intramural plexus

Mammalian airways receive their dominant afferent and efferent innervation from the vagus nerve. Most of the sensory afferent nerve fibres run *via* the vagus nerve to the brainstem. Cholinergic efferent nerves originate in the brainstem (motor nucleus of the vagus, nucleus ambiguus). Branches of the recurrent laryngeal and vagus nerves pass to the trachea and lung hila (bronchial branches) and synapse to the intramural multipolar ganglia which are located within the airway wall. Thus, the airway wall contains a complex neuronal network (ganglia, interneurons, postganglionic neurons) comparable to the small intestine [110, 111]. Within the central airways the ganglia are mainly placed in the posterior wall, but the location of ganglia varies in the smaller intrapulmonary bronchi. The neuronal plexus is composed of neurons exhibiting different electrical properties [112] and containing different transmitters/cotransmitters/modulators: acetylcholine, vasoactive intestinal polypeptide (VIP), peptide histidine isoleucine, galanin, tachykinines (substance P, neurokinine A, B), noradrenaline (very scanty in human airways), neuropeptide Y, somatostatin, enkephalins, NO and some other neuropeptides [113]. In guinea-pig airways the cell bodies of the intramural intrinsic neurons were devoid of positive VIP- and NO-immunoreactivity, but more or less all ganglia showed ChAT immunoreactivity [114, 115]. In contrast, in human airways, intrinsic ganglia co-expressed ChAT-, VIP- and NO-immunoreactivity, i.e., the innervation pattern of bronchoconstrictor and bronchodilator pathways was not separated anatomically [114].

The view is generally accepted that the parasympathetic nervous system represents the dominant neuronal bronchoconstrictor in animals and in humans. Efferent cholinergic neurons innervate the airway smooth muscle, mucous glands and pulmonary vessels [26, 29, 110, 111, 113, 116–118]. Experiments with electrical stimulation of the cervical vagus (cat, dog) and rapid lung freezing technique have revealed that the respiratory bronchioles and alveolar ducts were not constricted; i.e., extrinsic parasympathetic motor control does not include the terminal airways [119].

The vast majority of afferent sensory C-fibres originate as free endings within the airway wall (surface epithelium, submucosa, smooth muscle fibres) and lung parenchyma, run mainly *via* the vagus nerve to the ganglium nodosum and the medulla oblongata [25, 120]. Some afferent nerves show a non-vagal course and ter-

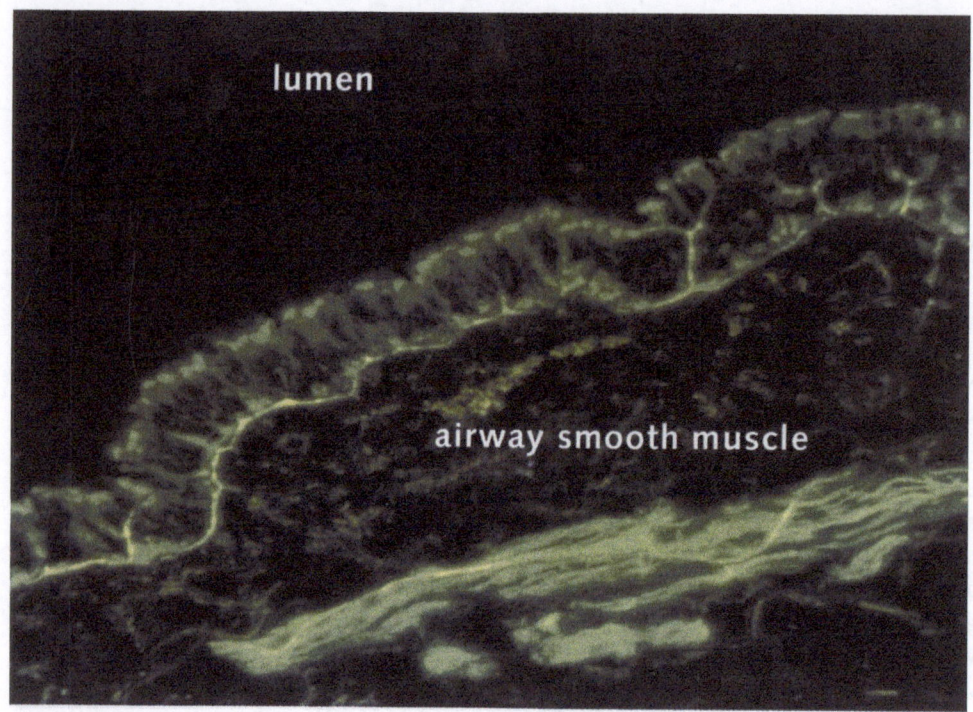

Figure 3
Positive anti-ChAT immunofluorescence (polyclonal rabbit anti-ChAT ab) in human bronchial airway smooth muscle (magnification ×200).

minate *via* dorsal roots (T1–T6) and sympathetic cervical nerves in the medulla oblongata [113, 121]. The afferent system is strongly involved in triggering reflex activity (chemoreceptors, mechanoreceptors, stretch receptors) and a retrograde release of proinflammatory neuropeptides. The efferent vagal neurons represent the motor innervation of the intramural plexus located within the airway wall. The intramural plexus, which has been regarded as a "mini brain" [29], consists of a highly complex network which integrates efferent and afferent reflex activity to maintain organ homeostasis. An exact quantification of the percentage of cholinergic neurons present in the airway wall does not exist, but cholinergic neurons (interneurons, postganglionic neurons) appear to dominate.

The co-transmitters/modulators are colocalized together with the classical transmitters in vagal (afferent, efferent) and sympathetic (few afferent and mainly efferent fibres) neurons which most likely represent the morphological counterpart of the non-adrenergic non-cholinergic (NANC) system. The exact organization of the

NANC system is still a matter of some debate, but it appears very unlikely that the NANC system is established by a separate, third neuronal network. VIP and NO are the most likely candidates to mediate inhibitory NANC responses (bronchodilation) and neuropeptides (neurokinins, calcitonin gene-related peptide (CGRP)) mediate excitatory NANC responses (bronchoconstriction, mucus secretion, enhanced microvascular leakage).

Neuronal acetylcholine is released from innervating extrinsic vagal neurons, from interneurons and short postganglionic intrinsic neurons. Neuronal acetylcholine released from extrinsic neurons and interneurons stimulates the activity of the intramural neurons *via* activation of nicotinic and muscarinic receptors (M1-subtype) [90]. In rat and guinea-pig trachea, release of newly synthesized radioactive acetylcholine evoked by stimulation of the preganglionic parasympathetic nerves was reduced to about 15% in the presence of hexamethonium or tubocurarine, which block nicotinic transmission [122]. This observation suggests that about 85% of the stimulated release of newly synthesized neuronal acetylcholine originates from interneurons and postganglionic cholinergic neurons and the rest from extrinsic vagal neurons.

Cholinergic neuroeffector junctions in the airway

Airway smooth muscle
Neuronal acetylcholine released from the postganglionic axons stimulates smooth muscle fibres and glands *via* activation of muscarinic receptors mainly of the M3-subtype. The pattern of neuronal control of airway smooth muscle differs between large and smaller airways and also between different species. The human trachealis muscle contains postganglionic axons whereas ganglia have not been found. The axons run parallel to the muscle fibres and the predominant type of varicosities are filled with small agranular vesicles (cholinergic), the closest contact between muscle membrane and varicosity varying between 100 and several 1000 nm in distance [123]. A substantial number of gap junctions is visible and the authors concluded that the human trachealis muscle may be controlled by a major myogenic mechanism [123]. The innervation of the smooth muscle in smaller bronchi (fourth to seventh order) is ten times more dense than that in the trachea [123], leading the authors to postulate a more neuronal control of the smooth muscles in the smaller than in the larger airways. Acetylcholine tissue content, however, did not differ between large (6–12 mm) and smaller (3–6 mm) human bronchi [124]. To explain this discrepancy it is suggested that the smaller airways may contain fewer intramural cholinergic ganglia or have a less dense cholinergic innervation of the glandular tissue than the central airways.

The cholinergic neuroeffector junctions at the airway smooth muscle do not represent specialized transmitting synapses. In humans airways most of the cholinergic varicosities terminate at a distance more than 1 μm away from the muscle membrane [123]. A similar distance has been reported for sympathetic neurons; i.e., the distance between adrenergic varicosities and vascular smooth muscle in the pulmonary artery is about 2 μm [125]. For comparison the synaptic cleft of the motor endplate, a specialized synapse at the skeletal muscle, is only 40 nm in width. The distance between cholinergic varicosities and their muscular target membrane in the airways is definitely larger than the distance between two neighbouring epithelial cells. Therefore it appears most likely that epithelial acetylcholine can mediate not only auto- but also paracrine effects, although one has to consider the different quantities of acetylcholine released. Thus, synchronized release of neuronal acetylcholine packed in vesicles can establish acetylcholine concentrations within the millimolar range, whereas only minute amounts of acetylcholine are released from non-neuronal cells.

Acetylcholine released from intrinsic cholinergic neurons causes depolarization and contraction of the smooth muscle and therewith bronchoconstriction. Acetylcholine-induced depolarization of airway smooth muscle is mediated by multiple cellular mechanisms. Binding of acetylcholine at M_3-receptors mediates phospholipase C activation *via* trimeric pertussis-toxin-sensitive G_p protein. Inositol-1,4,5-triphosphate (IP3)-mediated calcium release from internal calcium stores triggers the first component of muscle contraction [126, 127] but muscle membrane depolarization and contraction is sustained by several additional mechanisms [128]. An increase in a non-specific cation and chloride conduction, an increased entry of extracellular calcium through L-type calcium channels and a rapidly inactivating potassium conduction contribute to the acetylcholine-induced depolarization and contractile response [128]. Mechanisms which limit one of the components of the electric-contraction (ec) coupling may become useful in the treatment of obstructive airway disease, as more or less all spasmogenic mediators appear to operate in a comparable way to acetylcholine. It remains to be elucidated whether non-neuronal acetylcholine is involved in ec-coupling.

Secretory cells in the airway surface epithelium

Secretory cells are localized in the surface epithelium and submucosa. Surface secretory cells are goblet cells (serous and mucous) and Clara cells, the latter located mainly in the smaller airways [129]. Stimulation of the cut ends of the innervating parasympathetic nerve can cause mucus production. In cats the stimulated mucus originates mainly from submucosal glands [113], whereas in the goose epithelial mucus production has been observed. In principle, epithelial mucus secretion evoked by electrical stimulation of the vagus nerve can be mediated by efferent cholinergic innervation of the surface epithelium (rat trachea) or by antidromic activation of sensory afferents. In humans the surface epithelium does not contain varicosities filled with characteristic agranular 30-40 nm large cholinergic vesicles; i.e., the human surface epithelium is not innervated by efferent cholinergic neurons [26, 113, 130]. The mouse airway epithelium is also devoid of neuronal innervation, which may be related to the absence of cough reflex in this species [131]. Finally, the guinea-pig airway epithelium also appears to lack efferent innervation by cholinergic nerve fibres [115]. In contrast, the rat tracheal surface epithelium receives both efferent cholinergic and afferent vagal innervation [113, 116]. In agreement with this observation it has been shown that the surface epithelium of the rat trachea contains about 100-fold more acetylcholine than the epithelium of human bronchi or guinea-pig trachea [23, 132]. Thus, in rat airways, vagally induced epithelial mucus secretion can be mediated *via* innervating cholinergic efferent neurons. In human airways, cholinergic mediated epithelial mucus production may be triggered by neuronal acetylcholine originating from the subepithelial lamina propria which contains cholinergic varicosities and/or by an auto-/paracrine action of non-neuronal acetylcholine released from epithelial cells.

Submucosal glands

The submucosal glands (serous, mucous and duct cells) receive a dense cholinergic innervation in rabbit, dog and human airways [113]. Fine efferent cholinergic fibres are located within the gland acini. So far it is not clear whether serous or mucous cells differ in their pattern of cholinergic innervation. In close proximity to the submucosal glands, VIP-immunoreactivity has also been detected [113], which appears to be colocalized with neuronal acetylcholine and could modulate mucus secretion [133–135]. Efferent cholinergic innervation of the submucosal glands is involved in transmitting reflex activity when appropriate stimuli are applied to the nasal or nasopharyngeal surface [120, 136].

A baseline mucus secretion has been observed in different species and appears to be independent of neuronal input. This baseline activity may be mediated *via* auto-/paracrine effects of non-neuronal acetylcholine, which can be synthesized by the glandular cells because these cells express positive ChAT-immunoreactivity (Fig. 2B). For comparison, the eccrine sweat glands of human skin also show positive ChAT-immunoreactivity [105, 137].

Lung vascular tissue

Our present knowledge of the neuronal afferent and efferent innervation of bronchial and pulmonary vessels is still scanty. Innervation pattern and innervation density is species-dependent. Parasympathetic innervation appears very spare in contrast to the sympathetic innervation. In most species the innervation of pulmonary vasculature is less dense than that of bronchial vasculature, and veins receive a less dense innervation than arteries [138]. Cholinesterase staining technique, which is, however, not specific for cholinergic neurons, has given evidence of a fine network of fibres which surround the pulmonary vessels on the outer surface of the media [139, 140]. Applied acetylcholine causes substantial vasodilation, most likely by triggering the release of NO.

Taken together, neuronal acetylcholine released in response to vagus and intramural neuronal activity stimulates airway smooth muscle fibres and glandular cells. In human airways the surface epithelium does not receive efferent cholinergic innervation, thus surface secretory and ciliary cells are not directly controlled by neuronal acetylcholine.

Functional aspects of non-neuronal acetylcholine

Proliferation

Acetylcholine applied to primary cultures of human bronchial epithelial cells increased the proliferation rate in a concentration-related manner, whereby concentrations as low as 1 nM were effective [23]. The proliferative effect of acetylcholine could be prevented by blocking nicotinic and muscarinic receptors, and also by the ChAT-enzyme inhibitor bromoacetylcholine. These observations suggest that acetylcholine is involved in the regulation of the turnover of epithelial cells by an auto-/paracrine mode of action. Likewise, Grando and colleagues have demonstrated that in human epidermis, acetylcholine and nicotine promote proliferation and terminal differentiation to the different subtypes of keratinocytes and that acetylcholine is involved in wound healing [73–75]. Nicotine has also been found to facilitate cell proliferation in small cell lung carcinoma cell lines by stimulating the mitogen-activated protein kinase [141]. In other cell types also a proliferative activity of acetylcholine has been reported [142, 143].

A continuous shedding of surface epithelial cells exists under physiological conditions. The basal cell turnover balances this cell loss (less than 1% appear to proliferate within 24 h) [144]. Under pathological conditions (chronic airway inflammation, exposure to toxic material), cell proliferation rate increases substantially, so that about 17% of the surface cells show proliferation [144]. The enhanced cell turnover may possibly be linked to the proliferative property of acetylcholine, which is comparable to the role of acetylcholine in skin wound healing [21]. Taken together, epithelial acetylcholine may be involved in the regulation of epithelial turnover and repair mechanisms which contribute to the homeostasis of the airway epithelium.

Secretion/absorption of ions, water and mucus

Airway surface epithelium is exposed to a daily air volume of roughly 10,000 liters. This volume contains numerous inorganic and organic toxic and contaminating microbiological material. Secretion of ions, water and macromolecules as well as mucociliary clearance are very effective mechanisms in the local defense system. Secretion of chloride and reabsorption of sodium regulate the quantity of the airway tract fluid and therewith the efficiency of the mucociliary clearance [135, 144–147]. Ion channels, co-transporters and pumps at the basolateral and apical membrane of epithelial cells are involved in electrolyte and water transport. In order to understand a possible role of non-neuronal acetylcholine in the regulation of airway surface secretion, some basic mechanisms are summarized first.

At the basolateral membrane Na^+/K^+-ATPase maintains a sodium gradient which allows accumulation of Cl^- *via* a co-transporter most likely of the $Na^+/K^+/2Cl^-$ cotransporter. The basolateral membrane contains additional K^+ channels. Basolateral and apical ion transport is regulated in a coordinated manner. Stimulation of apical chloride secretion is accompanied by the activation of basolateral K^+ channels which can balance the raised intracellular K^+ mediated by increased activity of Na^+/K^+-ATPase and Cl^- cotransport [143]. Additionally, opening of basolateral K^+ channels causes hyperpolarization and thereby maintains the driving force for apical Cl^--secretion. Many secretagogues (β-adrenoceptor agonist, prostanoids, activators of protein kinase A or C (PKA, PKC)) which stimulate apical Cl^--secretion inhibit Na^+-reabsorption either directly (cAMP, increase in intracellular Ca) or *via* the reduction of the electrochemical gradient for Na^+ [146].

Some experimental evidence suggests that non-neuronal, epithelial acetylcholine may be directly involved in the regulation of described ion and water transport. In experiments with sheep tracheal epithelium it has been demonstrated that applied acetylcholine stimulates apical Cl^- channels and basolateral K^+ channels [148]. Moreover, acetylcholine suppresses net Na^+ absorption by decreasing the mucosal to serosal Na^+ flux, i.e., by inhibiting apical Na^+ channels [148]. These observations of acetylcholine-mediated effects on airway epithelial cells are not surprising. Similar effects have already been found in the gastrointestinal tract. Applied acetylcholine increases the short circuit current *via* the activation of Cl^-- and Na^+-secretion [149, 150]. Further references to a stimulatory effect of applied acetylcholine on salt-secreting cells are presented elsewhere [24]. Moreover, acetylcholine can increase potassium conductance *via* the activation of distinct K^+ channels. For example, in the *Necturus* gall bladder epithelium, applied acetylcholine causes hyperpolarization of the basolateral membrane *via* the stimulation of muscarinic receptors [151]. The mammalian heart (atrioventricular node) represents an exemplary tissue where acetylcholine increases potassium conductance.

Transepithelial transport of ions, water and macromolecules can, in addition to the transcellular route, be maintained *via* the paracellular pathway. This latter pathway is regulated by the activity of the tight junctions. Acetylcholine may also affect this target; for example, acetylcholine has already been found to increase tight junction permeability in the rabbit pancreas [152].

Mucus secretion from goblet cells and submucosal glands (mucous or serous cells) can be triggered by vagal nerve stimulation in some species, guinea-pig airways for example [153]. Thus, submucosal and epithelial secretory glands receive efferent cholinergic innervation in this species, although experimental evidence in favor of epithelial cholinergic varicosities is lacking in this species [115]. Alternatively, goblet cells may be stimulated by neuronal acetylcholine released within the lamina propria in close proximity to the surface epithelium. Hence, the cholinergic neuronal pathway can exert a strong regulatory function on the activity of the airway glands. Nevertheless, one cannot exclude that non-neuronal acetylcholine also affects mucus secretion. It has been shown that brief 30 s pulses of 30 nM acetylcholine cause transient glycoconjugate secretion from swine isolated submucosal glands [154]. In humans, topical nasal application of metacholine causes a complex secretory response (fluid, protein, lysozyme, lactoferrin) [155]. Submucosal serous cells of the ferret trachea, when exposed to metacholine, established intracellular vacuoles, which is indicative of ion and water transport [156]. It is possible that non-neuronal acetylcholine may target mucus secretion directly by an auto-/paracrine mechanism, in addition to the neuronal pathway.

In conclusion, non-neuronal acetylcholine may be directly involved in the regulation or fine-tuning of apical Cl^--secretion and Na^+-absorption/secretion by an auto-/paracrine mode of action. Thus epithelial acetylcholine may regulate the degree of water secretion/ absorption, the quantity/quality of mucus secretion and the efficiency of mucociliary clearance.

Mucociliary clearance

The airways are covered by a surface liquid which consists of two layers, the periciliary fluid and surface mucus layer. Mucociliary clearance represents a highly sophisticated mechanism to protect the airways against microbiological and toxic material. Mucociliary clearance depends on a synchronized ciliary activity (interaction between neighbouring ciliated cells), on the secretion of water, ions and mucus (cilia-mucus coupling) and on an appropriate organization of epithelial cell-cell contact (tight-junctions, gap junctions) [144]. Establishment of the physical barrier by closing tight-junctions and the increase in ciliary beating and reflex mucus secretion with the liberation of highly glucosylated proteins, which simulate the glycocalyx structure of epithelial cells to facilitate bacterial binding, contribute to the local defence mechanisms to protect against the invasion of infectious material [144].

It has been known for decades that acetylcholine affects ciliary and secretory activity. Evidence in favour of non-neuronal acetylcholine expressed in ciliated tissue comes from experiments with protozoa (*Paramecium caudatus*) or with the gill plates of the sea mussel, *Mytilis edulis* [9, 157]. Applied acetylcholine and physostigmine accelerated the ciliary movements [157]. From these observations the authors concluded that acetylcholine can be synthesized in a nerve-free and muscle-free tissue, and that acetylcholine is involved in controlling ciliary activity. In the rabbit trachea and in the mucous membrane of frog oesophagus, a low concentration of acetylcholine or of the cholinesterase inhibitor physostigmine increased and high concentrations diminished ciliary activity [158].

In human bronchial biopsies, applied agonists or antagonists at muscarinic receptors produce a stimulatory or inhibitory response on ciliary activity, respectively [159, 160]. Ciliary activity in mammalian airways can be reduced by atropine, but a basal activity still remains [19, 144]. This observation suggests that ciliary activity is generated by some basic mechanisms and additionally appears to be regulated by receptor-controlled pathways [161]. Meanwhile the expression of ChAT protein and the synthesis of acetylcholine have been demonstrated directly in isolated ciliated epithelial cells of human bronchi [23, 24, 97]. ChAT immunoreactivity was concentrated at the transition zone between the free cilia and the basal body in which the cilia are embedded (Fig. 2C) [97]. These histological findings suggest that epithelial acetylcholine is involved in initiating and maintaining the contractile events of ciliary motion.

Even at the present time the cellular events of initiating and maintaining synchronized ciliary movements within an individual cell and the coordinated activation of neighbouring ciliated cells remain obscure. Dynein and kinesin represent motor proteins (ATPase) which appear to play a key role in triggering ciliary activity. Upon activation and energy generation by ATP hydrolysis, the motor proteins induce bridge formation of the doublet microtubules and relative sliding of microtubules, comparable to actin-myosin interaction and myosin powerstroke in skeletal muscle fibres [144]. The powerstroke of the myosin head causes movement of the thin actin filament relative to the site of attachment and the repetition of the cycles mediates filament sliding, i.e., the contractile response. Similar motor proteins and microtubular filaments are involved in the organization of the cytoskeleton [162, 163]. It has been shown that the cytoskeleton is controlled by non-neuronal acetylcholine (see next paragraph) and, correspondingly, one might assume that the basic events of triggering ciliary activity are also linked to non-neuronal acetylcholine.

Airway cilia beat in a coordinated fashion; i.e., cell-cell interactions allow a coordinated movement of the ciliary activity [144]. Cilia still maintain operation in a coordinated fashion even when a 10 µm gap is interposed between neighbouring cells [164]. Wanner and colleagues have recently proposed that gap junctional communication and extracellular paracrine factors are responsible for triggering synchronization of cilia beats and of cilia-mucus coupling [144]. It is proposed here that cilia-mucus coupling and the coordinated transmission of ciliary activity to the neighbouring cells may be mediated *via* epithelial acetylcholine which can act intra- and extracellularly as a local signalling molecule to stimulate neighbouring ciliated cells, as well as fluid secretion and glandular activity. For example, in the intestine, gap junctional communication can be modulated *via* M_1-muscarinic receptors [165].

Organization of the cytoskeleton

Acetylcholine is involved in the control of the cytoskeleton and thereby in cell signalling and cell functions, such as ion transport, locomotion, cytokinesis, migration and cell-cell contact. The cytoskeleton represents the mechanical network between the subcellular compartments of an individual cell and its connection to neighbouring cells and the extracellular environment. The cytoskeleton consists of a filamentous network of F-actin, microtubules and intermediate filaments, providing a negatively charged surface which may facilitate interactions with the cation acetylcholine. The cytoskeleton affects the shape of the cell surface, the properties of ion channels and transporters, the activity of signalling molecules (protein and lipid kinases, phospholipases, GTPases), the intracellular transport of proteins, receptors and organelles, the nuclear signalling pathways, the organization of cell-cell contacts (tight junction, gap junctions, desmosomes) and finally properties like locomotion, proliferation and apoptosis.

Our knowledge that acetylcholine affects the function of the cytoskeleton originates from experiments with acetylcholine-receptor antagonists. Confluent cultures of rat or human airway epithelial cells, when exposed to nicotine receptor antagonists like tubocurarine, mecamylamine or κ-bungarotoxin, show a marked reduction in cell size, detachment from each other and retraction of their intermediate filaments [71, 166]. A similar observation has been reported for human skin keratinocytes [21, 75]. Very recently it has been shown that 1 µM atropine exposed to thin slices of human skin causes an impairment of the desmosomal cell-cell contact and an enlarged extracellular space [20]. All these observations can be interpreted as direct evidence that acetylcholine *via* an auto-/paracrine manner is involved in the organization of the cytoskeleton network. Whether this effect is mediated *via* extracellular membrane receptors or also *via* an intracellular mode of action remains to be elucidated. Nevertheless, the experiments suggest that non-neuronal, epithelial acetycholine can affect the shape, size and cell-cell contact of airway epithelial cells and thereby the quality of the mechanical barrier and aforementioned properties of the airway surface epithelium.

Airway smooth muscle

The biological function of the muscular acetylcholine is so far unknown. Nicotine has been found to stimulate the fusion of cultured human myoblasts [99]. Non-neuronal acetylcholine expressed by the muscle fibres may possibly be involved in maintaining the phenotypic functions of the muscle fiber that are, for example, the organization of the cytoskeleton and the contractile myofilaments as well as the generation of ATP. Therefore, acetylcholine may control the airway diameter by two different pathways, as a neurotransmitter released from the innervating postsynaptic varicosities at the neuroeffector junctions and as the non-neuronal cytomolecule synthesized in minute amounts directly within the muscle fibers. In this context, the high number of muscarinic M_2-receptors expressed on the muscle membrane is noteworthy. Muscarinic M_2-receptors may represent a target for non-neuronal acetylcholine, although possible muscular functions have not yet been identified. Muscarinic receptor antagonists applied in chronic obstructive airway disease can antagonize the actions of both neuronal and non-neuronal acetylcholine.

Immune functions

In the last few years, experimental evidence has been amassed that acetylcholine plays an important role in regulating immune functions, i.e., contributing to the unspecific and specific defense system and modulating inflammatory processes. For example, as already outlined in the two preceding paragraphs, epithelial acetylcholine can stimulate mucus secretion and may regulate the physical barrier function of the surface epithelium. Both functions serve as a defence system to prevent deeper bacterial invasion and to bind bacteria to mucus glycoconjugates. Acetylcholine detected in lymphocytes appears to be involved in the regulation of the activation and clonal expansion of these cells [51, 167]. For example, stimulation of T lymphocytes by muscarinic receptor agonists causes an increase in cGMP, gene expression and protein synthesis and the alteration of immune functions [107–109].

ChAT and acetylcholine have also been detected in B lymphocytes (rat), alveolar macrophages (human) and granulocytes [20, 168], but the function of acetylcholine in these cells remains to be elucidated in more detail. In the rat, B cells contain less acetylcholine than T cells and CD4+ T cells more than CD8+ cells [168]. It has also been found in the rat model that systemic application of cholinergic agonists increases the number of specific B-cell subsets [169–171]. Spleen and thymic cells (rat) also contain and release considerable levels of acetylcholine [168]. Mitogenic stimulation of spleen cells and leukocytes with phytohemagglutinin resulted in a marked increase in the content of acetylcholine [167]. The authors concluded that acetylcholine in the thymus or spleen modulates T-cell differentiation. Cell migration and phagocytosis are additional cellular functions which may be regulated by non-neuronal acetylcholine.

Acetylcholine can also modify immune functions indirectly *via* the release of chemo- or cytokines. For example, monolayers of bovine bronchial epithelial cells, exposed to acetylcholine, release neutrophil and monocyte chemotactic activity

[172]. Granulocyte macrophage-colony stimulation factor (GM-CSF) represents a pro-inflammatory cytokine which causes enhanced recruitment of T cells and eosinophils. It has recently been shown that acetylcholine, *via* stimulation of nicotinic receptors, increases the release of GM-CSF from primary cultures of human bronchial epithelial cells [173]. Thus, it appears possible that also under *in vivo* conditions non-neuronal acetylcholine may directly stimulate the release of GM-CSF from airway epithelial cells. Upregulated acetylcholine may facilitate GM-CSF release and thus contribute to chronic mucosal inflammation.

Finally, recent experimental evidence has indicated a strong inhibitory control by acetylcholine of the activity of mast cells which play a dominant role in triggering acute and chronic mucosal inflammatory processes. In conducting airways, mucosal mast cells are placed in a strategically prominent location to be among the first cells to interact with inhaled antigens or other reactive environmental material. A frequent activation of mast cells which liberate not only the classical mediators, such as histamine, prostanoids, leukotrienes and serotonin, but also proteases and chemo-/cytokines, may trigger acute and chronic inflammation. Thus, a regulatory pathway which limits mast cell mediator release activity represents an important pathway to prevent mucosal inflammation and thereby to maintain mucosal homeostasis.

Mast cells located in the human airway mucosa can be stimulated by the calcium-ionophore A23187 or by anti-human immunoglobulin E (IgE). Histamine release can be used as a marker to indicate the degree of activation of these mast cells. Applied acetylcholine, in a picomolar range, and physostigmine inhibited the stimulated histamine release substantially (see Fig. 4) [174]. The stable muscarinic receptor agonist oxotremorine (1 nM) suppressed A-23187-induced histamine release completely [174]. The acetylcholine-mediated inhibition was prevented by atropine, which itself facilitated the release kinetic of the anti-IgE-induced histamine release [174]. From all these observations the authors concluded that acetylcholine (non-neuronal and neuronal) exerts a strong inhibitory control on mast cell function [174]. Very recently a similar inhibitory pathway has been described in human skin. Human skin mast cells belong to a different mast cell population (tryptase- and chymase-positive) than the airway mucosal mast cells (tryptase-positive) [175]. In human skin, histamine release was significantly inhibited *via* nicotinic receptor stimulation [176]. Taken together, experimental evidence indicates that acetylcholine mediates an inhibitory control on the releasability of mast cells. This inhibitory pathway may limit mast cell activation in the skin and airway mucosa also under *in vivo* conditions. In fact, it had been shown that local histamine release triggered by nasal antigen challenge was increased after atropine pretreatment but reduced by metacholine [177]. The ineffectiveness of atropine-like drugs in the treatment of allergic asthma may, at least partly, be related to the elimination of the endogenous brake, which limits the facilitation of mast cell activity in the conducting airways.

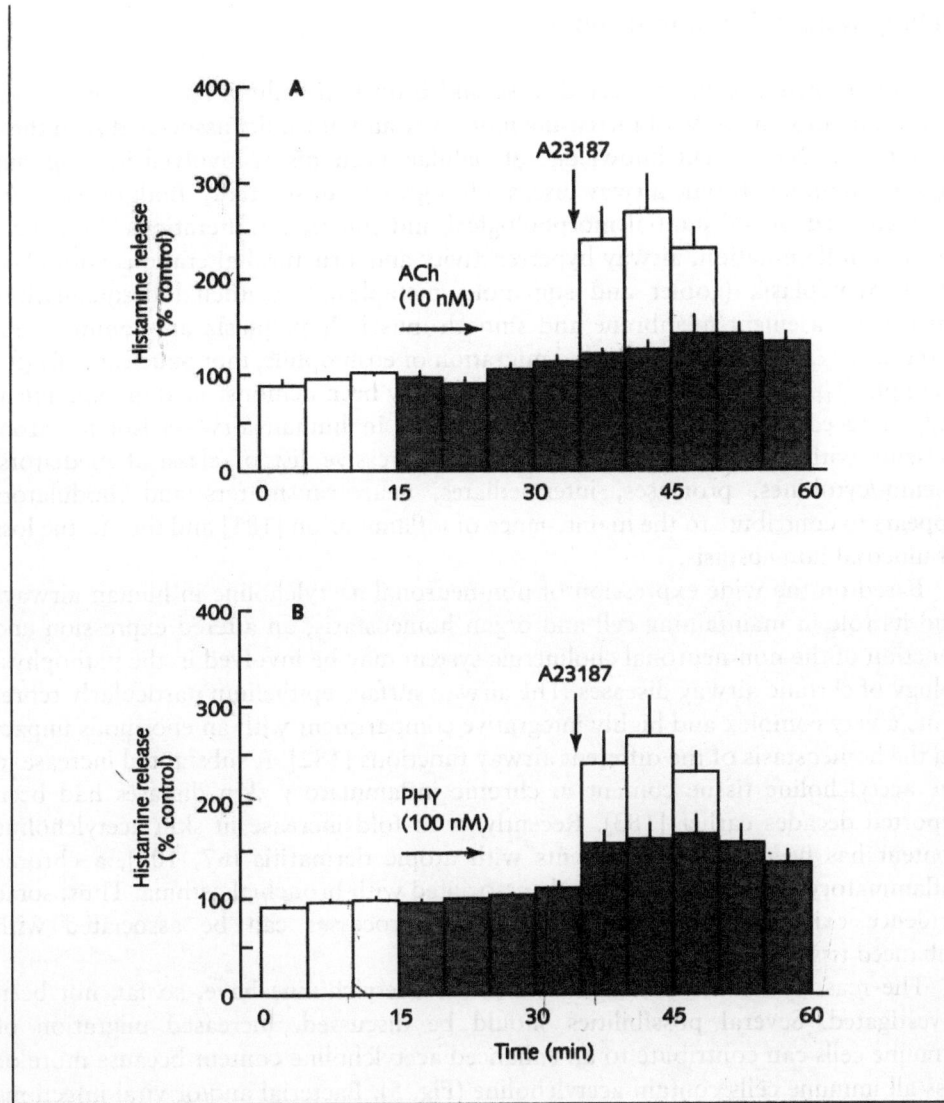

Figure 4
Inhibition of A23187-stimulated histamine release from human airway mucosal mast cells by 1 nM acetylcholine.
Basal and calcium-ionophore A23187-induced histamine release from isolated human bronchi was measured in 5 min intervals. A23187 present in the 34th min of incubation stimulated the histamine release 2.5-fold (open columns) under control conditions. In the presence of acetylcholine (ACh; A) or of the cholinesterase inhibitor physostigmine (PHY; B) the stimulated histamine release was substantially inhibited (values from [174]).

Pathophysiological implications

Chronic obstructive pulmonary disease and bronchial asthma have received wide public attention because of increasing morbidity and mortality associated with these conditions. Our current knowledge of cellular mechanisms involved in triggering and maintaining chronic airway diseases has grown substantially. Both diseases are characterized by substantial morphological and functional alterations. In asthma mucosal inflammation, airway hyperreactivity and structural alterations with glandular hyperplasia (goblet and squamous metaplasia), epithelial desquamation, thickened basement membrane and smooth muscle hyperplasia are common features [29, 178–180]. The increased migration of eosinophils, mononuclear cells (for example, T_H2 cells) and mast cells has repeatedly been demonstrated in experimentally induced airway inflammation, as well as in human airways isolated from patients with chronic airway disease. An impressive list of classical mediators, chemo-/cytokines, proteases, intermediates, neurotransmitters and modulators appears to contribute to the maintenance of inflammation [181] and thus to the loss of mucosal homeostasis.

Based on the wide expression of non-neuronal acetylcholine in human airways and its role in maintaining cell and organ homeostasis, an altered expression and function of the non-neuronal cholinergic system may be involved in the pathophysiology of chronic airway diseases. The airway surface epithelium particularly represents a very complex and highly integrative compartment with an enormous impact on the homeostasis of the different airway functions [182]. A substantial increase in the acetylcholine tissue content in chronic inflammatory skin diseases had been reported decades earlier [183]. Recently, a 15-fold increase in skin acetylcholine content has been found in patients with atopic dermatitis [67, 105], a chronic inflammatory skin disease frequently associated with bronchial asthma. Thus, some evidence exists that chronic inflammatory processes can be associated with enhanced tissue acetylcholine content.

The reasons for the increased content of acetylcholine have, so far, not been investigated. Several possibilities should be discussed. Increased migration of immune cells can contribute to an enhanced acetylcholine content because more or less all immune cells contain acetylcholine (Fig. 5). Bacterial and/or viral infections may cause upregulation of non-neuronal acetylcholine either *via* an increased synthesis or a reduced degradation of acetylcholine. Inflammatory and oxidative stress may likewise interfere with ChAT enzyme activity. In addition, prejunctional muscarinic receptors inhibiting evoked acetylcholine release from cholinergic neurons become impaired [184, 185] and more extracellular acetylcholine may be available. Whether tachykinins can enhance neuronal acetylcholine release in human airways remains to be elucidated [186, 187].

What could be the consequences of an enhanced tissue content of acetylcholine? Some consequences are diagrammatically illustrated in Figure 5. Raised levels of

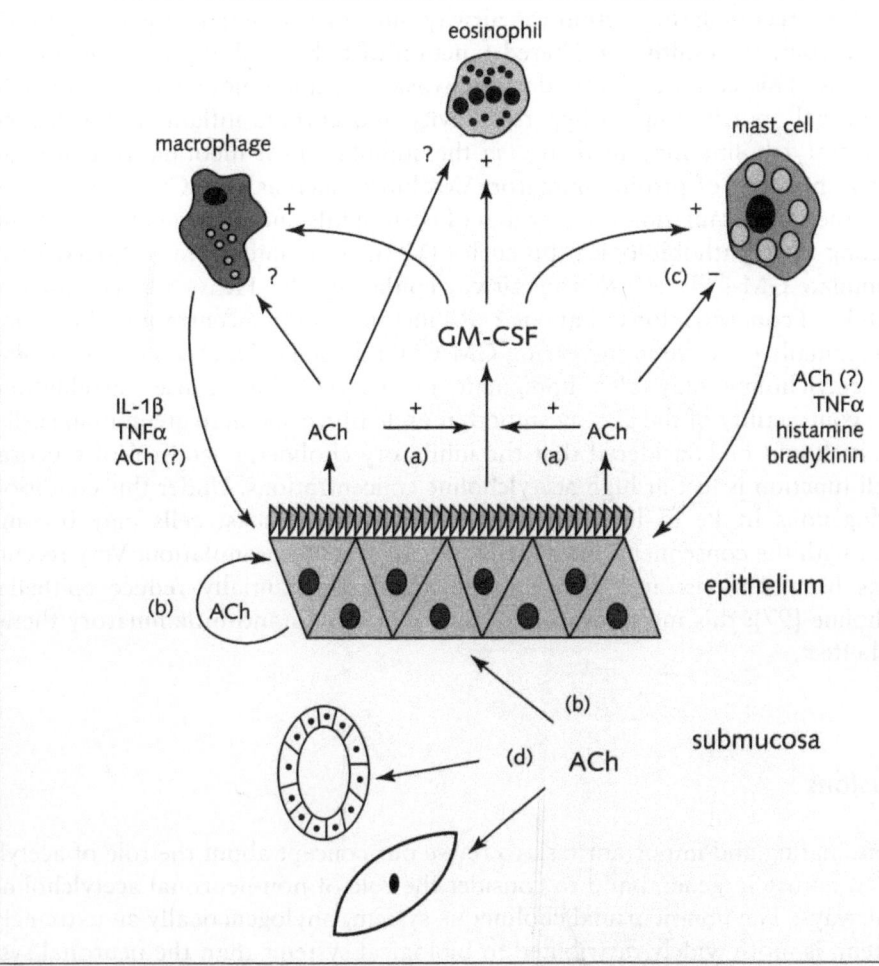

Figure 5
Pathophysiological implications of raised acetylcholine tissue content in chronic airway inflammation
Mucosal inflammation may cause increased levels of non-neuronal acetylcholine (ACh) which could result in the following consequences: a) increased mobilization of GM-CSF from epithelial cells and consequently an enhanced recruitment and activity of macrophages and eosinophils; b) loss of epithelial homeostasis in the regulation of the physical barrier (cell-cell contact) and mucociliary clearance; c) impairment of the inhibitory control of mucosal mast cells (in experiments with isolated human bronchi it was found that muscarinic inhibition of histamine release was lost at high acetylcholine concentrations [174]); d) increased acetylcholine levels in airway mucosa and submucosa may result in the change of proliferation, differentiation and phenotype function of cells.

acetylcholine accumulating within the airway mucosa may impair the homeostasis of cell-cell contact, resulting in altered function of tight- and gap junctional communications. This could facilitate deeper invasion of microbiological, allergic and toxic material contributing to hypersensitivity and chronic inflammation. Raised levels of acetylcholine may mediate, *via* the stimulation of nicotinic receptors an enhanced liberation of pro-inflammatory cytokines, such as GM-CSF from epithelial cells, thereby stimulating the invasion of eosinophils and mononuclear cells and heightening their pathobiological potencies [173]. In a similar way tobacco users may stimulate GM-CSF release from airway epithelial cells. Heavy tobacco use can establish local concentrations of about 1 µM nicotine, this concentration of nicotine being maximally effective in increasing GM-CSF release [173]. Increased tissue levels of acetylcholine may also upregulate mucus secretion, cause vasodilation, enhance contractility of the airway smooth muscle fibres and activate immune cells. Finally, it should be considered that the inhibitory cholinergic control of mucosal mast cell function is lost at high acetylcholine concentrations. Under this condition the endogenous brake to limit the secretory activity of mast cells may become impaired with the consequence of a facilitated mast cell degranulation. Very recently it has been demonstrated that glucocorticoids substantially reduce epithelial acetylcholine [97]; this mechanism may contribute to an anti-inflammatory therapeutical effect.

Conclusions

It is a fascinating and important task to revise our concept about the role of acetylcholine in nature in general and to consider the role of non-neuronal acetylcholine in the airways. The non-neuronal cholinergic system, phylogenetically an extremely old system, is more widely distributed in biological systems than the neuronal system. In the airways the vast majority of cells express ChAT and contain acetylcholine, for example, surface epithelial cells, glandular tissue, smooth muscle cells and migrated immune cells like alveolar macrophages, granulocytes and lymphocytes. Non-neuronal acetylcholine, a universal cytomolecule, is synthesized by the constitutively expressed ChAT and appears to be released continuously in minute quantities. Acetylcholine, *via* nicotinic and muscarinic receptors and possibly also *via* a direct protein interaction, can modulate most of the cellular signalling pathways and appears to be involved in the regulation of vital cell functions like proliferation, differentiation, organization of the cytoskeleton, cell-cell contact, ciliary activity, locomotion, migration, secretion and absorption of ions, water and mucus. Moreover, acetylcholine plays a role in the host defence system and in modifying or regulating various immune functions. It is important to clarify the role of the non-neuronal cholinergic system in airway disease in the immediate future.

Neuronal acetylcholine is expressed in extrinsic and intrinsic cholinergic neurons along the airway wall. Neuronal acetylcholine is highly concentrated in small axonal vesicles from which it can be released in a synchronized fashion to establish high threshold levels at synaptic communications and neuroeffector junctions within milliseconds. Neuronal acetylcholine serves as a nervous regulator of secretory and contractile activity and can be regarded as a reflex emergency system. The cytomolecule acetylcholine appears to regulate the cellular microenvironment (the so-called trophic property of acetylcholine), thus contributing to organ homeostasis.

References

1   Nadel JA, Widdicombe JG (1962) Reflex effects of upper airway irritation on total lung resistance and blood pressure. *J Appl Physiol* 17: 861–865
2   Nadel JA (1980) Autonomic regulation of airway smooth muscle. In: JA Nadel (ed): *Physiology and pharmacology of the airways*, Marcel Dekker, New York, 217–257
3   Gross NJ (1989) Cholinergic control. In: PJ Barnes, IW Rodger, NC Thomson (eds) : *Asthma, basic mechanisms and clinical management*. Academic Press, London, 381–393
4   Loewi O (1921) Über humorale Übertragbarkeit der Herznervenwirkung. *Pflügers Arch* 189: 239–242
5   Loewi O, Navratil E (1926) Über humorale Übertragbarkeit der Herznervenwirkung. X. Mitteilung. Über das Schicksal des Vagusstoff. *Pflügers Arch Gesamte Physiol* 214: 678–688
6   Ewins AJ (1914) Acetylcholine, a new active principle of ergot. *Biochem J* 8: 44–49
7   Burgen ASV (1995) The background of the muscarinic system. *Life Sci* 56: 801–806
8   Dale HH, Dudley HW (1929) The presence of histamine and acetylcholine in the spleen of the ox and the horse. *J Physiol (Lond)* 58: 97–123
9   Beyer G, Wense UT (1936) Über den Nachweis von Hormonen in einzelligen Tieren. Cholin und Acetylcholin in Paramecium. *Pflügers Arch Gesamte Physiol Menschen Tier* 237: 417–422
10   Comline RS (1946) Synthesis of acetylcholine by non-nervous tissue. *J Physiol (Lond)* 105: 6–7P.
11   Bülbring E, Lourie EM, Pardoe U (1949) Presence of acetylcholine in *Trypanosoma rhodesiense* and its absence from *Plasmodium gallinaceum*. *Br J Pharmacol* 4: 290–294
12   Lentz TL (1966) Histochemical localization of neurohumors in a sponge. *J Exp Zool* 162: 171–180
13   Saxena PR, Tangri KK, Bhargava KP (1966) Identification of acetylcholine, histamine, and 5-hydroxytryptamine in *Girardinia heterophylla* (DECNE). *Can J Physiol Pharmacol* 44: 621–627
14   Jaffe MJ (1970) Evidence for the regulation of phytochrome-mediated processes in bean roots by the neurohumor acetylcholine. *Plant Physiol* 46: 768–777

15   Hartmann E, Kilbinger H (1974) Occurrence of light-dependent acetylcholine concentrations in higher plants. *Experientia* 30: 1387–1388

16   Erzen I, Brzin M. (1979) Cholinergic mechanisms in *Planaria torva*. *Comp Biochem Physiol* 64C: 207 – 216

17   Whittaker VP (1963) Identification of acetylcholine and related esters of biological origin. In: O Eichler, A Farah, GB Koelle (eds): *Handbuch der Experimentellen Pharmakologie*, Bd. 15. Springer Verlag, Berlin, 1–39

18   Koelle GB (1963) Cytological distributions and physiological functions of cholinesterases. In: O Eichler, A Farah, GB Koelle (eds): *Handbuch der Experimentellen Pharmakologie*, Bd. 15. Springer Verlag, Berlin, 187–298

19   Sastry BVR, Sadavongvivad C (1979) Cholinergic systems in non-nervous tissues. *Pharmacol Rev* 30: 65–132

20   Wessler I, Kirkpatrick CJ, Racké K (1999) The cholinergic pitfall: acetylcholine, a universal cell molecule in biological systems, including humans. *Clin Exp Pharmacol Physiol* 26: 198–205

21   Grando SA (1997) Biological functions of keratinocyte cholinergic receptors. *J Invest Dermatol Symp Proc* 2: 41–48

22   Grando SA, Horton RM (1997) The keratinocyte cholinergic system: acetylcholine as an epidermal cytotransmitter. *Curr Opin Dermatol* 4: 262–268

23   Klapproth H, Reinheimer T, Metzen J, Münch M, Bittinger F, Kirkpatrick CJ, Höhle K-D, Schemann M, Racké K, Wessler I (1997) Non-neuronal acetylcholine, a signalling molecule synthesized by surface cells of rat and man. *Naunyn-Schmiedeberg's Arch Pharmacol* 355: 515–523

24   Wessler I, Kirkpatrick CJ, Racké K (1998) Non-neuronal acetylcholine, a locally acting molecule widely distributed in biological systems: expression and function in humans. *Pharmacol Ther* 77: 59–79

25   Widdicombe JG (1963) Regulation of tracheobronchial smooth muscle. *Physiol Rev* 43: 1–37

26   Richardson JB (1979) Nerve supply to the lungs. *Am Rev Respir Dis* 119: 785–802

27   Barnes PJ (1986) Neural control of human airways in health and Disease. *Am Rev Respir Dis* 134: 11289–1314

28   Barnes PJ (1991) Neural mechanisms in asthma. In: CP Page, PJ Barnes (eds): *Pharmacology of asthma*. Springer Verlag, Berlin, 143–159

29   Barnes PJ (1992) Modulation of neurotransmission in airways. *Physiol Rev* 72: 699–729

30   Nachmansohn D, Machado AL (1943) The formation of acetylcholine. A new enzyme choline acetylase. *J Neurophysiol* 6: 397–403

31   Tucek S (1988) Choline acetyltransferase and the synthesis of acetylcholine. In: VP Whittaker (ed): *Handbook of Experimental Pharmacology*, Vol. 86. Springer Verlag, Berlin, 129–131

32   Rossier J (1977) Acetyl-coenzyme A and coenzyme A analogues, their effects on rat brain choline acetyltransferase. *Biochem J* 165: 321–326

33 White HL, Wu JC (1973) Choline and carnitine acetyltransferase of heart. *Biochem* 12: 841–846

34 Tucek S (1982) The synthesis of acetylcholine in skeletal muscles of the rat. *J Physiol (Lond)* 322: 53–69

35 Tucek S (1978) Acetylcholine synthesis in neurons. Chapman & Hall, London

36 Adamic S (1972) Effects of quaternary ammonium compounds on choline entry into the rat diaphragm muscle fibre. *Biochem Pharmacol* 21: 2925–2929

37 Ducis I (1988) The high-affinity choline uptake systen. In: VP Whittaker (ed): *Handbook of Experimental Pharmacology*, Vol 86. Springer Verlag, Berlin, 409–437

38 Birks R, MacIntosh FC (1961) Acetylcholine metabolism of a sympathetic ganglion. *Can J Biochem Physiol* 39: 787–825

39 Jope RS (1979) High affinity choline transport and acetylCoA production in brain and their roles in the regulation of acetylcholine synthesis. *Brain Res Rev* 1: 313–344

40 Cohen-Haguenauer O, Brice A, Berrard S, Nguyen VC, Mallet J, Frezal J (1990) Localization of the choline acetyltransferase (ChAT) gene to human chromosome 10. *Genomics* 6: 374–378

41 Cervini R, Rocchi M, DiDonato S, Finocchiaro G (1991) Isolation and sub-chromosomal localization of a DNA fragment of the human choline acetyltransferase gene. *Neurosci Lett* 132: 191–194

42 Viegas-Pequignot E, Berrard S, Brice A, Apiou F, Mallet J (1991) Localization of a 900-bp-long fragment of the human choline acetyltransferase gene to 10q11.2 by nonradioactive *in situ* hybridization. *Genomics* 9: 210–212

43 Wu D, Hersh LB (1994) Choline acetyltransferase: celebrating its fiftieth year. *J Neurochem* 62: 1653–1663

44 Ibanez CF, Pelto-Huikko M, Soder O, Ritzen EM, Hersh LB, Hökfelt T, Persson H (1991) Expression of choline acetyltransferase mRNA in spermatogenic cells results in an accumulation of the enzyme in the postacrosomal region of mature spermatozoa. *Proc Natl Acad Sci USA* 88: 3676–3680

45 Benjanin S, Cervini R, Mallet J, Berrad S (1994) A unique gene organization for two cholinergic markers, choline acetyltransferase and a putative vesicular transporter of acetylcholine. *J Biol Chem* 269: 21944–57

46 Erickson JD, Varoqui H, Schäfer MK-H, Modi W, Diebler M-F, Weihe E, Rand J, Eiden LE, Bonner TI, Usdin TB (1994) Functional identification of a vesicular acetylcholine transporter and its expression from a "cholinergic" gene locus. *J Biol Chem* 269: 21929–21932

47 Berrard S, Varoqui H, Cervini R, Israel M, Mellet J, Diebler MF (1995) Coregulation of two embedded gene products, choline acetyltransferase and the vesicular acetylcholine transporter. *J Neurochem* 65: 939–942

48 Eiden LE (1998). The cholinergic gene locus. *J Neurochem* 70: 2227–2240

49 Mautner HG (1986) Choline acetyltransferase. In: AA Boulton, GB Baker, PH Yu PH (eds): *Neuromethods, Vol. 5, Neurotransmitter Enzymes*. Humana Press Inc. Clifton, New Jersey, USA, 273–317

50 Schäfer MKH, Eiden LE, Weihe E (1998) Cholinergic neurons and terminal fields revealed by immunohistochemistry for the vesicular acetylcholine transporter.II. The peripheral nervous system. *Neurosci* 84: 361–376

51 Kawashima K, Fujii T, Watanabe Y, Misawa H (1998) Acetylcholine synthesis and muscarinic receptors subtype mRNA expression in T-lymphocytes. *Life Sci* 62: 1701–1705

52 Fujii T, Tsuchiya T, Yamada S, Fujimoto K, Suzuki T, Kasahara T, Kawashima K (1996) Localization and synthesis of acetylcholine in human leukemic T cell lines. *J Neurosci Res* 44: 66–72

53 Vogel P, Reinheimer T, Racké K, Bittinger F, Kirkpatrick CJ, Wessler I (1998) Expression of the non-neuronal cholinergic system in human circulating and immune cells. *Naunyn-Schmiedeberg's Arch Pharmacol* (Suppl) 357: R21

54 Fuji T, Yamada S, Watanabe Y, Misawa H, Tajima S, Fujimoto K, Kasahara T, Kawashima K (1998) Induction of choline acetyltransferase mRNA in human mononuclear leukocytes stimulated by phytohemagglutinin, a T-cell activator. *J Neuroimmunol* 82: 101–107

55 Fritz S, Föhr KJ, Boddien S, Berg U, Bruckner C, Mayerhofer A (1999) Functional and molecular characterization of a muscarinic receptor type and evidence for expression of choline-acetyltransferase and vesicular acetylcholine transporter in human granulosa-luteal cells. *J Clin Endocrinol Metab* 84: 1744–1750

56 Anderson DC, King SC, Parsons SM (1983) Pharmacological characterization of the acetylcholine transport system in purified *Torpedo* electric organ synaptic vesicles. *Mol Pharmacol* 24: 46–54

57 Marshall IG, Parsons SM (1987) The vesicular acetylcholine transport system. *Trends Pharmacol Sci* 10: 174–177

58 Calakos N, Scheller R (1996) Synaptic vesicle biogenesis, docking, and fusion: a molecular description. *Physiol Rev* 76: 1–29

59 Jahn R, Südhof TC (1994) Synaptic vesicles and exocytosis. *Ann Rev Neurosci* 17: 219–246

60 Monck JR, Fernandez JM (1994) The exocytotic fusion pore and neurotransmitter release. *Neuron* 12: 707–716

61 van der Kloot W, Molgo J (1994) Quantal acetylcholine release at the vertebrate neuromuscular junction. *Physiol Rev* 74: 899–991

62 Volknandt W (1995) Commentary. The synaptic vesicle and its targets. *Neuroscience* 64: 277–300

63 Fisher LJ, Schinstine M, Salvaterra P, Dekker AJ, Thal L, Gage FH (1993) *In vivo* production and release of acetylcholine from primary fibroblasts genetically modified to express choline acetyltransferase. *J Neurochem.* 61: 1323–1332

64 Falk-Vairant J, Israel M, Bruner J, Stinnakre J, Meunier F-M, Gaultier P, Meunier FA, Lesbats B, Synguelakis M, Correges P et al (1996) Enhancement of quantal transmitter release and mediatophore expression by cyclic AMP in fibroblasts loaded with acetylcholine. *Neurosci* 75: 353–360

65 Rowell PR, Sastry BVR (1981) Human placental cholinergic system: depression of the

uptake of α-aminoisobutyric acid in isolated human placental villi by choline acetyltransferase inhibitors. *J Pharmacol Exp Ther* 216: 232–238

66 Sastry BVR (1997) Human placental cholinergic system. *Biochem Pharmacol* 53: 1577–1586

67 Reinheimer T, Vogel P, Bittinger F, Kirkpatrick CJ, Saloga J, Knop J, Wessler I (1998) Up-regulation of non-neuronal acetylcholine in patients with atopic dermatitis. *J Invest Dermatol* (Suppl) 110: 556

68 Zhang XY, Robinson NE, Wang ZW, Lu MC (1995) Catecholamine affects acetylcholine release in trachea: α_2-mediated inhibition and β_2-mediated augmentation. *Am J Physiol* 268: L368–L373

69 Edwards C, Dolezal V, Tucek S, Zemkova H, Vyskocil F (1985) Is an acetylcholine transport system responsible for nonquantal release of acetylcholine at the rodent myoneural junction? *Proc Natl Acad Sci USA* 82: 3514–3518

70 Dunant Y, Israel M (1993) Ultrastructure and biophysics of acetylcholine release: Central role of the mediatophore. *J Physiol (Paris)* 87: 179–192

71 Maus ADJ, Pereira EFR, Karachunski PI, Horton RM, Navaneetham D, Macklin K, Cortes WS, Albuquerque EX, Conti-Fine BM (1998) Human and rodent bronchial epithelial cells express functional nicotinic acetylcholine receptors. *Mol Pharmacol* 54: 779–788

72 Basbaum CB, Barnes PJ, Grillo M, Widdicombe JH, Nadel JA (1983) Adrenergic and cholinergic receptors in submucosal glands of the ferret trachea: autoradiographic localization. *Eur J Respir Dis* 64: 433–435

73 Grando SA, Crosby AM, Zelickson BD, Dahl MV (1993) Agarose gel keratinocyte outgrowth system as a model of skin re-epithelization: requirement of endogenous acetylcholine for outgrowth initiation. *J Invest Dermatol* 101: 804–810

74 Grando SA, Kist DA, Qi M, Dahl MV (1993) Human keratinocytes synthesize, secrete, and degrade acetylcholine. *J Invest Dermatol* 101: 32–36

75 Grando SA, Zelickson BD, Kist DA, Weinshenker D, Bigliardi PL, Wendelschafer-Crabb G, Kennedy WR, Dahl MV (1995) Keratinocyte muscarinic acetylcholine receptors: immunolocalization and partial characterization. *J Invest Dermatol* 104: 95–100

76 Grando SA, Horton RM, Mauro TM, Kist DA, Lee TX, Dahl, MV (1996) Activation of keratinocyte nicotinic cholinergic receptors stimulates calcium influx and enhances cell differentiation. *J Invest Dermatol* 107: 412–418

77 Wennerberg PA, Welch F (1977) Effects of cholinergic drugs on uptake of ^{14}C-α-aminoisobutyric acid by human term placenta fragments: Implication for acetylcholine recognition sites and observations on the binding of radioactive cholinergic ligands. *Fed Proc* 36: 980–988

78 Athweh A, Grayhack MS, Richman DP (1984) A cholinergic receptor site on murine lymphocytes with novel binding characteristics. *Life Sci* 35: 2459–2469

79 Brunner F, Kukovetz WR (1986) Muscarinic receptors of the vascular bed: radioligand binding studies on bovine splenic veins. *J Cardiovasc Pharmacol* 8: 712–721

80 Brunner F, Kühberger E, Brockmeier D, Kukovetz WR (1990) Evidence for muscarinic

receptors in endothelial cells from combined functional and binding studies. *Eur J Pharmacol* 187: 145–154

81 Costa LG, Kaylor G, Murphy D (1988) Muscarinic cholinergic binding sites on rat lymphocytes. *Immunopharmacol* 16: 139–149

82 Maslinski W (1989) Cholinergic receptors of lymphocytes. *Brain Behav Immunol* 3: 1–14

83 O'Malley KE, Farrell CB, O'Boyle KM, Baird AW (1995) Cholinergic activation of Cl-secretion in rat colonic epithelia. *Eur J Pharmacol* 275: 83–89

84 Hootman SR, de-Ondarza J (1995) Regulation of goblet cell degranulation in isolated pancreatic ducts. *Am J Physiol* 268: G24–G32

85 Hiemke C, Stolp M, Reuss S, Wevers A, Reinhardt S, Maelicke A, Schlegel S, Schröder H (1996) Expression of alpha subunit genes of nicotinic acetylcholine receptors in human lymphocytes. *Neurosci Lett* 214: 171–174

86 Lapied B, Le Corrone H, Hue B (1990) Sensitive nicotinic and mixed nicotinic-muscarinic receptors in insect neurosecretory cells. *Brain Res* 533:132–136

87 Shirvan MH, Pollard HB, Heldman E (1991) Mixed nicotinic and muscarinic features of cholinergic receptor coupled to secretion in bovine chromaffin cells. *Proc Natl Acad Sci USA* 88: 4860–4864

88 Minette PA, Barnes PJ (1990) Muscarinic receptor subtypes in lung. *Am Rev Respir Dis* 141: S162–S165

89 Mak JCW, Baraniuk JN, Barnes PJ (1992) Localization of muscarinic receptor subtype mRNAs in human lung. *Am J Respir Cell Mol Biol* 7: 344–348

90 Barnes PJ (1993) Muscarinic receptor subtypes in airways. *Life Sci* 52: 521–527

91 Singer-Lahat D, Rojas E, Felder CC (1996) Muscarinic receptor activated Ca^{2+} channels in non-excitable cells. In: J Klein, K Löffelholz K (eds): *Progress in Brain Research*, Vol 109. Elsevier, Amsterdam, 195–199

92 McDonald TV, Premack BA, Gardner P (1993) Flash photolysis of caged inositol-1,4,5,-trisphosphate activates plasma membrane calcium current in human T cells. *J Biol Chem* 268: 3889–3896

93 Alles GA, Hawes RC (1940) Cholinesterase in the blood of man. *J Biol Chem* 133: 375

94 Small RC, Good DM, Dixon JS, Kennedy I (1990) The effects of epithelium removal on the actions of cholinomimetic drugs in opened segments and perfused tubular preparations of guinea-pig trachea. *Br J Pharmacol* 100: 516–522

95 Adler M, Reutter SA, Moore DH, Filbert MG (1991) Regulation of acetylcholine hydrolysis in canine tracheal smooth muscle. *Eur J Pharmacol* 205: 73–79

96 Norel X, Angrisani M, Labat C, Gorenne I, Dulmet E, Rossi F, Brink C (1993) Degradation of acetylcholine in human airways: role of butyrylcholinesterase. *Br J Pharmacol* 108: 914–919

97 Reinheimer T, Münch M, Bittinger F, Racké K, Kirkpatrick CJ,Wessler I (1998) Glucocorticoids mediate reduction of epithelial acetylcholine in the airways of rat and man. *Eur J Pharmacol* 349: 277–284

98 Wessler I, Kirkpatrick CJ, Racké K (1999) Airway epithelium: more than just a barrier.

Airway epithelium:source of non-neuronal acetylcholine and modulator of neurotransmission. *Trends Pharmacol Sci* 20: 52–53

99 Krause RM, Hamann M, Bader CR, Liu J-H, Baroffio A, Bernheim L (1995) Activation of nicotinic acetylcholine receptors increases the rate of fusion of cultured human myoblasts. *J Physiol (Lond)* 489: 779–790

100 Miledi R, Molenaar PC, Polak RL, Tas JWM, van der Laaken T (1982) Neuronal and non-neuronal acetylcholine in the rat diaphragm. *Proc R Soc London B* 214: 153–168

101 Dolecal V,Tucek S (1983) The synthesis and release of acetylcholine in normal and denervated rat diaphragms during incubation *in vitro. J Physiol (Lond)* 334: 461–474

102 Parnavelas JG, Kelly W, Burnstock G (1985) Ultrastructural localization of choline acetyltransferase in vascular endothelial cells in rat brain. *Nature* 316: 724–725

103 Gonzalez JL, Santos-Benito FF (1987) Synthesis of acetylcholine by endothelial cells isolated from rat brain cortex capillaries. *Brain Res* 412: 148–150

104 Ikeda C, Morita I, Mori A, Fujimoto K, Suzuki T, Kawashima K, Murota S (1994) Phorbol ester stimulates acetylcholine synthesis in cultured endothelial cells isolated from porcine cerebral microvessels. *Brain Res* 655: 147–152

105 Reinheimer T, Vogel P, Racké K, Bittinger F, Kirkpatrick CJ, Saloga J, Knop J, Wessler I (1998) Non-neuronal acetylcholine is increased in chronic inflammation like atopic dermatitis. *Naunyn-Schmiedeberg's Arch Pharmacol* (Suppl) 358: R87

106 Fujii T, Yamada S, Misawa H, Tajima S, Fujimoto K, Suzuki T, Kawashima K (1995) Expression of choline acetyltransferase mRNA and protein in T-Lymphocytes. *Proc Japan Acad (Ser. B)* 71: 231–235

107 Strom TB, Sytkowski AT, Carpenter CB, Merrill JP (1974) Cholinergic augmentation of lymphocyte-mediated cytotoxicity. A study of the cholinergic receptor of cytotoxic T lymphocytes. *Proc Natl Acad Sci USA* 71: 1330–1334

108 Haddock AM, Patel KR, Alston WC, Kerr JW (1975) Response of lymphocyte guanyl cyclase to propranolol, noradrenaline, thymoxamine and acetylcholine in extrinsic bronchial asthma. *Br Med J* 2: 357–359

109 Masturzo P, Salmona M, Nordstrom O, Consolo S, Ladinski H (1985) Intact human lymphocyte membranes respond to muscarinic receptor stimulation by oxotremorine with marked changes in microviscosity and an increase in cyclic GMP. *FEBS Lett* 192: 194–198

110 Richardson JB, Ferguson CC (1979) Neuromuscular structure and function in the airways. *Federation Proc* 38: 202–208

111 Gabella G (1987) Innervation of airway smooth muscle: fine structure. *Ann Rev Physiol* 49: 583–594

112 Coburn RF (1987) Peripheral airway ganglia. *Ann Rev Physiol* 49: 573–582

113 Jeffery PK (1994) Innervation of the airways mucosa: structure, function and changes in airway disease. In: R Goldie (ed): *Immunopharmacology of epithelial barriers*, Academic Press, London, 85–118

114 Fischer A, Canning BJ, Kummer W (1996) Correlation of vasoactive peptide and nitric

oxide synthase with choline acetyltransferase in the airway innervation. *Ann NY Acad Sci* 805: 717–722

115 Canning BJ, Fischer A (1997) Localization of cholinergic nerves in lower airways of guinea pigs using antisera to choline acetyltransferase. *Am J Physiol Lung Cell Mol* 16: L731–L738

116 Jeffery P, Reid L (1973) Intra-epithelial nerves in normal rat airways: a quantitative electron microscopic study. *J Anat* 114: 35–45

117 Downing SE, Lee JC (1980) Nervous control of the pulmonary circulation. *Ann Rev Physiol* 42: 199–210

118 Andersson RGG, Grundström N (1987) Innervation of airway smooth muscle. Efferent mechanisms. *Pharmac Ther* 32: 107–130

119 Olsen CR, Colenbatch JH, Mebel PE, Nadel JA, Staub NC (1965) Motor control of pulmonary airways studied by nerve stimulation. *J Appl Physiol* 20: 202–208

120 Widdicombe JG (1954) Respiratory reflexes from the trachea and bronchi of the cat. *J Physiol (Lond)* 123: 55–70

121 Lundberg JM, Brodin E, Saria A (1983) Effects and distribution of vagal capsaicin-sensitive substance P neurons with special reference to the trachea and lungs. *Acta Physiol Scand* 119: 243–252

122 Wessler I, Klein A, Pohan D, Maclagan J, Racké K (1991) Release of [^{3}H]acetylcholine from the isolated rat or guinea-pig trachea evoked by preganglionic nerve stimulation; comparison to transmural stimulation. *Naunyn-Schmiedeberg's Arch Pharmacol* 344: 403–411

123 Daniel EE, Kannan M, Davis C, Posey-Daniel V (1986) Ultrastructural studies on the neuromuscular control of human tracheal and bronchial msucle. *Resp Physiol* 63: 109–128

124 Wessler I, Bender H, Härle P, Höhle K-D, Kirdorf G, Klapproth H, Reinheimer T, Ricny J, Schniepp-Mendelssohn KE, Racké K (1995) Release of [^{3}H]acetylcholine in human isolated bronchi: effect of indomethacin on muscarinic autoinhibition. *Am J Resp Crit Care Med* 151:1040–1046

125 Smith AD, Winkler H (1972) Fundamental mechanisms in the release of catecholamines. In: H Blaschko, E Musholl (eds): *Catecholamines, Handbuch der Experimentellen Pharmakologie*, Bd 33, Springer Verlag, Berlin, 500–617

126 Roffel AF, Meurs H, Elzinga CRS, Zaagsma J (1990) Characterization of the muscarinic receptor subtype involved in phosphoinositide metabolism in bovine tracheal smooth muscle. *Br J Pharmacol* 99: 293–296

127 Yang CM, Yo YY, Wang YY (1993) Intracellular calcium in canine cultured tracheal smooth muscle cells is regulated by M3 muscarinic receptors. *Br J Pharmacol* 110: 983–988

128 Daniel EE, Bourreau JP, Abela A, Jury J (1992) The internal calcium store in airway muscle: emptying, refilling and chloride. *Biochem Pharmacol* 43: 29–37

129 Jeffery PK (1990) Microscopic anatomy. In: RAL Brewis, GJ Gibson, DM Geddes DM (eds): *Respiratory medicine*, Bailliere Tindall, London, 57–78

130 Adriaensen D, Scheuermann DW (1993) Neuroendocrine cells and nerves of the lung. *The Anatomical Record* 236: 70–85

131 Pack RJ, Al-Ugaily LH, Widdicombe JG (1984) The innervation of the trachea and extrapulmonary bronchi of the mouse. *Cell Tissue Res* 238: 61–68

132 Reinheimer T, Bernedo P, Klapproth H, Oerlert B, Zeiske B, Racké K, Wessler I (1995) Synthesis and storage of acetylcholine in the isolated airways of rat, guinea-pig and man: species differences in the role of the airway mucosa. *Am J Physiol* 270: L722–L728

133 Webber SE, Widdicombe JG (1987) The effect of vasoactive intestinal peptide on smooth muscle tone and mucus secretion from the ferret trachea. *Br J Pharmacol* 91: 139–148

134 Tokuyama K, Kuo HP, Rohde JA, Barnes PJ, Rogers DF (1990) Neural control of goblet cell secretion in guinea-pig airways. *Am J Physiol* 259: L108–L115

135 Marin MG (1994) Update: Pharmacology of airway secretion. *Pharmacol Rev* 46: 35–65

136 Phipps RJ, Richardson PS (1976) The effects of irritation at various levels of the airway upon tracheal mucus secretion in the cat. *J Physiol (Lond)* 261: 563–581

137 Johansson O, Wang L (1993) Choline acetyltransferase-like immunofluorescence in epidermis of human skin. *Neurobiology* 1: 201–206

138 Downing SE, Lee JC (1980) Nervous control of the pulmonary circulation. *Ann Rev Physiol* 42: 199–210

139 Hepp C (1969) Motor innervation of the pulmonary blood vessels of mammals. In: AP Fishman, HH Hecht (eds): *The pulmonary circulation and interstitial space*, Univ Chicago Press, Chicago, 195

140 Waaler BA (1971) Physiology of the pulmonary circulation. *Angiologica* 8: 266–284

141 Cattaneo MG, Dátri F, Vicentini LM (1997) Mechanisms of mitogen-activated protein kinase activation by nicotine in small-cell lung carcinoma cells. *Biochem J* 328: 499–503

142 Guizzetti M, Cost LG (1996) Inhibition of muscarinic receptor-stimulated glial cell proliferation by ethanol. *J Neurochem* 67: 2236–2245

143 Qui Y, Peng Y, Wang J (1996) Immunoregulatory role of neurotransmitters. *Adv Neuroimmunol* 6: 223–231

144 Wanner A, Salathé M, O'Riordan TG (1996) Mucociliary clearance in the airways. *Am J Respir Crit Care Med* 154: 1868–1902

145 Olver REB, Davis B, Marin MG, Nadel JA (1975) Active transport of Na^+ and Cl^- across the canine tracheal epithelium *in vitro*. *Am Rev Respir Dis* 112: 811–815

146 Cullen JJ, Welsh MJ (1986) Regulation of sodium absorption by canine tracheal epithelium. *J Clin Invest* 79: 73–79

147 McCann JD, Welsh MJ (1990) Regulation of Cl^- and K^+ channels in airway epithelium. *Annu Rev Physiol* 52: 115–135

148 Acevedo M (1994) Effect of acetylcholine on ion transport in sheep tracheal epithelium. *Pflügers Arch* 427: 543–546

149 Stewart C P, Turnberg LA (1989) A microelectrode study of responses to secretagogues

by epithelial cells on villus and crypt of rat small intestine. *Am J Physiol* 257: G334–G343

150 Biagi B, Wang YZ, Cooke HJ (1990) Effects of tetrodotoxin on chloride secretion in rabbit distal colon: tissue and cellular studies. *Am J Physiol* 258: G223–G230

151 Altenberg GA, Subramanyam M., Bergmann JS, Johnson KM, Reuss L (1993) Muscarinic stimulation of gallbladder epithelium. I. Electrophysiology and signaling mechanisms. *Am J Physiol* 265: C1604–C1612

152 Jansen FW, Fleure-Jakobs AM, De-Pont JJ, Bonting SL (1980) Blocking by 2,4,6-triaminopyrimidine of increased tight junction permeability induced by acetylcholine in the pancreas. *Biochim Biophys Acta* 598: 115–126

153 Davis CW, Dowell ML, Lethem M, van Scott M (1992) Goblet cell degranulation in isolated canine tracheal epithelium: response to exogenous ATP, ADP, and adenosine. *Am J Physiol* 262: C1313–C1323

154 Dwyer TM, Szebeni A, Diveki K, Farley JM (1992) Transient cholinergic glycoconjugate secretion from swine tracheal submucosal gland cells. *Am J Physiol* 262: L418–L426

155 Gawin AZ, Emery BE, Baraniuk JN, Kaliner MA (1991) Nasal glandular secretory response to cholinergic stimulation in humans and guinea pigs. *J Appl Physiol* 71: 2460–2468

156 Basbaum CB, Ueki I, Brezina L, Nadel JA (1981) Tracheal submucosal gland serous cells stimulated *in vitro* with adrenergic and cholinergic agonists. *Cell Tissue Rev* 220: 481–498

157 Bülbring E, Burn, JH, Shelley HJ (1953) Acetylcholine and ciliary movement in the gill plate of *Mytilus edulis*. *Proc Roy Soc Ser Biol Sci* 141: 445–466

158 Kordik P, Bülbring E, Burn JH (1952) Ciliary movement and acetylcholine. *Br J Pharmacol* 7: 67–79

159 Wong LB, Miller IF, Yeates DB (1988) Regulation of ciliary beat frequency by autonomic agonists *in vivo*. *J Appl Physiol* 65: 971–981

160 Wong LB, Miller IF, Yeates DB (1988) Regulation of ciliary beat frequency by autonomic mechanisms *in vitro*. *J Appl Physiol* 65: 1895–1901

161 Eljamal M., Wong LB, Yeates DB (1994) Capsaicin-activated bronchial- and alveolar-initiated pathways regulating tracheal ciliary beat frequency. *J Appl Physiol* 77: 1239–1245

162 Hamm-Alvarez SF, Sheetz M (1998) Microtubule-dependent vesicle transport: modulation of channel and transporter activity in liver and kidney. *Physiol Rev* 78: 1109–1129

163 Janmey PA (1998) The cytoskeleton and cell signaling: component localization and mechanical coupling. *Physiol Rev* 78: 763–781

164 Gheber L, Priel Z (1994) metachronal activity of cultured mucociliary epithelium under normal and stimulated conditions. *Cell Motil Cytoskeleton* 28: 333–345

165 Ueda F, Ban K, Ishima T (1995) Irsoglandine activates gap-junctional intercellular com-

munication through M1 muscarinic acetylcholine receptor. *J Pharmacol Exp Ther* 274: 815–819

166 Klapproth H, Racké K, Wessler I (1994) Modulation of the airway smooth muscle tone by mediators released from cultured epithelial cells of rat trachea. *Naunyn-Schmiedeberg's Arch Pharmacol* (Suppl) 349: R72

167 Kawashima K, Fujii T, Misawa H, Yamada S, Tajima S, Suzuki T, Fujimoto K, Kasahara T (1996) Presence and synthesis of acetylcholine in the blood. *J Neurochem* (Suppl 6) 66: S73

168 Rinner I, Kawashima K, Schauenstein K (1998) Rat lymphocytes produce and secrete acetylcholine in dependence of differentiation and activation. *J Neuroimmunol* 81: 31–37

169 Rinner I, Schauenstein K (1991) The parasympathetic nervoues system takes part in the immuno-neuroendocrine dialogue. *J Neuroimmunol* 34: 165–172

170 Felsner P, Hofer D, Rinner I, Mangge H, Gruber M, Korsatko W Schauenstein K (1992) Continuous *in vivo* treatment with catecholamines suppresses *in vitro* reactivity of peripheral blood lymphocytes *via* α-adrenoceptor mediated mechanisms. *J Neuroimmunol* 37: 47–57

171 Felsner P, Hofer D, Rinner I, Porta S, Korsatko W Schauenstein K (1995) Adrenergic suppression of peripheral blood T-cell reactivity in the rat is due to activation of peripheral α2-receptors. *J Neuroimmunol* 57: 27–34

172 Koyama S, Rennard SI, Robbins RA (1992) Acetylcholine stimulates bronchial epithelial cells to release neutrophil and monocyte chemotactic activity. *Am J Physiol* 262: L466–L471

173 Klapproth H, Racké K, Wessler I (1998) Acetylcholine and nicotine stimulate the release of granulocyte-macrophage colony stimulating factor from cultured human bronchial epithelial cells. *Naunyn-Schmiedeberg's Arch Pharmacol* 357: 472–475

174 Reinheimer T, Baumgärtner D, Höhle KD, Racké K, Wessler I (1997) Acetylcholine *via* muscarinic receptors inhibits histamine release from human bronchi. *Am J Resp Crit Care Med* 156: 389–395

175 Metcalfe DD, Baram D, Mekori Y (1997) Mast cells. *Physiol Rev* 77: 1033–1079

176 Reinheimer T, Zimmermann S, Weikel W, Racké K, Wessler I (1999) Acetylcholine and nicotine inhibit histamine release from human skin. *Naunyn-Schmiedeberg's Arch Pharmacol* (Suppl.) 359: R22

177 Baroody FM, Ford S, Lichtenstein LM, Kagey-Sobotka, Naclerio RM (1994) Physiologic responses and histamine release after nasal antigen challenge. *Am J Respir Crit Care Med* 149: 1457–1465

178 Jeffery PK (1992) Pathology of asthma. In: PJ Barnes PJ (ed): *Asthma*, Br Med Bulletin 48, Churchill Livingstone, New York, 23–50

179 Arm JP, Lee TH (1992) The pathobiology of bronchial asthma. *Adv Immunol* 51: 323–382

180 Goldstein RA (1994) NIH Conference. Asthma. *Ann Inter Med* 121: 698–708

181 Hay DWP, Farmer SG, Goldie RG (1994) Inflammatory mediators and modulation of

epithelial/smooth muscle interactions. In: Goldie R (ed): *Immunopharmacology of epithelial barriers*, Academic Press, London, 119–146

182 Folkerts G, Nijkamp FP (1998) Airway epithelium: more than just a barrier. *Trends Pharmacol Sci* 19: 334–341

183 Scott A (1962) Acetylcholine in normal and diseased skin. *Brit J Derm* 74: 317–322

184 Fryer AD, Maclagan J (1984) Muscarinic inhibitory receptors in pulmonary parasympathetic nerves in the guinea-pig. *Br J Pharmacol* 83: 973–978

185 Fryer AD, Jacoby DB (1991) Parainfluenza virus infection damages inhibitory M2-muscarinic receptors on pulmonary parasympathetic nerves in the guinea-pig. *Br J Pharmacol* 102: 267–271

186 Watson N, Maclagan J, Barnes PJ (1993) Endogenous tachykinins facilitate transmission through parasympathetic ganglia in guinea-pig. *Br J Pharmacol* 109: 751–759

187 Belvisi MG, Patacchini R, Barnes PJ, Maggi CA (1994) Facilitatory effects of selective agonists for tachykinin receptors on cholinergic neurotransmission: evidence for species differences. *Br J Pharmacol* 111: 103–110

188 Ricny J, Höhle K-D, Racké K, Wessler I (1995) Long term application of inhalative steroids does not affect synthesis and content of acetylcholine in human bronchi. *Eur Resp J* 8: 589–589

Identification, localization and function of muscarinic receptor subtypes in the airways

Ad F. Roffel[1], Herman Meurs[2] and Johan Zaagsma[2]

[1]Department of Research Management, Pharma Bio-Research Group B.V., P.O. Box 200, 9470 AE Zuidlaren, The Netherlands; [2]Department of Molecular Pharmacology, University of Groningen, A. Deusinglaan 1, 9713 AV Groningen, The Netherlands

Introduction

Muscarinic receptors are prominently involved in the control of airway calibre, robustly mediating contraction of airway smooth muscle and secretion of mucus upon activation of the cholinergic nervous system. In line with these responses, autoradiography has visualized muscarinic receptors in extra- and intrapulmonary airway smooth muscle as well as in submucosal glands of both animal and man [1-6]. In addition, these studies identified muscarinic receptor binding sites on airway (parasympathetic) nerves and ganglia of cow and man [3, 4, 7] and on nerves of pig [6], on airway epithelium of ferret and guinea pig [1, 4] but not man [3, 4], on pulmonary blood vessels of ferret but not man and pig [1, 3, 6], as well as on alveolar walls of man, rabbit and pig [4–6] but not ferret and guinea pig [1, 4]. The relevance, with respect to the regulation of airway calibre, of muscarinic receptors in these additional locations has not been fully clarified yet, especially regarding the alveolar site.

Muscarinic receptor subtypes

Ever since muscarinic acetylcholine receptors were discovered to consist of more than one type (more potent blockade of ileal than atrial muscarinic receptors by specific receptor antagonists, 1976; radioligand displacement from more than one affinity binding site with pirenzepine, 1980; identification and cloning of five distinct muscarinic receptor genes, 1988) [8–10], the effort has been made to define which of the subtypes are present in airway and lung tissue and their specific cell types, and to delineate which of the well-known physiological responses produced by acetylcholine in the airways can be ascribed to individual receptor subtypes. The techniques used in this research include radioligand binding and autoradiography with highly specific, high-affinity radioligands and subtype-selective muscarinic

Muscarinic Receptors in Airways Diseases, edited by Johan Zaagsma, Herman Meurs and Ad F. Roffel

receptor antagonists, immunoprecipitation with subtype-selective antibodies, Northern blotting and *in situ* hybridization techniques with subtype-specific nucleic acid probes, and classical but elegant bioassays, again employing receptor antagonists. It should be noted that the subtype-selective muscarinic receptor antagonists that have been available over the past 20 years generally display relatively poor selectivity, i.e. only occasionally 100-fold or more, and very seldom for one subtype over all the others (see, e.g., [11]). However, the combination especially of the above techniques has produced important insights into muscarinic receptor subtype identity, localization and function in the airways.

Muscarinic receptors in airway smooth muscle

It has been clearly established that *extrapulmonary* airway smooth muscle expresses both M_2 and M_3 muscarinic receptor subtypes, with the former, sometimes referred to as the "cardiac" subtype, representing 70 to 90% of the total (see also the chapter by Eglen and Watson in this volume). This has been inferred from radioligand binding experiments on isolated cell membranes [12–16] as well as intact cells [17, 18] from bovine and dog trachealis muscle, and on cell membranes from rat, rabbit and pig trachealis [19–21]. In these studies, selective muscarinic receptor antagonists like AF-DX 116 and methoctramine competed with nonselective radioligands like the hydrophilic ^{3}H-N-methylscopolamine (^{3}H-NMS) or the more lipophilic ^{3}H-quinuclidinyl benzilate (^{3}H-QNB) or ^{3}H-dexetimide for high- (M_2) and low- (M_3) affinity binding sites. The identity of these binding sites has been underscored by the detection of high levels of M_2 and lower levels of M_3 receptor mRNAs in porcine [21, 22] and bovine trachea (Roffel, Mak, Zaagsma, Barnes, unpublished observations) and by the observation that 90% of muscarinic receptors in dog trachea immunoprecipitated with the M_2-antibody and 10% with M_3 [23]. M_1 type receptors were found to be absent in tracheal smooth muscle, as judged from the low affinity displayed by pirenzepine (see [24] and above references), as also was M_1 receptor mRNA in porcine trachea [21, 22]. The presence of M_4 receptors has not been broadly considered in these studies, probably because they are difficult to identify with the receptor antagonists available. Northern analysis, however, suggested that they are not expressed in rat and pig trachea [21, 22], and immunoprecipitation in dog airways showed that both M_1 and M_4 receptors are absent in tracheal as well as bronchial smooth muscle [23].

Levels of expression of M_2 type receptors in guinea pig and human airway smooth muscle may not be as high. Thus, guinea pig tracheal smooth muscle showed only 60% of M_2-type receptors in radioligand binding [25] and even less in autoradiography [4], with the remaining binding sites being of the M_3 subtype; M_1 receptors again appeared to be absent. Binding studies on human tracheal smooth muscle are virtually lacking, except for the demonstration of low affinity for pirenzepine [26]. Autoradiographic analysis was unable to detect methoctramine-sensitive, i.e., M_2 or M_4, binding sites in bronchial and bronchiolar smooth muscle [4], but *in situ* hybridization and Northern blotting demonstrated some M_2 in addition to M_3 receptor mRNA; M_4 was not detected [27]. Since M_2 receptor-mediated inhibition of adenylyl cyclase (one of the well-known biochemical responses produced by airway smooth muscle M_2 receptors – see the chapter by Eglen and Watson in this volume) has been demonstrated in cultured human bronchial smooth muscle cells [28], it appears that M_2 receptors are indeed present in human airway smooth muscle and may produce responses relevant to airway function (see the chapter by Eglen and Watson in this volume, and further below). Whether the relatively small proportion of M_2 receptors in human *intrapulmonary* airway smooth muscle represents a gradient compared to the large proportions found in tracheal smooth muscle of the animal species studied is at present unclear. Thus, although this might be inferred to be the case from immunoprecipitation experiments in dog airways, where the ratio of M_2:M_3 receptors was 89:11 in trachea but 44:56 in intrapulmonary bronchi [23], binding studies in rat, rabbit and pig, discussed below, found large proportions of M_2 receptors in peripheral lung tissue of these species, in addition to much smaller levels of M_3.

Finally, *in situ* hybridization has detected relatively abundant levels of M_4 receptor mRNA in bronchiolar airway smooth muscle in rabbit lung, together with lower levels of M_3 mRNA; a combination of direct (^{3}H-AF-DX 384) and indirect (^{3}H-QNB in the presence of telenzepine, methoctramine and 4-DAMP) autoradiography confirmed the expression of M_4 (and M_3) receptors [5]. These M_4 receptors may serve a function similar to M_2 receptors in other species, given the preferential (inhibitory) coupling of these receptor types to adenylyl cyclase.

Muscarinic receptors in peripheral lung

In addition to the above studies describing muscarinic receptor subtypes on defined airway smooth muscle, there is a wealth of experimental data on muscarinic receptor subtypes in lung tissue or lung parenchyma, without providing a completely clear-cut picture, however. In total rat lung, radioligand displacement and immunoprecipitation identified large (80–90%) proportions of M_2 and small proportions of M_3 receptors, probably located on the airway smooth muscle [19, 29–31]. M_4 receptors were not detected in rat lung [29], and M_1 receptors appeared completely absent in some studies [32, 33], but represented 10–70% of binding sites in two others, with the 70% proportion being found in isolated rat lung parenchyma rather than total lung [19, 31]. More or less similarly, no high-affinity pirenzepine (M_1) binding sites were detected by autoradiography or radioligand binding in guinea pig lung in two studies [4, 25]. However, the same investigators did find 40–50% of M_1-type binding sites in two other radioligand binding studies, using the same strains of animal [33, 34]. The remaining sites appeared to consist of mainly M_2 with some M_3 receptors according to the binding studies [25, 33, 34], but largely M_3, located on smooth muscle and epithelium, according to autoradiography as mentioned above [4].

The above discussion of muscarinic receptor subtypes in rat and guinea pig lung, and especially the putative presence of M_1 receptors, appears relevant since human peripheral lung tissue specimens (containing airways of ≤ 2 mm diameter) have rather consistently been shown to contain important levels of the M_1 subtype, i.e., some 55% of the total number of binding sites by direct ^{3}H-pirenzepine labeling and 65% by competition analysis [31, 33–37]. Remarkably, Northern blotting detected mRNA of the M_1 subtype only [27, 38]. The remaining binding sites appeared difficult to identify in these radioligand binding studies, showing antagonist binding profiles reminiscent of M_2 [33], M_3 [34] or M_2 plus M_3 [37]. Autoradiography suggested the complete absence of M_2 receptors in human lung [4], with M_3 receptors present in airway smooth muscle and M_1 on alveolar walls, while *in situ* hybridization indeed detected some M_2 (in addition to M_3) mRNA in smooth muscle and M_1 in alveoli [27]. M_3 receptor mRNA was also detected in airway epithelium, even though there was no autoradiographic labelling of this tissue in human lung. Finally, M_4 receptor mRNA and protein were not detected on any structure [4, 27].

Similar to man, rabbit peripheral lung tissue was reported to contain a large proportion (70–80%) of high-affinity pirenzepine binding sites, which were initially taken to be M_1 receptors [39]. However, since these binding sites showed relatively high affinity to M_2- and M_3-selective receptor antagonists as well, unlike classical M_1 receptors, and since only M_4 receptor mRNA was detected by Northern blotting, these sites were later characterized as M_4-type receptors [40]. Subsequent studies, both immunoprecipitation and *in situ* hybridization, corroborated the presence of important levels (50% of total) of M_4 receptors and their mRNA in rabbit lung [5, 29, 41], but also suggested the expression of M_1 (5%), M_2 (35%) and M_3 subtypes (4%). Interestingly, M_4 receptors were localized to alveolar walls, similar to the M_1 subtype in human lung, and co-localized with M_3 receptors in airway smooth muscle as discussed above [5].

M_4 receptors were also claimed to occur in porcine lung parenchyma, based on the observation in binding studies that the affinities of a number of muscarinic receptor antagonists in this tissue were very similar to those obtained in bovine adrenal medulla (which had previously been suggested to express M_4 type receptors), and with cloned human M_4 receptors expressed in COS cells [42]. Additional evidence for the expression of M_4 receptors in porcine lung, e.g., from Northern blotting or *in situ* hybridization, was not included in this study, however. This appears especially important since other studies found M_1, M_2 and M_3 receptor mRNA in porcine lung [6, 21] and also identified these receptor subtypes using binding analysis, but failed to detect M_4 receptor mRNA [21]. It appears that a well-controlled study is necessary to solve this question of M_4 receptor expression in porcine lung, since the above studies used similar tissue (peripheral lung free of bronchi), but from different age groups.

Taken together, M_2 and M_3 receptors appear to be present in extra- as well as in intrapulmonary airway smooth muscle of animal and man, and M_1 receptors are expressed in lung parenchyma, where they are putatively localized on alveolar walls. In rabbit, and possibly also in pig, parenchymal receptors may be of the M_4 subtype, and in rabbit intrapulmonary airway smooth muscle M_4 may replace M_2. Finally, parenchymal lung tissue from dog only appears to contain M_3 receptors [23].

Muscarinic receptor subtypes and airway smooth muscle contraction

Contraction of both extra- and intrapulmonary airway smooth muscle preparations by muscarinic receptor agonists appears to be mediated primarily *via* the M_3 receptor subtype, at least under normal conditions. This can be inferred from the high potencies with which M3-selective muscarinic receptor antagonists like 4-DAMP inhibit contraction induced by exogenous methacholine or carbachol (Tab. 1), or by endogenously released acetylcholine (as evoked by electrical field or nerve stimulation) [46, 50, 51, 56]. Lack of involvement of the M_2 receptor subtype follows from the low potencies found for M_2-selective antagonists like AF-DX 116, gallamine and methoctramine in these experiments (Tab. 1), combined with the observation that Schild analysis often produces straight lines with slopes not different from unity. Airway constriction in the isolated ventilated rat lung *in vitro* and in guinea pig lung *in vivo*, as well as resting airway calibre in mouse lung *in vivo*, similarly appear to be mediated through M_3 receptors [51, 57, 58]. This may not apply, however, to cholinergic contraction of the isolated guinea pig lung strip, which is sometimes regarded as an *in vitro* preparation better reflecting airway responsiveness than the extrapulmonary trachea. Thus, although initial studies concluded that contraction of this preparation is also mediated by M_3-type receptors indeed [25, 33], our own studies clearly showed that this receptor can best be designated as M_2-like [45, 59] (Tab. 2). When comparing these studies in detail, it appears that they all show the affinities for the M_2-selective antagonists to be higher in guinea pig lung strip compared to trachea, and that the reverse is true for M_3-selective compounds. The reason for arriving at different conclusions may reside in the number of antagonists investigated (3 in the former versus 9 in our studies), and in the larger differences observed between trachea and lung strip in the latter studies. Taken together, it would appear that the guinea pig lung strip may not provide a good model for peripheral airway responsiveness in the characterization of (subtype-selective) muscarinic receptor antagonists, the discrepancy possibly being due to the involvement of vascular elements in lung strip contraction.

The intracellular signalling pathway through which muscarinic M_3 receptors in airway smooth muscle produce contraction appears to be the activation of phos-

Table 1 - Antagonist affinities (pA$_2$ or pK$_B$ values) show the involvement of the muscarinic M$_3$ receptor subtype in the contraction of animal and human airway smooth muscle preparations.

		pirenzepine	AF-DX 116	gallamine	methoctramine	4-DAMP	Ref.
Bovine trachea		6.9	6.3	4.1[a]	6.2	9.0	[13]
Guinea pig trachea		6.5			5.8[a]	9.2	[25]
		6.9		<4.0	6.2	8.8	[43, 44]
	bronchus	6.8			6.3[a]	9.6	[25]
Rat	trachea	6.6		3.0			[31]
	bronchus	7.0			5.9	8.8	[54]
		6.4		3.3			[31]
Dog	trachea	7.2		<4.0		8.9	[46]
		7.1			6.1	9.2	[47]
	bronchus	6.9	6.7			8.6	[53]
Horse	trachea	6.8		<4.0		8.5	[48]
		6.7			6.5	8.9	[49]
Rabbit trachea		6.8	6.5			9.1	[20]
		7.0	6.4		6.0		[50]
Mouse trachea		6.5	6.3[a]			8.7	[51]
Pig	trachea	6.8	6.2		5.8	9.1	(b)
	bronchus	6.6	6.2		5.5[a]	8.9	(b)
Human trachea			5.7[a]			8.7	[52]
	bronchus	6.8	5.6			9.0	[45, 52]
		6.8		5.3		9.4	[55]

[a]*slope of Schild plot significantly different from unity*
[b]*Roffel et al., unpublished results*

phoinositide turnover, resulting in calcium release from intracellular stores and the activation of protein kinase C (see the chapter by Meurs et al. in this volume for detailed discussion). Thus, muscarinic receptor stimulation was shown to result in the production of inositol phosphates in human bronchus [62] and in guinea pig, bovine and canine trachea [47, 63, 64]. In the latter two species, the response was also shown to actually proceed *via* M3 type receptors, based on Schild analysis showing high affinity for the M3-selective antagonist 4-DAMP, intermediate affinity for the M1 receptor antagonist pirenzepine, and low affinity for the M2 antagonists AF-DX 116 and methoctramine [47, 64].

If cholinergic contraction of airway smooth muscle, whether induced by exogenous or endogenous agonist, and in animal and man alike, primarily involves M$_3$-

Table 2 - Affinities of muscarinic receptor antagonists at muscarinic receptors mediating contraction of the guinea pig lung strip correlate better to cardiac M_2 than to tracheal M_3 [45, 59] or to human m4 receptors expressed in Chinese hamster ovary cells (values taken from [60]). Functional affinities obtained by Haddad et al. [25] (in brackets) are shown for comparison (see text).

	Lung strip	Heart	Trachea	m4
pirenzepine	6.4	6.2	6.7	7.1
	(6.4	6.1	6.6	[25])
AF-DX 116	6.6	7.0	6.0	6.4
AQ-RA 741	7.5	8.3	6.6	8.0
gallamine	5.4	6.5	3.5	5.1
methoctramine	7.3	7.4	5.4	6.8
	(6.1[a]	7.5	5.8[a]	[25])
tripitramine	8.8	9.6	6.1	7.9[b]
AF-DX 474	6.4	6.2	7.1	7.0
AQ-RA 721	6.9	6.9	8.0	8.1
DAU 5884	6.8	6.6	8.7	8.5
UH-AH 371	7.0	7.4	8.2	7.8
(4-DAMP	8.4	7.7	8.9	[25])

[a]*slope of Schild plot significantly different from unity*
[b]*taken from [61] (binding affinity towards NG108-15 M_4 receptors)*

type receptors, it remains to be determined what function is served by the large population of smooth muscle M_2 receptors. As discussed in detail in the chapter by Eglen and Watson in this volume, a number of intracellular signalling responses have now been identified that may be activated by M_2 receptors in (human) airway smooth muscle cells, including the G_i-mediated inhibition of adenylyl cyclase and potassium channel activity, and activation of the small G-protein Rho and a nonselective cation channel. Apart from the attenuation of β-adrenoceptor-mediated airway smooth muscle relaxation or airway dilation that these biochemical responses may produce (in addition to M_3 receptor-mediated transductional cross-talk – see the chapter by Meurs et al. in this volume), it is suggested that M_2 receptors may indeed have some (additional) role in (fine-tuning) airway smooth muscle contraction. Another interesting possibility appears to be a role for M_2 receptors in (the modulation) of smooth muscle cell growth and proliferation (see the chapter by Meurs et al. in this volume).

As to the reason why the putative additional role of M_2 receptors in (fine-tuning) contraction has hitherto not been recognized by the use of M_2- and M_3-selective antagonists in the *in vitro* and *in vivo* contraction studies discussed above, interesting mathematical models have been recently devised to explain observations in guinea pig colon [65]. In this tissue, contractile responses to oxotremorine-M after irreversible blockade of the majority of M_3 receptors were sensitive to pertussis toxin, suggesting the involvement of M_2 receptors, yet they were relatively insensitive to AF-DX 116. One of the models proposed supposes that activation of M_3 receptors alone results in contraction, presumably *via* phosphoinositide signalling, and that activation of M_2 receptors can potentiate this response without generating contraction by itself. Alternatively, M_3 receptors may activate two parallel signalling pathways, of which one, putatively phosphoinositide turnover, can produce contraction by itself, and the other (possibly the non-selective cation current) only when simultaneously activated by M_2 plus M_3 receptors; again there is no contraction when M_2 would be activated alone. Both these models appeared to explain the observations made in guinea pig colon, with the latter model providing a somewhat better approximation. However, the contractile response mediated by M_2 together with M_3 receptors according to these models exhibited an M_3 receptor profile when probed with AF-DX 116 under normal (i.e., non-alkylated) conditions, questioning the physiological relevance of the potential contribution of the M_2 subtype [65].

Muscarinic receptors in mucus glands

From the relatively limited number of studies identifying muscarinic receptor subtype(s) in airway submucosal glands, it appears quite clear that M_1 and M_3 are present, the latter putatively in higher numbers. Thus, pirenzepine showed high (M_1-type) affinity for 27% and low (M_3-type) affinity for the remaining 73% of [^{3}H]-NMS binding sites in isolated porcine tracheal submucosal glands, and autoradiography identified M_1 and M_3 receptors in human bronchial and ferret tracheal mucus glands in 1:2 and 1:1 ratios, respectively [4, 66]. A binding study on cat tracheal submucosal glands demonstrated monophasic and low-affinity binding for AF-DX 116, indicating the absence of M_2 type receptors, but shallow and high-affinity pirenzepine binding [67], putatively due to the co-localisation of M_1 and M_3.

The functional response of the submucosal gland, i.e., the secretion of mucus, appears to be mediated primarily by the M_3 receptor subtype. Indirect evidence for this was obtained in porcine tracheal submucosal gland cells treated for 7 days with an irreversible inhibitor of acetylcholinesterase; such treatment down-regulated M3 receptors much more than M_1 receptors and almost abolished mucus secretion [68]. More direct evidence was obtained in the intact isolated cat trachea, where the antagonist affinities for 4-DAMP (high) and AF-DX 116 (low) clearly indicated the involvement of M_3 [69]. A similar conclusion was reached in isolated cat tracheal submucosal glands, where cholinergic stimulation of both intracellular calcium, as the signalling pathway underlying the cellular response, and production of mucus were inhibited with a potency order 4-DAMP > pirenzepine > AF-DX 116 [67]. Finally, a recent study showed that M_3 receptors mediate mucus secretion in ferret trachea, both vagally or exogenously induced *in vitro* and vagally induced *in vivo* [66] (see also the chapter by Rogers in this volume). A role for M_1 receptors in mucosal gland function, in addition to the primary involvement of M_3 receptors, has been suggested based on the rather indirect finding in porcine tracheal gland cells that both M_1 receptor expression and chloride secretion were resistant to the chronic treatment with an acetylcholinesterase-inhibitor [70], in contrast to M_3 receptors and mucus secretion as discussed above. Intermediate-to-high pirenzepine potencies in inhibiting agonist-induced mucus secretion in cat trachea also left room for some involvement of M_1 receptors [67, 69], but low telenzepine potency does not support a role for M_1 receptors in ferret trachea [66].

Functional responses of ganglionic, neuronal and alveolar muscarinic receptors

Muscarinic receptors in airway ganglia, as detected by autoradiography, may represent postsynaptic M_1 receptors involved in the fine-tuning of ganglionic neurotransmission that is primarily mediated *via* nicotinic receptors. Such ganglionic M_1 receptors have been reported in rabbit [71, 72], dog [73] and guinea pig airways [74] *in vivo* and *in vitro*, based on pirenzepine's preference to inhibit vagal nerve stimulation-induced airway constriction, compared to either vagally induced bradycardia (M_2 receptor effect) or electrical field stimulation-induced airway constriction (M_3 receptor effect). However, the involvement of ganglionic M_1 receptors in vagal bronchoconstriction as measured *in vivo* as well as *in vitro* has been disputed in both rabbit [50, 75] and guinea pig [76, 77]. As a putative solution to this dispute, we have found that under normal conditions of nerve stimulation, M_1 receptors do not appear to be involved in vagally induced contraction of the guinea pig main bronchus, but that a facilitatory role, albeit of limited magnitude, can be demonstrated under very specific conditions. Thus, when nerve stimulation was prolonged to produce plateau contractions, with pre- and postganglionic inhibitory M_2 receptors selectively blocked using AQ-RA 741 and nicotinic ganglionic receptors partially inhibited using hexamethonium, M_1-selective concentrations of pirenzepine inhibited the resulting contractions by approximately 30% [78].

Vagally mediated bronchoconstriction in man, induced by the inhalation of SO_2, has also been reported to be especially sensitive to (inhaled) pirenzepine, suggesting that facilitatory M_1 receptors exist in airway parasympathetic ganglia in man [79]. M_1 receptors also appeared to be involved in vagally mediated tracheal constriction in hyperreactive cats [80]. As a consequence, a number of studies have sought to establish to what extent selective blockade of M_1 receptors may affect airway tone in humans, or may be of therapeutic importance in patients suffering from airway obstruction that results from increased vagal drive. In normal subjects, inhaled pirenzepine was shown to have a significant but small (25–30% of ipratropium bromide) bronchodilator effect [81], purportedly produced by acting on M_1 receptors in peripheral airways [82]. A bronchodilator effect was also observed in patients with reversible or partially reversible airway obstruction (i.e., putatively asthma and COPD) when pirenzepine was given in relatively high intravenous doses [83, 84], but not when given by inhalation [85]. The more potent M_1 receptor antagonist telenzepine, given orally in a single 5 mg dose, was reported to produce a significant and important bronchodilation in patients with COPD [86], but this effect could not be reproduced with a slightly lower dose given once daily for 5 days [87], nor in patients with nocturnal asthma [88]; dry mouth was reported as a side-effect in all three studies. Taken together, these studies have failed to clearly demonstrate a role for the putative ganglionic M_1 receptors in controlling bronchomotor tone in humans, either with or without obstructive airways disease. It remains to be seen whether this is due to the lack of involvement of such receptors, or rather to the relatively weak selectivity for M_1 over M_3 receptors of the compounds available for use in man.

Muscarinic receptors on (airway parasympathetic) nerves, as identified using autoradiography, have a clear functional correlate as prejunctional (auto)inhibitory receptors serving as a feedback mechanism to control excess neurotransmitter release. Since such receptors, and especially their dysfunction in airways disease (see also below), are the subject of the chapter by Adamko et al. in this book, it may suffice to mention here that they have been demonstrated in every mammalian species studied to date (Tab. 3), *in vivo* as well as *in vitro*, and by indirect (vagally or electrical field stimulation-induced airway constriction or contraction) as well as by direct (vagally or electrically induced release of acetylcholine) techniques.

The functional importance of muscarinic receptors in parenchymal lung tissue, especially those identified as M_1 or M_4 and expressed on alveolar walls as detected by autoradiography, is far from clear. Initial reports [4, 71, 72] suggested that these receptors might be expressed in the smooth muscle of small bronchioli, but contraction experiments as discussed above do not support this explanation. The alternative suggestion of localization to blood vessels [19] has been denied by autoradiography [4, 6]. There is, however, some evidence that muscarinic receptors may mediate surfactant production from type II alveolar cells in rat lung, although this is not without dispute (see [5] and [115]); a more recent report suggests that muscarinic receptors on type II cells may be involved in fluid reabsorption in fetal guinea pig lung [116]. Given the vast population of muscarinic receptors on alveolar walls, more research seems warranted.

Table 3 - Experimental evidence for the presence of prejunctional muscarinic autoreceptors on cholinergic and adrenergic nerve endings where they serve as a negative feedback mechanism inhibiting acetylcholine and noradrenaline release. VNS, vagal nerve stimulation; EFS, electrical field stimulation.

Species, tissue		Parameter measured	Autoreceptor type (antagonist used)	Refs.
Guinea pig				
	lung *in vivo*	VNS, constriction	M_x (gallamine)	[89]
		VNS, constriction	M_2 (methoctramine)	[90]
			M_2 (gallamine)	[91]
	trachea *in vivo* and *in vitro*	VNS, EFS, constriction	M_x (gallamine)	[92]
	trachea *in vitro*	EFS, contraction	M_2 (gallamine)	[93]
			M_2 (various)	[56]
		EFS, contraction/release	M_2-like (tripitramine)	[59]
		EFS, release	M_x (atropine)	[94, 95]
			M_2-like (various)	[96]
			M_4 (various)	[97]
			M_x (pilocarpine)	[98]
		EFS, release (NA)	M_2 (various)	[99]
Cat	lung *in vivo*	VNS, constriction	M_2 (gallamine)	[100]
	trachea *in vivo*	VNS, contraction	No autoinhibition (gallamine)	[80]
	trachea/bronchus *in vitro*	EFS, contraction	M_2 (gallamine)	[101]
Rat	bronchus *in vitro*	EFS, release	M_x (scopolamine)	[102]
	trachea *in vitro*	EFS, constriction	M_2 (gallamine, methoctramine)	[103]
		VNS, release	M_x (scopolamine)	[104]
		EFS, release (NA)	M_x (scopolamine)	[105]
	lung *in vivo*	VNS, constriction	M_2 (AF-DX 116)	[106]
Dog	lung *in vivo*	VNS, constriction	M_2 (gallamine)	[73]
	bronchus *in vitro*	EFS, release	M_x (atropine)	[107]
	trachea *in vitro*	EFS, contraction	M_2 (gallamine)	[46]
Man	bronchus *in vitro*	EFS, contraction	M_2 (gallamine)	[93]
	lung *in vivo*	SO_2, constriction	M_x (pilocarpine)	[108]
	trachea *in vitro*	EFS, release	M_2 (methoctramine)	[109]
	bronchus *in vitro*	EFS, contraction	M_2 (AQ-RA 741, gallamine)	[110]

Table 3 (continued)

Species, tissue	Parameter measured	Autoreceptor type (antagonist used)	Refs.
Rabbit lung *in vivo*	VNS, constriction	No autoinhibition (gallamine)	[75]
trachea *in vitro*	EFS, release	M_2 (various)	[111]
	EFS, release (NA)	M_2-like (various)	[112]
bronchus *in vitro*	EFS, contraction	M_2 (not M_4) (various)	[50]
Horse trachea *in vitro*	EFS, contraction	No autoinhibition (gallamine)	[48]
	EFS, release	M_x (atropine)	[113]
Mouse lung *in vivo*	Baseline lung resistance	M_2 (AF-DX116)	[51]
trachea *in vitro*	EFS, release	M_2 (gallamine)	[114]
Ferret trachea *in vivo* and *in vitro*	VNS, mucus secretion EFS, mucus secretion	M_2 (methoctramine)	[66]

Muscarinic receptors in disease and drug treatment

Changes in the expression of muscarinic receptors may be expected to occur in obstructive airways diseases, based on the frequent observation of bronchial hyper-responsiveness to methacholine in asthma and the robust bronchodilator responses to ipratropium and tiotropium bromide in COPD (see the chapters by Chapman and by Disse in this volume). However, total muscarinic receptor numbers were only slightly (by 8-25%) and not significantly increased in central and segmental airways smooth muscle of patients suffering from chronic and severe airflow obstruction (emphysema, chronic obstructive bronchitis) as compared to controls, and agonist binding was actually somewhat weaker [117]. In peripheral lung tissue from a COPD-like group of patients, total muscarinic receptor number was decreased rather than increased (by 70%) and agonist binding affinity marginally increased [118]. A more recent study, comparing peripheral lung tissue from normal and asthmatic individuals, also observed only non-significant changes in total muscarinic receptor number (−28%) and mRNA expression (+25%) [38]. It therefore appears that in obstructive airways diseases, changes in muscarinic receptors either do not occur at all, or only with respect to specific receptor subtypes and/or their functional responses. Concerning these possibilities, it should be noted that cholinergic contraction of isolated airway preparations was not increased in COPD [119, 120], and only non-specifically (i.e., together with other spasmogens) in some asthmatic preparations [121–123], suggesting that M_3 receptor function may not be enhanced in obstructive airways disease. In contrast, prejunctional M_2 autoreceptors, inhibiting acetylcholine release as a negative feedback mechanism under normal conditions, indeed appear to be dysfunctional in human asthma [38, 108] and in animal models thereof, and this may contribute to bronchial hyperreactivity. Thus, in our laboratory it was found that ovalbumin-sensitised guinea pigs developed M_2 receptor dysfunction as well as bronchial hyperreactivity to histamine already 6 h after a single allergen challenge, i.e., after the early allergic response in this model [91, 124], and that this dysfunction contributed to the observed hyperreactivity [125]. It is interesting that M_2 receptor dysfunction had recovered virtually completely after the late reaction, when histamine bronchial hyperreactivity, although still detectable, had also decreased [91, 124]. The reader is referred to the chapter by Adamko et al. in this book for a more extensive discussion on this subject.

Ganglionic M_1 receptors may be involved in increased tracheal reactivity in hyperreactive cats [80], but their importance is highly questionable in man (see above) and could not be demonstrated in vagal hyperreactivity in antigen-challenged guinea pigs [126]. Finally, increased expression and/or activity of postjunctional M_2 receptors may be involved in decreased β-adrenoceptor relaxant function, as has been suggested in Basenji-Greyhound dogs and in isolated rabbit trachea exposed to pro-inflammatory cytokines or human atopic serum (see the chapter by Eglen and Watson in this volume for elaborate discussion and references).

Concerning the effects of drug therapy on airway muscarinic receptors, a number of different observations have now been made. Treatment of patients suffering from COPD and undergoing surgery for lung carcinoma with subcutaneous terbutaline for 24–72 h resulted in non-significant changes in muscarinic receptor number (+20%) and antagonist affinity in peripheral lung tissue; significant changes were not observed either when isolated lung tissue was exposed to terbutaline for up to 36 h *in vitro* [37]. In apparent contrast, chronic β-agonist treatment of rabbits (28 day albuterol from s.c. osmotic minipumps) and guinea pigs (6 weeks fenoterol per inhalation) resulted in increased muscarinic receptor function, expressed as a 40–70% increase in maximum cholinergic contraction of isolated smooth muscle preparations *in vitro* or bronchoconstriction *in vivo* [127, 128]. Although the mechanism of this hyperreactivity was not investigated in these studies, a number of suggestions were put forward, including increased muscarinic receptor number and/or G-protein coupling, changes in G-protein levels, and increased MLCK activity; similar mechanisms have indeed been observed in other tissues and species (see the chapter by Eglen and Watson in this volume).

Muscarinic receptor number and function were (also) found to be increased in rabbits after chronic anticholinergic treatment (28 day atropine from s.c. osmotic minipumps); total receptor number increased by 40–60% in trachea, main bronchus and peripheral lung tissue (and 90% in heart), and maximum contractility by 30–60% in isolated trachea and bronchus. Immunoprecipitation showed that receptor increase included both M_2 and M_3, but not M_4 receptors [129]. This observation was taken to explain why chronic use of anticholinergics in obstructive airways disease may be accompanied by (increased) cholinergic hypersensitivity.

Finally, the effects of glucocorticosteroids on airway muscarinic receptors have not been fully established yet. Thus, *in vivo* treatment of rats has been reported to result in 40% upregulation [130] or 40–70% downregulation [131] of total receptor number, and no change or 50% upregulation in guinea pigs [132, 133]. Downregulation in rat lung was accompanied by decreased contractile sensitivity in airway smooth muscle, but upregulation in guinea pig lung did not affect lung strip sensitivity [131, 133]. A recent study in Basenji-Greyhound dogs showed that both tracheal M2 and M3 receptors were decreased, by some 45 and 70%, respectively, after *in vivo* but not after *in vitro* exposure to methylprednisolone; contractile function was not assessed [134]. Human studies on this subject are completely lacking (cf. [135]).

Concluding remarks

The presence and localization of muscarinic receptor subtypes in the airways have been established extensively over the past 15 years, with the possible exception of the M_4 subtype. Concerning the functional responses mediated by these receptor subtypes, much has been learned, but even more awaits delineation. Thus, it has been made very clear that M_3 receptors play a prominent role in smooth muscle contraction and mucus secretion in probably all animal species as well as in man, and that the same applies to prejunctional M_2 receptors in the autoinhibition of neurotransmitter release. It remains to be established, however, to what extent smooth muscle M_2 receptors are of actual importance in the fine-tuning of airway smooth muscle contraction and in the induction and/or modulation of proliferation. Even more uncertainty exists regarding the function of alveolar M_1 receptors. Moreover, changes in the presence, localization and function of the above muscarinic receptor subtypes in (chronic) airways diseases, and during and after drug therapy have only begun to be investigated. Thus, prejunctional M_2 receptors have been shown to be dysfunctional in human asthma and in various animal models of this disease. Most of the observations made in animals, however, such as the role of dysfunctional M_2 receptors in acute allergen-induced asthma, and the roles of eosinophil major basic protein, viral infections, cytokines and tachykinins in the development of M_2 receptor dysfunction (see the chapter by Adamko et al. in this volume), have not yet been extrapolated to man. It appears that these mechanisms, if indeed found to be important in human asthma, may provide a number of new targets for drug therapy, aimed at decreasing bronchial hyperresponsiveness and shortness of breath. Finally, important progress in our understanding of the role of muscarinic receptor subtypes may also be expected regarding the regulation of these receptors during and after drug therapy. This appears especially interesting now that long-acting β-adrenoceptor agonists such as formoterol and salmeterol have obtained a prominent position in bronchodilator therapy, and the long-acting and putatively M_3-selective anticholinergic bronchodilator tiotropium bromide may be introduced in the treatment of COPD.

References

1   Barnes PJ, Nadel JA, Roberts JM, Basbaum CB (1983) Muscarinic receptors in lung and trachea: autoradiographic localization using [³H]quinuclidinyl benzilate. *Eur J Pharmacol* 86: 103–106

2   Barnes PJ, Basbaum CB, Nadel JA (1983) Autoradiographic localization of autonomic receptors in airway smooth muscle. *Am Rev Respir Dis* 127: 758–762

3   Van Koppen CJ, Blankesteijn WM, Klaassen ABM, Rodrigues de Miranda JF, Beld AJ, Van Ginneken CAM (1988) Autoradiographic visualization of muscarinic receptors in human bronchi. *J Pharmacol Exp Ther* 244: 760–764

4   Mak JCW, Barnes PJ (1990) Autoradiographic visualization of muscarinic receptor subtypes in human and guinea pig lung. *Am Rev Respir Dis* 141: 1559–1568

5   Mak JCW, Haddad E-B, Buckley NJ, Barnes PJ (1993) Visualization of muscarinic m4 mRNA and M4 receptor subtype in rabbit lung. *Life Sci* 53: 1501–1508

6   Hislop AA, Mak JCW, Reader JA, Barnes PJ, Haworth SG (1998) Muscarinic receptor subtypes in the porcine lung during postnatal development. *Eur J Pharmacol* 359: 211–221

7   Van Koppen CJ, Blankesteijn WM, Klaassen ABM, Rodrigues de Miranda JF, Beld AJ, Van Ginneken CAM (1987) Autoradiographic visualization of muscarinic receptors in pulmonary nerves and ganglia. *Neurosci Lett* 83: 237–240

8   Barlow RB, Berry KJ, Glenton PAM, Nikolaou NM, Soh KS (1976) A comparison of affinity constants for muscarine-sensitive acetylcholine receptors in guinea pig atrial pacemaker cells at 29 °C and in ileum at 29 °C and 37 °C. *Br J Pharmacol* 58: 613–620

9   Hammer R, Berrie CP, Birdsall NJM, Burgen ASV, Hulme EC (1980) Pirenzepine distinguishes between different subclasses of muscarinic receptors. *Nature* 283: 90–92

10   Bonner TI (1989) New subtypes of muscarinic acetylcholine receptors. *Trends Pharmacol Sci* 10 (Suppl. Subtypes of Muscarinic Receptors IV):11–15

11   Eglen RM, Hegde SS, Watson N (1996) Muscarinic receptor subtypes and smooth muscle function. *Pharmacol Rev* 48: 531–565

12   Roffel AF, In 't Hout WG, De Zeeuw RA, Zaagsma J (1987) The M_2 selective antagonist AF-DX 116 shows high affinity for muscarine receptors in bovine tracheal membranes. *Naunyn-Schmiedeberg's Arch Pharmacol* 335: 593–595

13   Roffel AF, Elzinga CRS, Van Amsterdam RGM, De Zeeuw RA, Zaagsma J (1988) Muscarinic M_2 receptors in bovine tracheal smooth muscle: discrepancies between binding and function. *Eur J Pharmacol* 153: 73–82

14 Roffel AF, Elzinga CRS, Meurs H, Zaagsma J (1989) Allosteric interactions of three muscarine antagonists at bovine tracheal smooth muscle and cardiac M_2 receptors. *Eur J Pharmacol* 172: 61–70

15 Lucchesi PA, Scheid CR, Romano FD, Kargacin ME, Mullikin-Kilpatrick D, Yamaguchi H, Honeyman TW (1990) Ligand binding and G protein coupling of muscarinic receptors in airway smooth muscle. *Am J Physiol* 258: C730–C738

16 Fernandes LB, Fryer AD, Hirshman CA (1992) M_2 muscarinic receptors inhibit isoproterenol-induced relaxation of canine airway smooth muscle. *J Pharmacol Exp Ther* 262: 119–126

17 Yang CM (1991) Characterization of muscarinic receptors in dog tracheal smooth muscle cells. *J Auton Pharmacol* 11: 51–61

18 Schaefer OP, Ethier MF, Madison JM (1995) Muscarinic regulation of cyclic AMP in bovine trachealis cells. *Am J Respir Cell Mol Biol* 13: 217–226

19 Fryer AD, El-Fakahany EE (1990) Identification of three muscarinic receptor subtypes in rat lung using binding studies with selective antagonists. *Life Sci* 47: 611–618

20 Mahesh VK, Nunan LM, Halonen M, Yamamura HI, Palmer JD, Bloom JW (1992) A minority of muscarinic receptors mediate rabbit tracheal smooth muscle contraction. *Am J Respir Cell Mol Biol* 6: 279–286

21 Haddad E-B, Mak JCW, Hislop A, Haworth SG, Barnes PJ (1994) Characterization of muscarinic receptor subtypes in pig airways: radioligand binding and Northern blotting studies. *Am J Physiol* 266: L642–L648

22 Maeda A, Kubo T, Mishina M, Numa S (1988) Tissue distribution of mRNAs encoding muscarinic acetylcholine receptor subtypes. *FEBS Lett* 239: 339–342

23 Emala CW, Aryana A, Levine MA, Yasuda RP, Satkus SA, Wolfe BB, Hirshman CA (1995) Expression of muscarinic receptor subtypes and M_2-muscarinic inhibition of adenylyl cyclase in lung. *Am J Physiol* 268: L101–L107

24 Madison JM, Jones CA, Tom-Moy M, Brown JK (1987) Affinities of pirenzepine for muscarinic receptors isolated from bovine tracheal mucosa and smooth muscle. *Am Rev Respir Dis* 135: 719–724

25 Haddad E-B, Landry Y, Gies J-P (1991) Muscarinic receptor subtypes in guinea pig airways. *Am J Physiol* 261: L327–L333

26 Van Koppen CJ, Rodrigues de Miranda JF, Beld AJ, Hermanussen MW, Lammers JWJ, Van Ginneken CAM (1985) Characterization of the muscarinic receptor in human tracheal smooth muscle. *Naunyn-Schmiedeberg's Arch Pharmacol* 331: 247–252

27 Mak JCW, Baraniuk JN, Barnes PJ (1992) Localization of muscarinic receptor subtype mRNAs in human lung. *Am J Respir Cell Mol Biol* 7: 344–348

28 Widdop S, Daykin K, Hall IP (1993) Expression of muscarinic M2 receptors in cultured human airway smooth muscle cells. *Am J Respir Cell Mol Biol* 9: 541–546

29 Yasuda RP, Ciesla W, Flores LR, Wall SJ, Li M, Satkus SA, Weisstein JS, Spagnola BV, Wolfe BB (1993) Development of antisera selective for m4 and m5 muscarinic cholinergic receptors: distribution of m4 and m5 receptors in rat brain. *Mol Pharmacol* 43: 149–157

30 Scott T, McMahon KK (1988) Antagonists and agonists interactions with the muscarinic receptor of the rat large airways. *J Pharmacol Exp Ther* 247: 136–142

31 Gardier RW, Blaxall HS, Killian LN, Cunningham J (1991) Reserpine-induced post-receptor reduction in muscarinic-mediated airway smooth muscle contraction. *Life Sci* 48: 1705–1713

32 Wall SJ, Yasuda RP, Li M, Wolfe BB (1991) Development of an antiserum against m3 muscarinic receptors: distribution of m3 receptors in rat tissues and clonal cell lines. *Mol Pharmacol* 40: 783–789

33 Gies J-P, Bertrand C, Vanderheyden P, Waeldele F, Dumont P, Pauli G, Landry Y (1989) Characterization of muscarinic receptors in human, guinea pig and rat lung. *J Pharmacol Exp Ther* 250: 309–315

34 Mak JCW, Barnes PJ (1989) Muscarinic receptor subtypes in human and guinea pig lung. *Eur J Pharmacol* 164: 223–230

35 Casale TB, Ecklund P (1988) Characterization of muscarinic receptor subtypes on human peripheral lung. *J Appl Physiol* 65: 594–600

36 Bloom JW, Halonen M, Yamamura HI (1988) Characterization of muscarinic cholinergic receptor subtypes in human peripheral lung. *J Pharmacol Exp Ther* 244: 625–632

37 Böhm M, Gengenbach S, Hauck RW, Sunder-Plassmann L, Erdmann E (1991) Beta-adrenergic receptors and m-cholinergic receptors in human lung. *Chest* 100: 1246–1253

38 Haddad E-B, Mak JCW, Belvisi MG, Nishikawa M, Rousell J, Barnes PJ (1996) Muscarinic and β-adrenergic receptor expression in peripheral lung from normal and asthmatic patients. *Am J Physiol* 270: L947–L953

39 Bloom JW, Halonen M, Lawrence LJ, Rould E, Seaver NA, Yamamura HI (1987) Characterization of high affinity [^{3}H]pirenzepine and (–)-[^{3}H]quinuclidinyl benzilate binding to muscarinic cholinergic receptors in rabbit peripheral lung. *J Pharmacol Exp Ther* 240: 51–58

40 Lazareno S, Buckley NJ, Roberts FF (1990) Characterization of muscarinic M_4 binding sites in rabbit lung, chicken heart, and NG108-15 cells. *Mol Pharmacol* 38: 805–815

41 Dörje F, Levey AI, Brann MR (1991) Immunological detection of muscarinic receptor subtype proteins (m1-m5) in rabbit peripheral tissues. *Mol Pharmacol* 40: 459–462

42 Chelala JL, Kilani A, Miller MJ, Martin RJ, Ernsberger P (1998) Muscarinic receptor binding sites of the M4 subtype in porcine lung parenchyma. *Pharmacol Toxicol* 83: 200–207

43 Eglen RM, Whiting RL (1988) Comparison of the muscarinic receptors of the guinea-pig oesophageal muscularis mucosae and trachea *in vitro*. *J Auton Pharmacol* 8: 181–189

44 Eglen RM, Cornett CM, Whiting RL (1990) Interaction of p-F-HHSiD (p-fluoro-hexahydrosila-difenidol) at muscarinic receptors in guinea-pig trachea. *Naunyn-Schmiedeberg's Arch Pharmacol* 342: 394–399

45 Roffel AF, Elzinga CRS, Zaagsma J (1993) Cholinergic contraction of the guinea pig lung strip is mediated by muscarinic M_2-like receptors. *Eur J Pharmacol* 250: 267–279

46   Brichant J-F, Warner DO, Gunst SJ, Rehder K (1990) Muscarinic receptor subtypes in canine trachea. *Am J Physiol* 258: L349–L354

47   Yang CM, Chou S-P, Sung T-C (1991) Muscarinic receptor subtypes coupled to generation of different second messengers in isolated tracheal smooth muscle cells. *Br J Pharmacol* 104: 613–618

48   Yu M, Robinson NE, Wang Z, Derksen FJ (1992) Muscarinic receptor subtypes in equine tracheal smooth muscle. *Vet Res Commun* 16: 301–310

49   Van Nieuwstadt RA, Henricks PAJ, Hajer R, Van der Meer Van Roomen WA, Breukink HJ, Nijkamp FP (1997) Characterization of muscarinic receptors in equine tracheal smooth muscle *in vitro*. Vet Q 19: 54–57

50   Eltze M, Galvan M (1994) Involvement of muscarinic M_2 and M_3, but not of M_1 and M_4 receptors in vagally stimulated contractions of rabbit bronchus/trachea. *Pulmon Pharmacol* 7: 109–120

51   Garssen J, Van Loveren H, Gierveld CM, Van der Vliet H, Nijkamp FP (1993) Functional characterization of muscarinic receptors in murine airways. *Br J Pharmacol* 109: 53–60

52   Roffel AF, Elzinga CRS, Zaagsma J (1990) Muscarinic M_3 receptors mediate contraction of human central and peripheral airway smooth muscle. *Pulm Pharmacol* 3: 47–51

53   Itabashi S, Aikawa T, Sekizawa K, Ohrui T, Sasaki H, Takishima T (1991) Pre- and postjunctional muscarinic receptor subtypes in dog airways. *Eur J Pharmacol* 204: 235–241

54   Chiba Y, Sakai H, Misawa M (1998) Characterization of muscarinic receptors in rat bronchial smooth muscle *in vitro*. *Res Commun Mol Pathol Pharmacol* 102: 205–208

55   Watson N, Magnussen H, Rabe KF (1995) Pharmacological characterization of the muscarinic receptor subtype mediating contraction of human peripheral airways. *J Pharmacol Exp Ther* 274: 1293–1297

56   Ten Berge REJ, Roffel AF, Zaagsma J (1993) The interaction of selective and non-selective antagonists with pre- and postjunctional muscarinic receptor subtypes in the guinea pig trachea. *Eur J Pharmacol* 233: 279–284

57   Post MJ, Te Biesebeek JD, Doods HN, Wemer J, Van Rooij HH, Porsius AJ (1991) Functional characterization of the muscarinic receptor in rat lungs. *Eur J Pharmacol* 202: 67–72

58   Howell RE, Laemont K, Gaudette R, Raynor M, Warner A, Noronha-Blob L (1991) Characterization of the airway smooth muscle muscarinic receptor *in vivo*. *Eur J Pharmacol* 197: 109–112

59   Roffel AF, Davids JH, Elzinga CRS, Wolf D, Zaagsma J, Kilbinger H (1997) Characterization of the muscarinic receptor subtype(s) mediating contraction of the guinea-pig lung strip and inhibition of acetylcholine release in the guinea-pig trachea with the selective muscarinic receptor antagonist tripitramine. *Br J Pharmacol* 122: 133–141

60   Doods HN, Willim KD, Boddeke HWGM, Entzeroth M (1993) Characterization of muscarinic receptors in guinea-pig uterus. *Eur J Pharmacol* 250: 223–230

61   Melchiorre C, Bolognesi ML, Chiarini A, Minarini A, Spampinato S (1993) Synthesis

and biological activity of some methoctramine-related tetraamines bearing a 11-acetyl-5,11-dihydro-6H-pyrido[2,3-b][1,4]-benzodiazepin-6-one moiety as antimuscarinics: a second generation of highly selective M_2 muscarinic receptor antagonists. J Med Chem 36: 3734–3737

62   Meurs H, Timmermans A, Van Amsterdam RGM, Brouwer F, Kauffman HF, Zaagsma J (1989) Muscarinic receptors in human airway smooth muscle are coupled to phosphoinositide metabolism. Eur J Pharmacol 164: 369–371

63   Langlands JM, Rodger IW, Diamond J (1989) The effect of M&B 22948 on methacholine- and histamine-induced contraction and inositol 1,4,5-trisphosphate levels in guinea-pig tracheal tissue. Br J Pharmacol 98: 336–338

64   Roffel AF, Meurs H, Elzinga CRS, Zaagsma J (1990) Characterization of the muscarinic receptor subtype involved in phosphoinositide metabolism in bovine tracheal smooth muscle. Br J Pharmacol 99: 293–296

65   Sawyer GW, Ehlert FJ (1999) Muscarinic M_3 receptor inactivation reveals a pertussis toxin-sensitive contractile response in the guinea pig colon: evidence for M_2/M_3 receptor interactions. J Pharmacol Exp Ther 289: 464–476

66   Ramnarine SI, Haddad E-B, Khawaja AM, Mak JCW, Rogers DF (1996) On muscarinic control of neurogenic mucus secretion in ferret trachea. J Physiol 494.2: 577–586

67   Ishihara H, Shimura S, Satoh M, Masuda T, Nonaka H, Kase H, Sasaki T, Sasaki H, Takishima T, Tamura K (1992) Muscarinic receptor subtypes in feline tracheal submucosal gland secretion. Am J Physiol 262: L223–L228

68   Yang CM, Farley JM, Dwyer TM (1988) Muscarinic stimulation of submucosal glands in swine trachea. J Appl Physiol 64: 200–209

69   Gater PR, Alabaster VA, Piper I (1989) A study of the muscarinic receptor subtype mediating mucus secretion in the cat trachea in vitro. Pulm Pharmacol 2: 87–92

70   Yang CM, Farley JM, Dwyer TM (1988) Acetylcholine-stimulated chloride flux in tracheal submucosal gland cells. J Appl Physiol 65: 1891–1894

71   Bloom JW, Yamamura HI, Baumgartener C, Halonen M (1987) A muscarinic receptor with high affinity for pirenzepine mediates vagally induced bronchoconstriction. Eur J Pharmacol 133: 21–27

72   Bloom JW, Baumgartener-Folkerts C, Palmer JD , Yamamura HI, Halonen M (1988) A muscarinic receptor subtype modulates vagally stimulated bronchial contraction. J Appl Physiol 65: 2144–2150

73   Beck KC, Vettermann J, Flavahan NA, Rehder K (1987) Muscarinic M1 receptors mediate the increase in pulmonary resistance during vagus nerve stimulation in dogs. Am Rev Respir Dis 136: 1135–1139

74   Yang Z-J, Biggs DF (1991) Muscarinic receptors and parasympathetic neurotransmission in guinea-pig trachea. Eur J Pharmacol 193: 301–308

75   Maclagan J, Faulkner D (1989) Effect of pirenzepine and gallamine on cardiac and pulmonary muscarinic receptors in the rabbit. Br J Pharmacol 97: 506–512

76   Maclagan J, Fryer AD, Faulkner D (1989) Identification of M_1 muscarinic receptors in

pulmonary sympathetic nerves in the guinea-pig by use of pirenzepine. *Br J Pharmacol* 97: 499–505

77 Undem BJ, Myers AC, Barthlow H, Weinreich D (1990) Vagal innervation of guinea pig bronchial smooth muscle. *J Appl Physiol* 69: 1336–1346

78 Ten Berge REJ, Roffel AF, Zaagsma J (1995) Conditional involvement of muscarinic M$_1$ receptors in vagally mediated contraction of guinea-pig bronchi. *Naunyn-Schmiedeberg's Arch Pharmacol* 352: 173–178

79 Lammers J-WJ, Minette P, McCusker M, Barnes PJ (1989) The role of pirenzepine-sensitive (M$_1$) muscarinic receptors in vagally mediated bronchoconstriction in humans. *Am Rev Respir Dis* 139: 446–449

80 Killingsworth CR, Robinson NE (1992) The role of muscarinic M$_1$ and M$_2$ receptors in airway constriction in the cat. *Eur J Pharmacol* 210: 231–238

81 Fujimura M, Kamio Y, Matsuda T (1992) Effect of a M1-selective muscarinic receptor antagonist (pirenzepine) on basal bronchomotor tone in young women. *Respiration* 59: 102–106

82 Cazzola M, Russo S, De Santis D, Principe P, Marmo E (1987) Respiratory responses to pirenzepine in healthy subjects. *Int J Clin Pharmacol Ther Toxicol* 25: 105–109

83 Sertl K, Meryn S, Graninger W, Laggner A, Schlick W, Rameis H (1986) Acute effects of pirenzepin on bronchospasm. *Int J Clin Pharmacol Ther Toxicol* 24: 655–657

84 Cazzola M, Matera MG, D'Amato G, De Santis D, Maione S, Lisa M, Cenicola ML, Marmo E (1989) Evidence of muscarinic receptor subtypes in airway smooth muscle of normal volunteers and of chronic obstructive pulmonary disease patients. *Int J Clin Pharmacol Res* 9: 65–70

85 Ceyhan B, Celikel T, Simsir S, Kandemir B (1993) Comparison of the bronchodilator efficacy of nebulized pirenzepine and ipratropium bromide in patients with airway obstructive lung disease. *Int J Clin Pharmacol Ther Toxicol* 31: 510–513

86 Cazzola M, D'Amato G, Guidetti E, Staudinger H, Steinijans VW, Kilian U (1990) An M1-selective muscarinic receptor antagonist telenzepine improves lung function in patients with chronic obstructive bronchitis. *Pulmon Pharmacol* 3: 185–189

87 Ukena D, Wehinger C, Engelstätter R, Steinijans V, Sybrecht GW (1993) The muscarinic M$_1$-receptor-selective antagonist, telenzepine, had no bronchodilator effects in COPD patients. *Eur Respir J* 6: 378–382

88 Cazzola M, Matera MG, Liccardi G, Sacerdoti G, D'Amato G, Rossi F (1994) Effect of telenzepine, an M1-selective muscarinic receptor antagonist, in patients with nocturnal asthma. *Pulmon Pharmacol* 7: 91–97

89 Fryer AD, Maclagan J (1984) Muscarinic inhibitory receptors in pulmonary parasympathetic nerves in the guinea-pig. *Br J Pharmacol* 83: 973–978

90 Watson N, Barnes PJ, Maclagan J (1992) Actions of methcotramine, a muscarinic M$_2$ receptor antagonist, on muscarinic and nicotinic cholinoceptors in guinea-pig airways *in vivo* and *in vitro*. *Br J Pharmacol* 105: 107–112

91 Ten Berge REJ, Krikke M, Teisman ACH, Roffel AF, Zaagsma J (1996) Dysfunctional

muscarinic M_2 autoreceptors in vagally induced bronchoconstriction of conscious guinea pigs after the early allergic reaction. *Eur J Pharmacol* 318: 131–139

92   Faulkner D, Fryer AD, Maclagan J (1986) Postganglionic muscarinic inhibitory receptors in pulmonary parasympathetic nerves in the guinea-pig. *Br J Pharmacol* 88: 181–187

93   Minette PAH, Barnes PJ (1988) Prejunctional inhibitory muscarinic receptors on cholinergic nerves in human and guinea pig airways. *J Appl Physiol* 64: 2532–2537

94   D'Agostino G, Chiari MC, Grana E, Subissi A, Kilbinger H (1990) Muscarinic inhibition of acetylcholine release from a novel preparation of the guinea-pig trachea. *Naunyn-Schmiedeberg's Arch Pharmacol* 342: 141–145

95   Baker DG, Don HF, Brown JK (1992) Direct measurement of acetylcholine release in guinea pig trachea. *Am J Physiol* 263: L142–L147

96   Kilbinger H, Schneider R, Siefken H, Wolf D, D'Agostino G (1991) Characterization of prejunctional muscarinic autoreceptors in the guinea-pig trachea. *Br J Pharmacol* 103: 1757–1763

97   Kilbinger H, Von Bardeleben RS, Siefken H, Wolf D (1995) Prejunctional muscarinic receptors regulating neurotransmitter release in airways. *Life Sci* 56: 981–987

98   Ten Berge REJ, Weening EC, Roffel AF, Zaagsma J (1996) Differences in the prejunctional effects of methacholine and pilocarpine on the release of endogenous acetylcholine from guinea-pig trachea. *Naunyn-Schmiedeberg's Arch Pharmacol* 354: 606–611

99   Racké K, Hey C, Wessler I (1992) Endogenous noradrenaline release from guinea-pig isolated trachea is inhibited by activation of M_2 receptors. *Br J Pharmacol* 107: 3–4

100 Blaber LC, Fryer AD, Maclagan J (1985) Neuronal muscarinic receptors attenuate vagally-induced contraction of feline bronchial smooth muscle. *Br J Pharmacol* 86: 723–728

101 Killingsworth CR, Yu M, Robinson NE (1992) Evidence for the absence of a functional role for muscarinic M_2 inhibitory receptors in cat trachea *in vivo*: contrast with *in vitro* results. *Br J Pharmacol* 105: 263–270

102 Aas P, Fonnum F (1986) Presynaptic inhibition of acetylcholine release. *Acta Phsyiol Scand* 127: 335–342

103 Aas P, Maclagan J (1990) Evidence for prejunctional M_2 muscarinic receptors in pulmonary cholinergic nerves in the rat. *Br J Pharmacol* 101: 73–76

104 Wessler I, Klein A, Pohan D, Maclagan J, Racké K (1991) Release of [^{3}H]acetylcholine from the isolated rat or guinea-pig trachea evoked by preganglionic nerve stimulation; a comparison with transmural stimulation. *Naunyn-Schmiedeberg's Arch Pharmacol* 344: 403–411

105 Racké K, Bähring A, Brunn G, Elsner M, Wessler I (1991) Characterization of endogenous noradrenaline release from intact and epithelium-denuded rat isolated trachea. *Br J Pharmacol* 103: 1213–1217

106 Belmonte KE, Jacoby DB, Fryer AD (1997) Increased function of inhibitory neuronal M_2 muscarinic receptors in diabetic rat lungs. *Br J Pharmacol* 121: 1287–1294

107 Deckers IA, Rampart M, Bult H, Herman AG (1989) Evidence for the involvement of prostaglandins in modulation of acetylcholine release from canine bronchial tissue. *Eur J Pharmacol* 167: 415–418

108 Minette PAH, Lammers J-WJ, Dixon CMS, McCusker MT, Barnes PJ (1989) A muscarinic agonist inhibits reflex bronchoconstriction in normal but not in asthmatic subjects. *J Appl Physiol* 67: 2461–2465

109 Patel HJ, Barnes PJ, Takahashi T, Tadjkarimi S, Yacoub MH, Belvisi MG (1995) Evidence for prejunctional muscarinic autoreceptors in human and guinea pig trachea. *Am J Respir Crit Care Med* 152: 872–878

110 Ten Berge REJ, Zaagsma J, Roffel AF (1996) Muscarinic inhibitory autoreceptors in different generations of human airways. *Am J Respir Crit Care Med* 154: 43–49

111 Loenders B, Rampart M, Herman AG (1992) Selective M_3 muscarinic receptor antagonists inhibit smooth muscle contraction in rabbit trachea without increasing the release of acetylcholine. *J Pharmacol Exp Ther* 263: 773–779

112 Hey C, Wessler I, Racké K (1994) Muscarinic inhibition of endogenous noradrenaline release from rabbit isolated trachea: receptor subtype and receptor reserve. *Naunyn-Schmiedeberg's Arch Pharmacol* 350: 464–472

113 Wang Z, Robinson NE, Yu M (1993) Ach release from horse airway cholinergic nerves: effects of stimulation intensity and muscle preload. *Am J Physiol* 264: L269–L275

114 Larsen GL, Fame TM, Renz H, Loader JE, Graves J, Hill M, Gelfand EW (1994) Increased acetylcholine release in tracheas from allergen-exposed IgE-imunne mice. *Am J Physiol* 266: L263–L270

115 Brown LAS, Longmore WJ (1981) Adrenergic and cholinergic regulation of lung surfactant secretion in the isolated perfused rat lung and in the alveolar type II cell in culture. *J Biol Chem* 256: 66–72

116 Woods BA, Ng W, Thakorlal D, Liu AL, Perks AM (1996) Effects of acetylcholine on lung liquid production by *in vitro* lungs from fetal guinea pigs. *Can J Physiol Pharmacol* 74: 918–927

117 Van Koppen CJ, Lammers J-W J, Rodrigues de Miranda JF, Beld AJ, Van Herwaarden CLA, Van Ginneken CAM (1989) Muscarinic receptor binding in central airway musculature in chronic airflow obstruction. *Pulmon Pharmacol* 2: 131–136

118 Raaijmakers JAM, Terpstra GK, Van Rozen AJ, Witter A, Kreukniet J (1983) Muscarinic cholinergic receptors in peripheral lung tissue of normal subjects and of patients with chronic obstructive lung disease. *Clin Sci* 66: 585–590

119 De Jongste JC, Mons H, Block R, Bonta IL, Frederiks AP, Kerrebijn KF (1987) Increased *in vitro* histamine responses in human small airways smooth muscle from patients with chronic obstructive pulmonary disease. *Am Rev Respir Dis* 135: 549–553

120 Van Koppen CJ, Rodrigues de Miranda JF, Beld AJ, Van Ginneken CAM, Lammers J-W J, Van Herwaarden CLA (1988) Muscarinic receptor sensitivity in airway smooth muscle of patients with obstructive airway disease. *Arch Int Pharmacodyn* 295: 238–244

121 De Jongste JC, Mons H, Bonta IL, Kerrebijn KF (1987) *In vitro* responses of airways from an asthmatic patient. *Eur J Respir Dis* 71: 23–29

122 Bai TR (1990) Abnormalities in airway smooth muscle in fatal asthma. *Am Rev Respir Dis* 141: 552–557

123 Bai TR (1991) Abnormalities in airway smooth muscle in fatal asthma. A comparison between trachea and bronchus. *Am Rev Respir Dis* 143: 441–443

124 Ten Berge REJ, Santing RE, Hamstra JJ, Roffel AF, Zaagsma J (1995) Dysfunction of muscarinic M_2 receptors after the early allergic reaction: possible contribution to bronchial hyperresponsiveness in allergic guinea-pigs. *Br J Pharmacol* 114: 881–887

125 Santing RE, Pasman Y, Olymulder CG, Roffel AF, Meurs H, Zaagsma J (1995) Contribution of a cholinergic reflex mechanism to allergen-induced bronchial hyperreactivity in permanently instrumented, unrestrained guinea-pigs. *Br J Pharmacol* 114: 414–418

126 Yang Z-J, Biggs DF (1991) Muscarinic-receptor functioning in tracheas from normal and ovalbumin-sensitive guinea pigs. *Can J Physiol Pharmacol* 69: 871–876

127 Witt-Enderby PA, Yamamura HI, Halonen M, Palmer JD, Bloom JW (1993) Chronic exposure to a β_2-adrenoceptor agonist increases the airway response to methacholine. *Eur J Pharmacol* 241: 121–123

128 Wang Z-L, Bramley AM, McNamara A, Pare PD, Bai TR (1994) Chronic fenoterol exposure increases *in vivo* and *in vitro* airway responses in guinea pigs. *Am J Respir Crit Care Med* 149: 960–965

129 Witt-Enderby PA, Yamamura HI, Halonen M, Lai J, Palmer JD, Bloom JW (1995) Regulation of airway muscarinic cholinergic receptor subtypes by chronic anticholinergic treatment. *Mol Pharmacol* 47: 485–490

130 Marquardt DL, Motulsky HJ, Wasserman SI (1982) Rat lung cholinergic receptor: characterization and regulation by corticosteroids. *J Appl Physiol* 53: 731–736

131 Nabishah BM Morat PB, Kadir BA, Khalid BAK (1991) Effect of steroid hormones on muscarinic receptors of bronchial smooth muscle. *Gen Pharmac* 22: 389–392

132 Suzuki R, Takagi K, Satake T (1985) Changes in muscarinic acetylcholine receptors in guinea-pig lung: effect of aging, inhalation of an allergen, administration of drugs, and vagotomy. *Lung* 163: 173–182

133 Scherrer D, Lach E, Landry Y, Gies J-P (1997) Glucocorticoid modulation of muscarinic and β-adrenergic receptors in guinea pig lung. *Fundam Clin Pharmacol* 11: 111–116

134 Emala CW, Clancy J, Hirshman CA (1997) Glucocorticoid treatment decreases muscarinic receptor expression in canine airway smooth muscle. *Am J Physiol* 272: L745–L751

135 Hirst SJ, Lee TH (1998) Airway smooth muscle as a target of glucocorticoid action in the treatment of asthma. *Am J Respir Crit Care Med* 158: S201–S206

Functional roles of postjunctional muscarinic M₂ receptors in airway smooth muscle

Richard M. Eglen[1] and Nikki Watson[2]

[1]DiscoveRx Corp, 42501 Albrae St., Fremont, CA 94538, USA; [2]Department of Pharmacology, University of Virginia, 5316 Jordan Hall, 1300 Jefferson Park Ave., Charlottesville, VA 22908, USA

Introduction

Muscarinic receptors

Acetylcholine, the main neurotransmitter of the parasympathetic nervous system, acts by binding to two major receptor types, the nicotinic and the muscarinic receptor classes. Muscarinic receptors are composed of five subtypes, M_1–M_5, encoded by intronless genes, with endogenously expressed correlates in several tissues, including the respiratory tract [1]. Given this diversity, it has been a challenge to define the physiological roles for each subtype. As discussed below, a paucity of selective ligands remains for use as defining pharmacological tools, but gene-targeting techniques, such as receptor antisense and transgenic animals, could assist in this respect. While the antisense techniques have not met with much success in the examination of smooth muscle function, it is fair to state that this has not been extensively studied. Alternatively, transgenic mice, lacking muscarinic M_1, M_2, M_4 or M_5 receptor genes, have now been constructed but, as yet, no studies have been reported relating to airway smooth muscle function.

All muscarinic receptors conform to the archetypal motif suggested for many G protein-coupled receptors, in that they possess an extracellular amino terminus and an intracellular carboxyl terminal, between which there are seven transmembrane domains involved in ligand binding. In terms of cellular signaling, muscarinic M_2 and M_4 receptors preferentially couple to $G_{\alpha o/i}$, resulting in inhibition of adenylyl cyclase and closure of K^+ channels. Muscarinic M_1, M_3 and M_5 preferentially couple to G_{q11}, resulting in the stimulation of phospholipase C_β and an increase in intracellular calcium concentration [1]. Therefore, defining the role of a subtype on the basis of intracellular signaling mechanisms *per se* is problematic.

Characterization of the function of muscarinic receptor subtypes is generally restricted to pharmacological approaches. A rigorous approach is to define the muscarinic receptor by measuring ligand affinities under equilibrium conditions. Currently the subject of extensive investigation, muscarinic agonists have only limited

Muscarinic Receptors in Airways Diseases, edited by Johan Zaagsma, Herman Meurs and Ad F. Roffel

use in muscarinic receptor characterization, as none possess acceptable selectivity for a singular subtype. Several agonists have been developed that exhibit a functional muscarinic receptor subtype selectivity but, as this phrase implies, their selectivity relates to a differential efficacy rather than affinity [1, 2]. Consequently, their potency relies upon several factors such as receptor density and efficiency of the stimulus response coupling, as well as differences in receptor affinity or intrinsic efficacy.

Several antagonists have now been identified as displaying selectivity between the muscarinic receptors, although no compound has been identified that displays selectivity of more than two orders of magnitude for one subtype over the remaining four [2]. Compounds that display preferential selectivity for each of the five subtypes, although currently lacking, will provide invaluable tools in the classification of muscarinic receptors. Until then, however, it is critical to generate a profile of antagonist affinities to unambiguously characterize the functional response. Collectively, these classification issues render analysis of the physiological role of muscarinic receptors difficult; therefore, the combined use of several approaches, molecular, biochemical and pharmacological, is preferable. This paradigm must be remembered in the foregoing discussion defining the roles of the subtypes in airway smooth muscle. Careful reading of the literature reveals that extensive pharmacological analysis is rarely done and the identity of a subtype is frequently defined on minimal evidence.

Nonetheless, a consensus is now emerging on muscarinic receptor heterogeneity and its role in airway function. The interplay of the sympathetic and parasympathetic nervous system in maintenance of airway smooth muscle tone is well characterized and involves activation of the muscarinic M_3 receptor to elicit contraction and β-adrenoceptors to induce relaxation (Fig. 1a). This implies that the level of smooth muscle tone is a balance between those cellular events leading to contraction and those causing relaxation. The thesis of the present chapter is that the situation is more complex, involving participation of both muscarinic M_2 and M_3 receptors in the parasympathetic control of muscle tone.

Airway smooth muscle and receptor heterogeneity

Smooth muscle preparations from the respiratory tract have been historically employed as bioassays for studying the structure-activity relationships of many novel compounds at the muscarinic M_3 receptor [3]. However, of the five muscarinic receptor subtypes, at least four (M_1, M_2, M_3 and M_4) are expressed in airway tissue. In most species, contraction of airway smooth muscle is mediated by activation of muscarinic M_3 receptors; only recently has a contractile role of post-junctional muscarinic M_2 receptors been argued [4]. Although airway contractile responses are clearly mediated by activation of muscarinic M_3 receptors, competition radioligand-

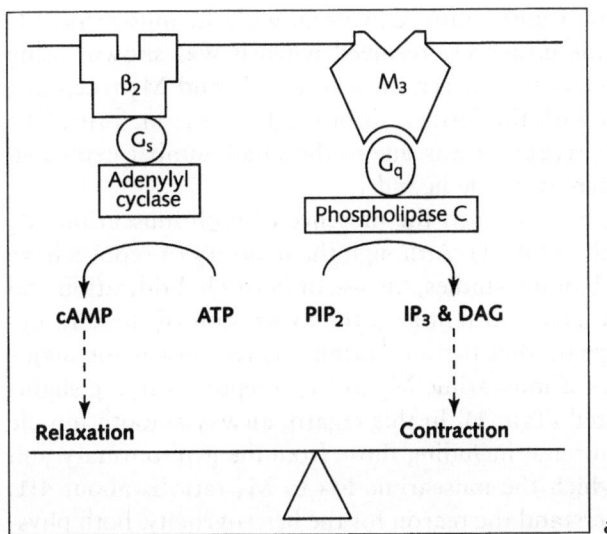

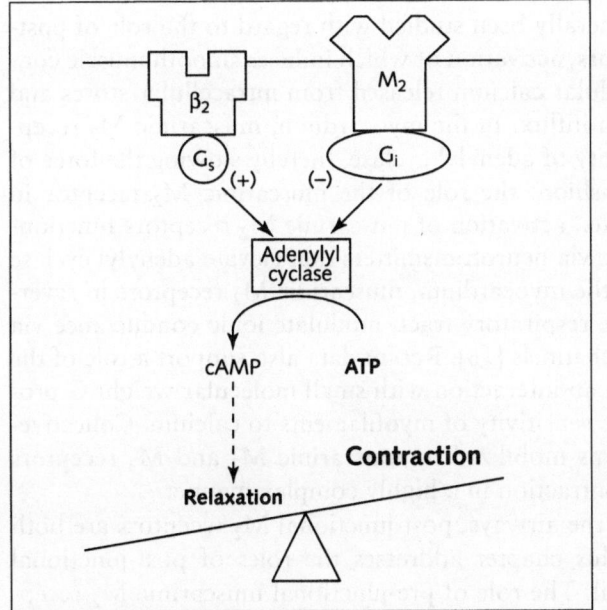

Figure 1

(a) Schematic representation of the traditional view of the regulation of smooth muscle tone.
(b) Inhibition by the muscarinic M₂ receptors of the β-adrenoceptor-mediated stimulation of
adenylyl cyclase. Overall contractile tone is schematically represented as a balance. cAMP,
cyclic adenosine monophosphate; PIP₂, Phosphatidyl inositol bisphosphate; IP₃, inositol
trisphosphate; DAG, diacylglycerol.

binding studies reveal a profile of affinities more consistent with the muscarinic M_2 receptor [5, 6]. This paradox was ultimately resolved when it was shown, using antagonists with a degree of selectivity between muscarinic M_2 and M_3 receptors, that both receptors were present with the former, surprisingly, in the majority. The inability to detect muscarinic M_3 receptors was due to the small number expressed and the poor selectivity of the then available ligands.

Since then, several studies have confirmed the presence of both muscarinic M_2 and M_3 receptors post-junctionally (Tab. 1). Although the majority of reports have utilized competition radioligand binding studies, the use of *in situ* hybridization and immunoprecipitation techniques have substantiated the expression of these receptors. In all of these studies it appears that the muscarinic M_2 receptor is the major subtype expressed, with the ratio of muscarinic M_2 to M_3 receptors varying slightly according to species investigated (Tab. 1). In this regard, airway smooth muscle resembles many other smooth muscles, including those from the genitourinary and gastrointestinal tracts [15], in which the muscarinic M_2 to M_3 ratio is about 4:1. Therefore, it is important to understand the reason for the heterogeneity, both physiologically and pathophysiologically, in the context of airway function.

Smooth muscle tissue has generally been studied with regard to the role of post-junctional muscarinic M_3 receptors, activation of which induces smooth muscle contraction via elevation of intracellular calcium released from intracellular stores and subsequent activation of calcium influx. In the myocardium, muscarinic M_2 receptor activation decreases the activity of adenylyl cyclase, thereby slowing the force of contraction. In an analogous fashion, the role of the muscarinic M_2 receptor in smooth muscle may be similar, i.e., activation of muscarinic M_2 receptors functionally opposes relaxation occurring via neurotransmitters that elevate adenylyl cyclase activity. Again by analogy with the myocardium, muscarinic M_2 receptors in several smooth muscles, including the respiratory tract, modulate ionic conductance via a direct interaction with several channels [16]. Recent data also support a role of the muscarinic M_2 receptor, through an interaction with small molecular weight G proteins and the enhancement of the sensitivity of myofilaments to calcium. Collectively, the signal transduction systems mobilized by muscarinic M_2 and M_3 receptors appear to act to cause muscle contraction in a highly complex manner.

It is therefore evident that in the airways, post-junctional M_2 receptors are both predominant and functional. This chapter addresses the roles of post-junctional muscarinic M_2 receptors in detail. The role of pre-junctional muscarinic M_2 receptors is covered elsewhere in this edition. It will be argued that muscarinic M_2 receptors can regulate smooth muscle tone via several intracellular mechanisms. However, for clarity, this chapter will be divided into discrete sections, including the relationship of muscarinic M_2 receptors to intracellular signaling molecules, ion channels and their role in the etiology of disease and aging. It should be noted, however, that this is an artificial separation since these mechanisms probably act in concert to regulate tone.

Table 1 - Heterogeneity of airway smooth muscle identified by radioligand binding studies using a range of muscarinic antagonists and immunoprecipitation studies.

Species	AF-DX 116 M$_2$:M$_3$	Methoctramine M$_2$:M$_3$	4-DAMP M$_2$:M$_3$	HHSiD M$_2$:M$_3$	Ref.
Cow	74%:26%	83%:17%	100%	100%	[5]
	85%:15%	N/D	85%:15%	N/D	[7]
Calf	N/D	N/D	60%:40%	N/D	[8]
Dog	N/D	72%:28%	55%:45%	56%:44%	[9]
	89%:11%	N/D	N/D	100%	[10]
Guinea-pig	52%:48%	64%:36%	100%	N/D	[11]
Rabbit	83%:17%	N/D	76%:24%	72%:28%	[12]
Pig	N/D	70%:30%	100%	N/D	[13]

AF-DX 116, [11-((((dimethylamino)-methyl)-1-piperidinyl)acetyl)-5,11-dihydro-6H-pyrido(2,3-b)(1,4)benzodiazepine-6-one]; 4-DAMP, 4-diphenylacetoxy-N-methyl piperidine methiodide; HHSiD, hexahydrosiladifenidol; N/D, not determined. Immunoprecipitation techniques in canine tracheal smooth muscle have revealed a ratio of M$_2$:M$_3$ receptors of approximately 8:1 [14].

Muscarinic M$_2$ receptors and signal transduction

Adenylyl cyclase

Classically, the activation of β-adrenoceptors relaxes airway smooth muscle via stimulation of adenylyl cyclase activity (Fig. 1a). Furthermore, the relaxant potency of β-adrenoceptor agonists, such as isoprenaline, depends not only upon the initial contractile state, but also on the type of spasmogen used to induce contraction [17–19]. It is important to note that muscarinic agonist-induced tone is more resistant to relaxation by β-adrenoceptor agonists than equivalent levels of tone generated by non-cholinergic agonists, including histamine, 5-HT or leukotriene D$_4$ [17, 18]. Two explanations have been advanced to account for this finding. One hypothesis centers on the muscarinic M$_3$ receptor subtype and phosphoinositide metabolism [19, 20] and is covered elsewhere in this book. The second explanation involves activation of the muscarinic M$_2$ receptor subtype and this is discussed below.

Augmentation of adenylyl cyclase activity generates cAMP, causing relaxation of smooth muscle. Since muscarinic M$_2$ receptors inhibit adenylyl cyclase activity and thus reduce intracellular cAMP, they may functionally oppose airway smooth muscle relaxation induced by this process (Fig. 1b). Substantial evidence now exists to support this model in smooth muscle from a number of different species and organs [21]. Muscarinic receptor-mediated inhibition of adenylyl cyclase has been demonstrated in airway smooth muscle [22–25] and since this inhibition is pertussis toxin-sensitive, the involvement of muscarinic M_2 receptors has been presumed [23]. Furthermore, both pertussis toxin treatment [26] and muscarinic M_2 receptor antagonists [10, 27] augment the relaxant potency of β-adrenoceptor agonists in airways pre-contracted with muscarinic agonists, further implicating a role of muscarinic M_2 receptors. This may also be the case in human airways, as direct inhibition of adenylyl cyclase activity has been demonstrated in cultured human tracheal smooth muscle cells [28]. Moreover, under certain conditions of contraction, activation of muscarinic M_2 receptors functionally antagonizes β-adrenoceptor-mediated relaxations of human bronchial smooth muscle [29].

There are, however, some contradictory reports in the literature that may argue against a functional role of the muscarinic M_2 receptor in opposing adenylyl cyclase-mediated relaxations. Notably in bovine trachea, smooth muscle muscarinic M_2 receptor antagonism does not affect the relaxant potency of isoprenaline [30, 31]. Nonetheless, in this tissue, muscarinic M_2 receptor antagonism alters the relaxant potency of forskolin [31]. The involvement of the muscarinic M_2 receptor appears to depend upon the amount of cAMP formed. Thus the ability of the muscarinic M_2 receptors to oppose β-adrenoceptor-mediated relaxation is markedly less than that seen with forskolin, a direct activator of adenylyl cyclase.

Furthermore, a portion of the relaxant response to isoprenaline in bovine trachea is mediated through a non-cAMP-dependent mechanism, and this mechanism is largely unaffected by M_2 receptor activation [21, 31]. This finding raises an interesting issue with respect to the mechanism of β-adrenoceptor-mediated relaxation. It is now established that, in addition to cAMP-dependent mechanisms, β-adrenoceptor agonists induce smooth muscle relaxation by cAMP-independent mechanisms, specifically by direct coupling of the receptor to calcium-dependent potassium channels [32] (Fig. 2). Discrepancies have also been reported in studies with guinea-pig trachea [27, 33] and human bronchi [29, 34]. However, in tissue from these species it remains to be determined whether differences in the level of adenylyl cyclase stimulation or non-cAMP-dependent mechanisms of β-adrenoceptor-mediated relaxation account for these differences.

The nature of the muscarinic receptor subtype mediating inhibition of β-adrenoceptor-mediated relaxation cannot be conclusively established from the functional studies discussed above, because the muscarinic antagonist affinity cannot be determined directly. This problem can be circumvented by muscarinic M_3 receptor alkylation studies [35, 36]. In these experiments selective alkylation of the muscarinic M_3 receptor population provides conditions (i.e., contracted with histamine and relaxed with isoprenaline) for studying muscarinic M_2 receptor function in isolation [37]. Consequently, the muscarinic M_2 receptor function appears as a "re-contraction", i.e., a reversal of the relaxation [35, 36]. This approach has revealed a func-

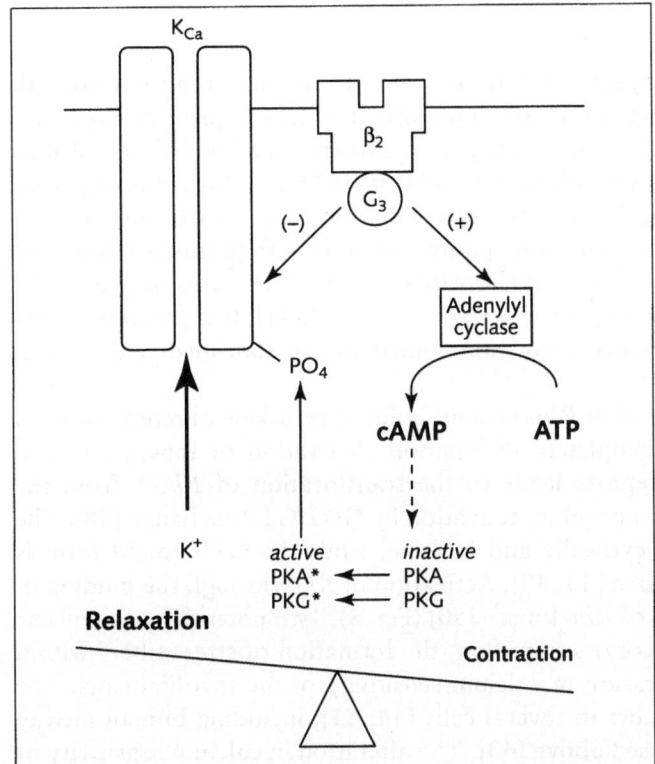

Figure 2
Schematic representation of the interaction of the β-adrenoceptors with both cyclic AMP-dependent relaxant pathways and the cAMP-independent maxi K channel (K_{Ca}). The efflux of potassium from the cell leads to membrane hyperpolarization and relaxation. PKA and PKG, protein kinase A and G, respectively; PO_4, site of phosphorylation.

tional role for muscarinic M_2 receptors in several isolated smooth muscle preparations, including the oesophagus, urinary bladder and fundus of the rat [38–40] and guinea-pig ileum [35, 36]. For guinea-pig isolated oesophagus or trachea the data using this approach are, however, contradictory [40–42]. These apparent differences may relate to the level of cAMP generated. That is, where muscarinic M_2 receptor-mediated "re-contraction" was not seen [41, 42], isoprenaline was used to stimulate adenylyl cyclase, while forskolin was used to stimulate adenylyl cyclase when "re-contractions" were seen [40]. Once again, it is possible that the greater the concentration of cAMP generated, the greater the role of muscarinic M_2 receptors in opposing adenylyl cyclase-induced relaxation.

Rho proteins

Recently, muscarinic M_2 receptors have been shown to interact in airway smooth muscle cells via G_i with a monomeric, small molecular weight G-protein, *Rho*, [43, 44]. As a subfamily of the *Ras* superfamily, *Rho* proteins are involved in various actin-dependent functions in smooth muscle cells, including cytoskeletal organization and contraction [45, 46]. In human airway smooth muscle, carbachol induces stress fiber formation through a signaling pathway that is both pertussis toxin- and *Clostridium botulinium* C3 exoenzyme-sensitive. Collectively, this suggests the involvement of both G_i and *Rho* proteins, respectively [43, 44]. It is possible, therefore, that muscarinic M_2 receptor activation contributes to *Rho*-mediated smooth muscle contraction [47].

Currently, the steps involved in *Rho* protein-induced cytoskeletal reorganization and cell contraction are incompletely understood. Activation of muscarinic and other G-protein coupled receptors leads to the translocation of *Rho*A from the cytosol to the plasma membrane and its activation by GDT/GTP exchange [48]. The GDP-bound form of *Rho* is cytosolic and inactive, while the GTP-bound form is membrane-associated and active [44, 49]. Activation of *Rho* through the binding of GTP results in the generation of *Rho* kinase [50] (Fig. 3). Two potential mechanisms contribute to *Rho*-mediated contraction: first, the formation of stress fibers within the cell and, second, an alteration in calcium sensitivity of the myofilaments. The formation of stress fibers occurs in several cells [44, 51], including human airway smooth muscle cells as discussed above [43]. The alteration in calcium sensitivity of the myofilaments is also inhibited by pertussis toxin, implying that muscarinic M_2 receptors induce contraction by sensitizing the muscle cell to intracellular calcium. This may be achieved by the activation of *Rho* kinase as a result of *Rho* activation. *Rho* kinase itself inactivates the myosin-binding subunit of myosin phosphatase. In this respect, activation of the muscarinic M_2 receptor acts in parallel with the activation of the muscarinic M_3 receptor and consequent mobilization of inositol phospholipids. Indeed, it has been argued that this is an explanation for the low Schild slope of the muscarinic antagonist, methoctramine, in airway smooth muscle, since activation of both muscarinic receptors results in contraction [47]. Against this, however, several studies have demonstrated both Schild slopes of unity [52] for selective muscarinic M_2 antagonists, as well as a lack of effect of pertussis toxin on the contraction [53].

Muscarinic M_2 receptors and ion channel function

Activation of muscarinic receptors modulates ion channel activity in two fundamental ways: directly, by an interaction with G-proteins and indirectly, via alterations in the activity of intracellular enzymes, ultimately leading to channel phos-

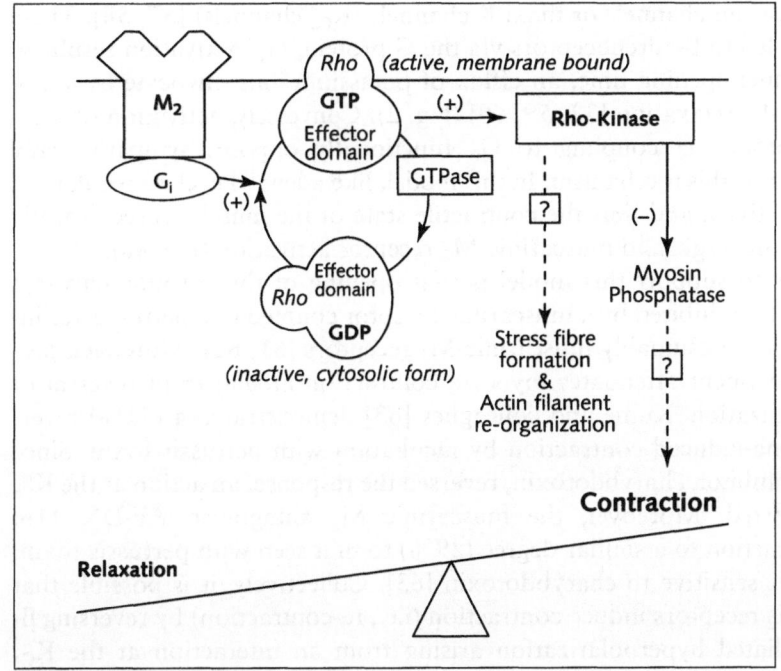

Figure 3
Schematic representation of the events leading to muscarinic M_2 receptor-mediated activation of Rho and consequent stress fiber formation and cell contraction.

phorylation [54]. Much of the early work identifying a role for muscarinic receptors in the regulation of ion channel function comes from studies in neurons and cardiac muscle. In neurons, activation of muscarinic M_2 receptors can increase potassium conductance, resulting in hyperpolarizing of the nerve terminals, a reduction in action potential duration and ultimately pre-synaptic inhibition. Alternatively, activation of muscarinic M_2 receptors can decrease potassium conductance (M current), suppressing a calcium-activated potassium channel [54]. The evidence for the interaction of the muscarinic M_2 receptor with ion channels in smooth muscle is discussed below.

Potassium channels

Several types of potassium channels exist in smooth muscles and regulate membrane potential and consequently smooth muscle tone [55, 56]. Of relevance to airway smooth muscle are the calcium-activated potassium channels, also referred to as large conductance potassium channels or maxi K channels (K_{Ca} channels) [57, 58]. These channels are coupled to β-adrenoceptors via the G protein, G_s. Activation results in an increased channel opening time, an efflux of potassium ions, myocyte hyperpolarization and finally relaxation [32, 59, 60] (Fig. 2). Conversely, activation of muscarinic M_2 receptors, via coupling to G_i, functionally opposes sympathetically induced relaxation by this mechanism. In this model, like adenylyl cyclase regulation, the K_{Ca} channel activity, and thus the contractile state of the muscle, is reciprocally modulated by β-adrenergic and muscarinic M_2 receptor activation (Fig. 4a).

Some evidence to support this model is that opening of the calcium-activated potassium channel is inhibited by a muscarinic receptor coupled to a pertussis toxin-sensitive G protein – presumably muscarinic M_2 receptors [61, 62]. Moreover, pertussis toxin pretreatment attenuates myocyte contraction arising from reversal of muscle hyperpolarization. Kume and colleagues [63] demonstrated a (31%) reversal of methacholine-induced contraction by incubation with pertussis toxin. Since the K_{Ca} channel inhibitor, charybdotoxin, reversed the response, an action at the K_{Ca} channel was inferred. Moreover, the muscarinic M_2 antagonist, AF-DX 116, reversed the contraction to a similar degree (29%) to that seen with pertussis toxin; again in a manner sensitive to charybdotoxin [63]. Collectively, it is possible that post-junctional M_2 receptors induce contraction (i.e., re-contraction) by reversing β-adrenoceptor-mediated hyperpolarization arising from an interaction at the K_{Ca} channel.

A further point is that elevations of intracellular calcium also inhibit potassium channel activity, thus eliminating the calcium release events that drive spontaneous channel opening. In airway myocytes, this occurs via activation of the muscarinic M_3 receptor. Consequently, both subtypes may play a role in functionally opposing relaxations induced by opening of potassium channels. This model provides a rationale for understanding why these two receptors are expressed in airway smooth muscle, since it allows for dual inhibitory control of membrane hyperpolarization.

Non-selective cation

Another means by which muscarinic M_2 and M_3 receptors regulate contraction by depolarization lies in the control of non-selective cation channels. The interplay of the regulation of intracellular calcium and opening of membrane- and sarcolemmal-associated channels is complex [64]. Activation of muscarinic receptors increases intracellular calcium concentration, as a result of both extracellular cation entry (*via* a non-selective cation channel) and liberation of calcium from intracellular stores. The influx of extracellular calcium occurs via opening of calcium channels on the muscle cell membrane (Fig. 4b). In airway smooth muscle, as in tissues such as the guinea-pig ileum, muscarinic receptor activation opens a pertussis toxin-sensitive, non-selective cation channel implicating the involvement of the post-junctional mus-

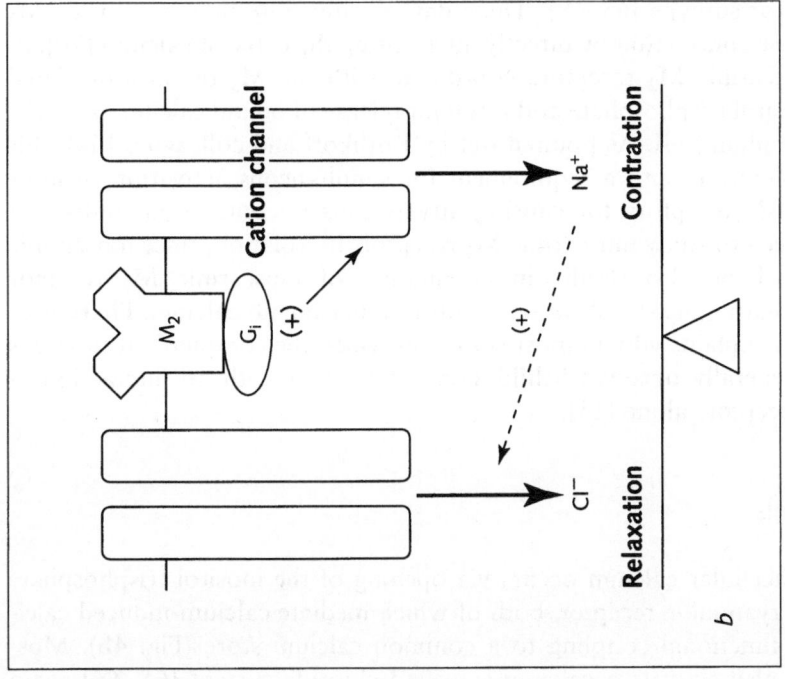

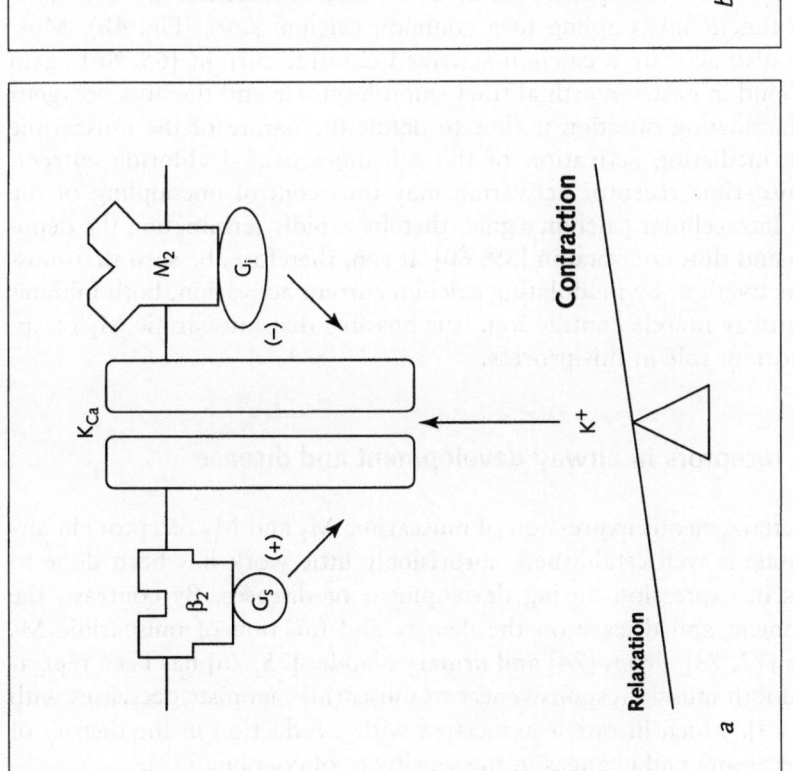

Figure 4

(a) Schematic representation of how muscarinic M$_2$ receptors interact with the maxi K channel to inhibit potassium efflux and there-fore tip the balance toward contraction.

(b) Illustration of the muscarinic M$_2$ receptor-mediated influx of calcium through non-selective cation channels promoting contrac-tion. Also shown is the calcium-dependent stimulation of chloride influx, tending to oppose contraction.

carinic M_2 receptor subtype [65–69]. These data demonstrate that muscarinic M_2 receptors can cause contraction by directly augmenting the entry of calcium [70]. In this manner, muscarinic M_2 receptors coordinate with the M_3 receptor-mediated generation of inositol trisphosphate and resultant release of bound calcium from the sarcoplasmic reticulum [64]. As pointed out by Kotlikoff and colleagues [64], this translates physiologically into a requirement for simultaneous activation of muscarinic M_2 and M_3 receptors for causing airway muscle contraction. Indeed, it emphasizes the need to study muscarinic M_2 receptors in isolation, since muscarinic M_3 receptor blockade also results in attenuation of muscarinic M_2 receptor response, by preventing the mobilization of intracellular bound calcium. Pharmacologically, this also explains why in most studies in which the contractile response is measured, one generally observes Schild slopes consistent with an interaction at muscarinic M_3 receptors alone [15].

Chloride channels

Elevation of intracellular calcium occurs via opening of the inositol trisphosphate receptor and the ryanodine receptor, both of which mediate calcium-induced calcium release by a functional coupling to a common calcium store (Fig. 4b). Muscarinic receptors also activate a calcium-activated chloride current [65, 66] again similar to that found in gastrointestinal tract smooth muscle and the anococcygeus muscle [71]. A fascinating question is thus to define the nature of the muscarinic receptor subtype mediating activation of the calcium-activated chloride current. Furthermore, muscarinic receptor activation may thus control uncoupling of the current from the intracellular calcium signal, thereby rapidly terminating the depolarizing stimulus and thus contraction [59, 60]. It can, therefore, be seen that muscarinic receptor activation, by modulating calcium current activation, both initiates and terminates airway muscle contraction. It is possible that muscarinic M_2 receptors play an important role in this process.

Muscarinic M_2 receptors in airway development and disease

Given that the heterogeneous expression of muscarinic M_2 and M_3 receptors in airway smooth muscle is well established, surprisingly little work has been done to measure changes in expression during development or diseases. By contrast, the effect of development and disease on the density and function of muscarinic M_2 receptors in atria [72, 73], ileum [74] and urinary bladder [75, 76] has been reported. In cardiac smooth muscle responsiveness to muscarinic agonists decreases with maturation [72, 73], which in rats is associated with a reduction in the density of muscarinic M_2 receptors and changes in the sensitivity of coupling [72].

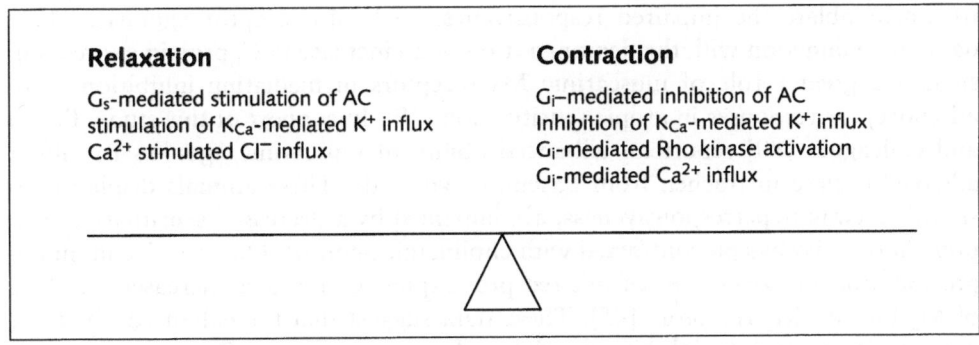

Relaxation	Contraction
G_s-mediated stimulation of AC	G_i-mediated inhibition of AC
stimulation of K_{Ca}-mediated K^+ influx	Inhibition of K_{Ca}-mediated K^+ influx
Ca^{2+} stimulated Cl^- influx	G_i-mediated Rho kinase activation
	G_i-mediated Ca^{2+} influx

Figure 5
An illustration of factors promoting relaxation and M_2 receptor-mediated contraction of airway smooth muscle. Overall contractile tone is schematically represented as a balance.

In the airways, studies of age-related changes in muscarinic M_2 receptor distribution are limited [77]. Moreover, discrimination between preparations of smooth muscle (airways) and parenchymal (lung tissue) is often not made. This is a concern, because of the regional differences associated with the distribution of muscarinic receptor subtypes in the airways [14]. In canine trachea, immunoprecipitation studies indicate that muscarinic M_2 receptors predominate over muscarinic M_3 receptors, in agreement with the majority of radioligand binding studies (Tab. 1). However, in bronchi there are approximately equal numbers of both muscarinic M_2 and M_3 subtypes, while in parenchymal tissue only the M_3 receptor subtype is detected [14]. Clearly, performing radioligand binding studies on membranes prepared from whole lung preparations may not detect subtle changes in receptor number associated with specific regions of the respiratory tract.

In rabbit tracheal smooth muscle, increased muscarinic functional antagonism of β-adrenoceptor-mediated relaxation correlates with age [78]. However, the inhibitory influence of muscarinic M_2 receptor activation on β-adrenoceptor-mediated relaxation is not the cause of this age-related effect, since differences in relaxant responsiveness persisted after M_2 receptor blockade. Furthermore, no significant changes in $G_{\alpha i}$ subunit expression were detected using Northern blot analysis [78]. These findings thus argue against a role for changes in muscarinic M_2 receptor function in maturation. However, this may not be the case in disease. In tissues from two models of airways hyperresponsiveness (atopic sensitized airways and the Basenji-greyhounds), alterations in muscarinic M_2 receptors may contribute to the models' pathology [79, 80]. Both models exhibit hyporesponsiveness to β-adrenoceptor agonist stimulation, which is also a characteristic feature of airways from asthmatic patients [81]. In atopic sensitized rabbit airway smooth muscle, Hakonarson and colleagues [79] report that muscarinic M_2 receptor antagonists and pertussis toxin treatment ablate the impaired responsiveness to β-adrenoceptor agonists. These data, in conjunction with the demonstration of an increase in G_i protein expression, suggest a greater role of muscarinic M_2 receptors in mediating inhibition of β-adrenoceptor responses in atopic sensitization [79]. In support of this study, Emala and colleagues [80] report an enhanced ability of muscarinic agonists to inhibit adenylyl cyclase in trachea from Basenji-greyhounds. These animals display non-specific airway hyperresponsiveness, accompanied by a decreased sensitivity to iso-prenaline in airways precontracted with cholinergic agonists. Quantitative immuno-precipitation analysis of muscarinic receptor expression revealed increased numbers of M_2 but not M_3 receptors [85]. These data suggest that the enhanced ability of muscarinic agonists to inhibit adenylyl cyclase in airways of Basenji-greyhounds relates to a greater density of muscarinic M_2 receptors [85]. Moreover, this may account, in part, for the impaired relaxation responses to β-adrenoceptor agonists. Unfortunately, no equivalent studies have been undertaken using airways from asth-matics, even though they too display reduced responsiveness to β-adrenoceptor ago-nists [81]. Therefore, it currently remains unclear whether changes in muscarinic M_2 receptors might contribute to asthmatic β-adrenoceptor hyporesponsiveness.

Chronic exposure to muscarinic antagonists and β-adrenoceptor agonists results in heterologous regulation of β-adrenoceptor and muscarinic receptor expression, respectively [82]. This study suggests that cross-talk exists between these receptors. Indeed, in chick isolated heart cells, chronic exposure to the β-adrenoceptor agonist, isoprenaline, results in a protein kinase A-dependent increase in muscarinic M_2 receptor expression [83]. In contrast, in human embryonic lung 299 cells, short-term exposure to the selective $β_2$-adrenoceptor agonist, procaterol, caused seques-tration and a consequent decrease in muscarinic M_2 receptors, with a significant increase in m2 mRNA. Twenty-four hour treatment leads to a down regulation and functional desensitization of the muscarinic M_2 receptor, a process that involves cAMP-dependent kinase and protein kinase C mechanisms [84]. Given that current treatment strategies for asthma involve repeated exposures to either muscarinic antagonists, long-acting β-adrenoceptor agonists or combinations thereof, studies such as these have clinical relevance. It is possible that the therapeutic efficacy of combination treatment regimens [85] of these agents in acute exacerbation of juve-nile asthma result from the inhibition of post-junctional muscarinic M_2, as well as muscarinic M_3, receptor function.

Conclusion

Substantial evidence supports a role of post-junctional muscarinic M_2 receptors in airway smooth muscle function. The ability of the muscarinic M_2 receptor to oppose adenylyl cyclase activity and ultimately relaxation is synergistic with a direct con-tractile role of the muscarinic M_3 receptor. Coupled with the ability of the muscarinic M_2 receptor to sensitize the contractile myofilaments to calcium, one can envisage a control over the contractile state of airway smooth muscle that varies according to the prevailing intracellular calcium concentration. Although the roles of muscarinic receptors in controlling potassium and calcium channel activity are not fully elucidated, details are emerging. Conceivably, muscarinic M_2 and M_3 receptors operate in parallel to reverse hyperpolarization and thus relaxation. Similarly, both subtypes act to regulate depolarization and thus contraction. This sophisticated model argues for a fine coordinate control over airway smooth muscle tone by the simultaneous activation of both muscarinic receptor subtypes. Since both are expressed in airway smooth muscle, this is likely to occur during parasympathetic nerve stimulation.

Relatively unexplored, then, is the influence of disease, such as asthma or chronic obstructive pulmonary disease (COPD), or development, on muscarinic M_2 receptor function in airway smooth muscle. This has considerable relevance, considering the role of the M_2 receptor in modulating the function of both the M3 receptor and the β-adrenoceptor. Since the M_2 receptor impairs β-adrenoceptor relaxation in atopically sensitized airways, the development of highly selective M_3 receptor antagonists as therapeutics for respiratory disease is arguable. Consequently, clinical efficacy data, from a direct comparison of a non-selective muscarinic antagonist against selective M_3 antagonists, are important in this respect. Future research will undoubtedly address these issues regarding the role of the muscarinic M_2 receptor in smooth muscle activity. Already it is evident that the role is subtle and diverse – exactly how diverse and subtle remains to be established.

References

1   Caulfield MO, Birdsall NJM (1990) International Union of Pharmacology. XVII Classification of muscarinic acetylcholine receptors. *Pharmacol Rev* 50: 279–290

2   Eglen RM, Watson N (1996) Selective muscarinic receptor agonists and antagonists. *Pharmacol Toxicol* 78: 59–68

3   Barlow RB, Franks FM, Pearson JDM (1972) A comparison of the affinities of antagonist for acetylcholine receptors in the ileum, bronchial muscle and iris of the guinea pig. *Br J Pharmacol* 46: 300–312

4   Eglen RM, Reddy H, Watson N, Challis RAJ (1994) Muscarinic acetylcholine receptor subtypes in smooth muscle. *Trends Pharmacol Sci* 15: 114–119

5   Roffel AF, Elzinga CRS, Van Amsterdam RGM, De Zeeuw RA, Zaagsma J (1988) Muscarinic M_2 receptors in bovine tracheal smooth muscle: discrepancies between binding and function. *Eur J Pharmacol* 153: 73–82

6   Roffel AF, Meurs H, Elzinga CRS, Zaagsma J (1989) Characterization of the muscarinic receptor subtype involved in phosphoinositide metabolism in bovine tracheal smooth muscle. *Br J Pharmacol* 99: 293–296

7   Lucchesi PA, Scheid CR, Romano FD, Kargacin ME, Mullikin-Kilpatrick D, Yamaguchi H, Honeyman TW (1990) Ligand binding and G protein coupling of muscarinic receptors in airway smooth muscle. *Am J Physiol* 258: C730–C738

8   Roets E, Burvenich C, Roberts M (1992) Muscarinic receptor subtypes, β-adrenoceptors and cAMP production in the trachealis smooth muscle of conventional and double-muscled calves. *Vet Res Commun* 16: 465–467

9   Yang CM (1991) Characterization of muscarinic receptors in dog tracheal smooth muscle cells. *J Auton Pharmacol* 11: 51–61

10 Fernandes LB, Fryer AD, Hirshman CA (1992) M_2 muscarinic receptors inhibit isoproterenol-induced relaxation of canine airway smooth muscle. *J Pharmacol Exp Ther* 262: 119–126

11 Haddad EB, Landry Y Gies J-P (1991) Muscarinic receptor subtypes in guinea-pig airways. *Am J Physiol.* 261: L327–L333

12 Mahesh VK, Nunan LM, Halonen M, Yamamura HI, Palmer JD, Bloom JW (1992) A minority of muscarinic receptors mediated rabbit tracheal smooth muscle contraction. *Am J Respir Cell Mol Biol* 6: 279–286

13 Haddad EB, Mak JCW, Hislop A, Haworth SG, Barnes PJ (1994) Characterization of muscarinic receptor subtypes in pig airways: radioligand binding and northern blotting studies. *Am J Physiol* 266: L642–L648

14 Emala CW, Aryana A, Levine MA, Yasuda RP, Satkus SA, Wolfe BB, Hirshman CA (1995) Expression of muscarinic receptor subtypes and M_2-muscarinic inhibition of adenylyl cyclase in lung. *Am J Physiol* 268: L101–L107

15 Eglen RM, Hegde SS, Watson N (1996) Muscarinic receptor subtypes and smooth muscle function. *Pharmacol Rev* 48: 531–565

16 Wang Y-X, Kotlikoff, MI (1998) Calcium release and calcium-activated chloride channels in airway smooth muscle. *Am J Respir Crit Care Med* 158: S109–S114

17 Koenig SM, Mitchel RW, Kelly E, White SR, Leff AR, Popovich KJ (1989) β-adrenergic relaxation of dog trachealis: contractile agonist-specific interaction. *J Appl Physiol* 67: 181–185

18 Torphy TJ (1988) Differential relaxant effects of isoproterenol on methacholine-versus leukotriene D_4-induced contraction in the guinea-pig trachea. *Eur J Pharmacol* 102: 549–553

19 Van Amsterdam RGM, Meurs H, Brouwer F, Posterma JB, Timmermans A, Zaagsma J (1989) Role of phosphoinositol metabolism in functional antagonism of airway smooth muscle contraction by b-adrenoceptor agonists. *Eur J Pharmacol* 172: 175–183

20 Offer GJ, Chilvers ER, Nahorski SR (1991) β-adrenoceptor-induced inhibition of muscarinic receptor-stimulated phosphoinositide metabolism is agonist-specific in bovine trachea smooth muscle. *Eur J Pharmacol.* 207: 243–248

21 Ehlert FJ, Sawyer GW, Esqueda EE (1999) Contractile role of M_2 and M_3 muscarinic receptors in gastrointestinal smooth muscle. *Life Sci* 64: 375–380

22 Jones CA, Madison JM, Tom-Moy M, Brown JK (1987) Muscarinic cholinergic inhibition of adenylate cyclase in smooth muscle. *Am J Physiol* 235: C97–C104

23   Sankary RM, Jones CA, Madison JM, Brown JK (1988) Muscarinic cholinergic inhibition of cyclic AMP in airway smooth muscle: Role of pertussis toxin-sensitive protein. *Am Rev Respir Dis* 138:145–150

24   Yang CM, Chou S-P, Sung T-C (1991) Muscarinic receptor subtypes coupled to generation of different second messengers in isolated tracheal smooth muscle cells. *Br J Pharmacol* 104: 613–618

25   Pyne NJ, Grady MW, Shehnaz D, Stevens PA, Pyne S, Rodger IW (1992) Muscarinic blockade of β-adrenoceptor-stimulated adenylyl cyclase; the role of stimulatory and inhibitory guanine-nucleotide binding regulatory proteins (G_s and G_i). *Br J Pharmacol* 107: 881–887

26   Mitchell RW, Koenig SM, Popovich KJ, Kelly E, Tallet, Leff AR (1993) Pertussis toxin augments β-adrenoceptor relaxation of muscarinic contraction canine trachealis. *Am Rev Respir Dis* 147: 327–331

27   Watson N, Eglen RM (1994) Effect of muscarinic M_2 and M_3 receptor stimulation and antagonism on responses to isoprenaline of guinea-pig trachea *in vitro*. *Br J Pharmacol* 112: 179–187

28   Widdop S, Daykin K, Hall IP (1993) Expression of muscarinic M_2 receptors in cultured human airway smooth muscle cells. *Am J Respir Cell Mol Biol* 9: 541–546

29   Naline E, Sarria B, Blanc M, Molimard M, Advenier C, Morcillo EJ (1997) Influence of muscarinic M_2 receptors on the acetylcholine-isoprenaline functional antagonism in the human isolated bronchus. *Am J Respir Crit Care Med* 155: A57

30   Meurs H, Elzinga CRS, De Boer REP, Van Amsterdam RGM, Brouwer F, Zaagsma J (1992) Muscarinic receptor mediate inhibition of adenylyl cyclase and its role in the functional antagonism of cholinergic airway smooth muscle contraction by β-agonist. *Am Rev Respir Dis* 145: A438

31   Ostrom RS, Ehlert FJ (1998) M_2 muscarinic receptors inhibit forskolin- but not isoproterenol-mediated relaxation in bovine tracheal smooth muscle. *J Pharmacol Exp Ther* 286: 234–242

32   Torphy TJ (1994) β-adrenoceptors, cAMP and airway smooth muscle relaxation: challenges to the dogma. *Trends Pharmacol Sci* 15: 370–374

33   Roffel AF, Meurs H, Elzinga CRS, Zaagsma J (1993) Muscarinic M2 receptors do not participate in the functional antagonisms between methacholine and isoprenaline in guinea-pig tracheal smooth muscle. *Eur J Pharmacol* 249: 235–238

34   Watson N, Magnussen H, Rabe KF (1995) Antagonism of β-adrenoceptor-mediated relaxations of human bronchial smooth muscle by carbachol. *Eur J Pharmacol* 275: 307–310

35   Reddy H, Watson N, Ford APDW, Eglen RM (1995) Characterization of the interaction between muscarinic M_2 receptors and β-adrenoceptor subtypes in guinea-pig isolated ileum. *Br J Pharmacol* 114: 49–56

36   Thomas EA, Baker S, Ehlert FJ (1996) Functional role of the M_2 muscarinic receptor in smooth muscle of guinea-pig ileum. *Mol Pharmacol* 4: 102–110

37   Eglen RM, Reddy H, Watson N (1994) Selective inactivation of muscarinic receptor subtypes. *Int J Biochem* 26: 1357–1368

38 Eglen RM, Peele B, Pulido-Rios MT, Leung E (1996) Functional interactions between muscarinic M_2 receptors and 5-hydroxytryptamine $(5HT)_4$ and β_3-adrenoceptors in isolated eosophageal muscularis mucosae of the rat. *Br J Phrmacol* 119: 595–601

39 Hegde SS, Choppin A, Bonhaus S, Briaud S, Loeb, TM Moy, Loury D, Eglen RM (1997) Functional role of M_2 and M_3 muscarinic receptors in the urinary bladder of rats *in vitro* and *in vivo*. *Br J Pharmacol* 120: 1409–1418

40 Thomas EA, Ehlert FJ (1996) Involvement of the M_2 muscarinic receptor in contractions of guinea-pig trachea, guinea-pig esophagus and rat fundus. *Biochem Pharmacol* 51: 779–788

41 Watson N, Reddy H, Eglen RM (1995) Characterization of muscarinic receptor and b-adrenoceptor interactions in guinea-pig oesophageal muscularis mucosae. *Eur J Pharmacol* 275: 307–310

42 Watson N, Reddy H, Eglen RM (1995) Role of muscarinic M_2 and M_3 receptors in guinea-pig trachea: effects of receptor alkylation. *Eur J Pharmacol* 278: 195–201

43 Togashi H, Emala CW, Hall IP, Hirshman CA (1998) Carbachol-induced actin reorganization involves G_i activation of Rho in human airway smooth muscle cells. *Am J Physiol* 274: L803–L809

44 Hirshman CA, Togashi H, Shao D, Emala CW (1998) $G_{\alpha i}$-2 is required for carbachol-induced stress fiber formation in human airway smooth muscle cells. *Am J Physiol* 275: L911–L916

45 Wang P, Bitar KN (1998) Rho A regulates sustained smooth muscle contraction through cytoskeletal reorganization of HSP27. *Am J Physiol* 275: G1454–G1462

46 Hall A (1990) The cellular functions of small GTP-binding proteins. *Science* 249: 635–640

47 Hirshman C, Lande B, Croxtomn TL (1999) Role of M_2 muscarinic receptors in airway smooth muscle contraction. *Life Sci* 64: 443–448

48 Keller J, Schmidt M, Hussein B, Rümenapp U, Kakobs KH (1997) Muscarinic receptor-stimulated cytosol-membrane translocation of RhoA. *FEBS Letts* 403: 299–302

49 Hirata K, Kikuchi A, Sasaki T, Kuroda S, Kaibuchi K, Matsuura Y, Seki H, Saida K, Takai Y (1992) Involvement of rho p21 in the GTP-enhanced calcium ion sensitivity of smooth muscle contraction. *J Biol Chem* 267: 8719–8722

50 Kimura K, Ito M, Amano M, Chihara K, Fukata Y, Nakafuku M, Yamamori B, Feng J, Nakano T, Okawa K et al (1996) Regulation of myosin phosphatase by Rho and Rho-associated kinase (Rho-Kinase). *Science* 273: 245–248

51 Bussey H (1996) Cell shape determination: A pivotal role for Rho. *Science* 272: 224–225

52 Eglen RM, Montgomery WW, Dainty IA, Dubuque LK, Whiting RL (1988) The interaction of methoctramine and himbacine at atrial, smooth muscle and endothelial muscarinic receptors *in vitro*. *Br J Pharmacol* 95: 1031–1038

53 Eglen RM, Huff MM, Montgomery WW, Whiting RL (1988) Differential effects of pertussis toxin on muscarinic responses in isolated atria and smooth muscle. *J Auton Pharmacol* 8: 29–37

54   Christie MJ, North RA (1988) Control of ion conductances by muscarinic receptors. *Trends Pharmacol Sci Suppl*: 30–34

55   Edwards G, Weston AH (1990) Potassium channel openers and vascular smooth muscle relaxation. *Pharmac Ther* 48: 237–258

56   Yamade M, Inanobe A, Kurchi Y (1998) G protein regulation of potassium ion channels. *Pharmacol Rev* 50: 723–757

57   Jones TR, Charette L, Garcia ML, Kaczorowski GJ (1990) Selective inhibition of relaxation of guinea-pig trachea by charybdotoxin, a potent Ca^{2+}-activated K^+ channel inhibitor. *J Pharmacol Exp Ther* 255: 697–706

58   Miura M, Belvisi MG, Stretton CD, Yacoub MH, Barnes PJ (1992) Role of potassium channels in bronchodilator responses in human airways. *Am Rev Respir Dis* 146: 132–136

59   Wang Y-X, Fleishmann BK, Kotlikoff, MI (1997) Modulation of maxi-K^+ channels by voltage dependent Ca2+ channels and methacholine in single airway myocytes. *Am J Physiol* 272: C1151–C1159

60   Wang Y-X, Kotlikoff MI (1997) Muscarinic signaling pathway for calcium release and calcium-activated chloride current in smooth muscle. *Am J Physiol* 273: C509–C519

61   Kume H, Kotlikoff MI (1991) Muscarinic inhibition of single K_{Ca} channels in smooth muscle cells by a pertussis-sensitive G protein. *Am J Physiol.* 261: C1204–C1209

62   Kume H, Hall IP, Washabau RJ, Takagi K, Kotlikoff MI (1994) Beta-adrenergic agonists: regulate K_{Ca} channels in airway smooth muscle by cAMP-dependent and -independent mechanisms. *J Clin Invest* 93: 371–379-C

63   Kume H, Mikawa K, Takagi K, Kotlikoff MI (1995) Role of G proteins and K_{Ca} channels in the muscarinic regulation and β-adrenergic of tracheal smooth muscle. *Am J Physiol* 286: L221–L229

64   Kotlikoff MI, Wang Y-X (1998) Calcium release and calcium-activated chloride channels in airway smooth muscle cells. *Am J Respirt Crit Care Med* 158: S109–S114

65   Janssen LJ, Sims SM (1992) Acetylcholine activates non-selective cation and chloride conductances in canine and guinea pig tracheal myocytes. *J Physiol* 453: 197–218

66   Janssen LJ, Sims SM (1993) Emptying and refilling of Ca^{2+} store in tracheal myocytes as indicated by Ach-evoked currents and contraction. *Am J Physiol* 265: C877–C886

67   Wade GR, Barbera J, Sims SM (1996) Cholinergic inhibition of Ca^{2+} current in guinea pig gastric and tracheal smooth muscle cells. *J Physiol* 491: 307–319

68   Pucovsky V, Zholos AV, Bolton TB (1998) Muscarinic cation current and suppression of Ca^{2+} current in guinea-pig ileal smooth muscle cells. *Eur J Pharmacol* 346: 323–330

69   Zholos AV, Bolton TB (1997) Muscarinic receptor subtypes controlling the cation current in guinea-pig ileal smooth muscle. *Br J Pharmacol* 122: 885–893

70   Wang Y-X, Fleishmann BK Kotlikoff MI (1997) M₂ receptor activation of nonselective cation channels in smooth muscle cells: calcium and G_i/G_o requirements. *Am J Physiol* 273: 42: C500–C508

71   Byrne NG, Large WA (1987) Membrane mechanism associated with muscarinic recep-

tor activation in single cells freshly dispersed from the rat anococcygeus muscle. *Br J Pharmacol* 92: 371–379

72   Borda ES, Leiros CP, Camusso JJ, Bacman S, Sterin-Borda L (1997) Differential cholinoceptor subtype-dependent activation of signal transduction pathways in neonatal versus adult rat atria. *Biochem Pharmacol* 53: 959–967

73   Poller U, Nedelka G, Radke J, Pönicke K, Brodde O-E (1997) Age-dependent changes in cardiac muscarinc receptor function in healthy volunteers. *JACC* 29:187–193

74   Michalek H, Fontana S, Pintor A (1993) Age-related changes in muscarinic receptor and post-receptor mechanisms in brain and ileum strips of rats. *Acta Neurobiol Exp* 53: 93–101

75   Latifpour A, Kondo S, O'Hollaren B, Morita T, Weiss RM (1990) Autonomic receptors in urinary tract: sex and age differences. *J Pharmacol Exp Ther* 253: 661–667

76   Braverman A, Legos J, Young W, Luthin G, Ruggieri M (1999) M_2 receptors in genitourinary smooth muscle pathology. *Life Sci* 64: 429–436

77   Hislop AA, Mak JCW, Reader JA, Barnes PJ, Haworth SG (1998) Muscarinic receptor subtypes in the porcine lung during postnatal development. *Eur J Pharmacol* 359: 211–221

78   Schramm CM, Arjona NC, Grunstein MM (1995) Role of muscarinic M_2 receptors in regulation β-adrenergic responsiveness in maturing rabbit airway smooth muscle. *Am J Physiol* 269: L783–L790

79   Hakonarson H, Herrick DJ, Grunstein MM (1995) Mechanism of impaired β-adrenoceptor responsiveness in atopic sensitized airway smooth muscle. *Am J Physiol* 269: L645–L652

80   Emala CW, Aryana A, Levine MA, Yasuda RP, Satkus SA, Wolfe BB, Hirshman CA (1995) Basenji-greyhound dog; increased m2 muscarinic receptor expression in trachealis muscle. *Am J Physiol* 268: L935–L940

81   Goldie RG, Spina D, Henry PJ, Lulich KM, Paterson JW (1986) *In vitro* responsiveness of human asthmatic bronchus to carbachol, histamine β-adrenoceptor agonist and theophyline. *Br J Clin Pharmacol* 22: 669–676

82   Haddad E-B, Rousell J (1998) Regulation of the expression and function of the M_2 muscarinic receptor. *Trends Pharmacol Sci* 19: 322–327

83   Jackson DA, Nathanson NM (1995) Subtype-specific regulation of muscarinic receptor expresssion and function by heterologous receptor activation. *J Biol Chem* 270: 22374–22377

84   Rousell J, Haddad E-B, Mak JCW, Webb BLJ, Giembycz MA, Barnes PJ (1996) β-Adrenoceptor-mediated down-regulation of M_2 muscarinic receptors: Role of cyclic adenosine 5'-monophosphate-dependent protein kinase and protein kinase C. *Mol Pharmacol* 49: 629–635

85   Plotnick LH, Ducharme FM (1998) Should inhaled anticholinergics be added to β_2-agonists for treating acute childhood and adolescent asthma? *BMJ* 317: 971–977

Dysfunction of prejunctional muscarinic M$_2$ receptors: role of environmental factors

Darryl J. Adamko, Allison D. Fryer and David B. Jacoby

Johns Hopkins School of Medicine and Johns Hopkins School of Public Health, Johns Hopkins Asthma and Allergy Center, 5501 Hopkins Bayview Circle, Baltimore, MD 21224, USA

Introduction

Asthma is characterized by periods of quiescence interrupted by exacerbations. As many as 20% of patients with asthma require hospitalization at some point for these exacerbations.

In most cases, the increase in difficulty with breathing is caused by an environmental factor. Asthma attacks may be precipitated by viral infections of the airways [1] as well as by inhalation of allergens [2]. Associations of air pollution with increases in hospitalization rates for asthma suggest that irritation of the airways by pollutants may also exacerbate asthma [3–5]. In all cases, increased vagally mediated reflex bronchoconstriction may be important.

Inhibitory muscarinic receptors on airway parasympathetic nerves

Our studies suggest that loss of function of an inhibitory receptor, the M$_2$ muscarinic receptor, on the vagus nerves may account for this increase in vagally mediated bronchoconstriction. In the lungs, the vagus nerves provide the dominant autonomic control. The release of acetylcholine from postganglionic parasympathetic fibers stimulates smooth muscle contraction by binding to M$_3$ receptors on the smooth muscle. At the same time, acetylcholine binds to inhibitory M$_2$ receptors on the nerve endings themselves (Fig. 1). This provides a negative feedback, decreasing further release of acetylcholine.

When these inhibitory neuronal M$_2$ receptors are blocked using M$_2$ selective antagonists such as gallamine, the bronchoconstrictor response to electrical stimulation of the vagus nerve can be increased as much as tenfold. When M$_2$ receptors are stimulated with the muscarinic agonist pilocarpine, bronchoconstrictor response to vagal stimulation is inhibited by as much as 85%. These inhibitory neuronal M$_2$ receptors have been demonstrated in the airways of guinea pigs [6], dogs [7], cats [8, 9] and rats [10]; they are also present in man [11].

Muscarinic Receptors in Airways Diseases, edited by Johan Zaagsma, Herman Meurs and Ad F. Roffel

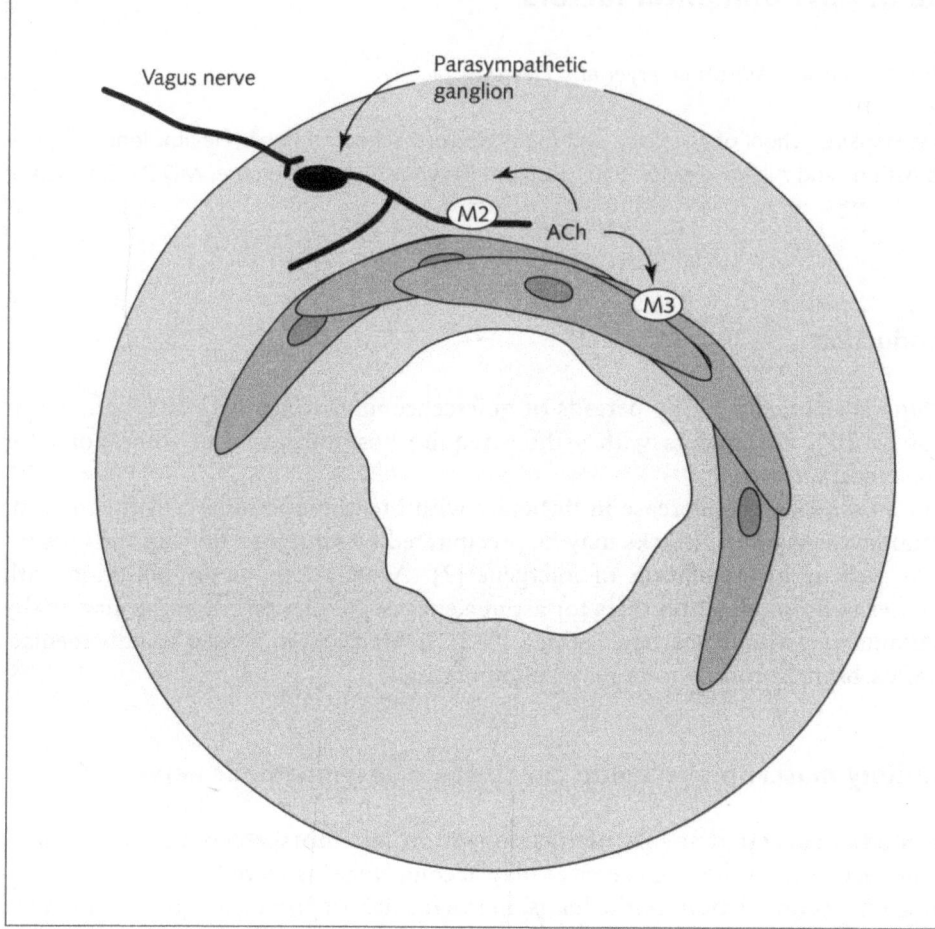

Figure 1
Acetylcholine is released from postganglionic nerve fibers in the airways. Binding of acetyl-
choline to M_3 muscarinic receptors on the airway smooth muscle causes contraction and
bronchoconstriction. Binding of acetylcholine to M_2 muscarinic receptors on the nerve end-
ing provides a negative feedback, limiting the further release of acetylcholine.

The function of these receptors is also impaired in some [12, 13], but not all [14], patients with asthma We have shown that inhibitory neuronal M_2 receptors are dysfunctional in guinea pigs after viral infections [15], inhalation of allergens [16, 17] and exposure to ozone [18]. In each of these conditions, and in patients with asthma, parasympathetic drive to the airway smooth muscle is increased [19–21]. Loss of M_2 receptor function increases release of acetylcholine from the parasympathetic nerves, potentiating vagally mediated bronchoconstriction. This may contribute to the airway hyperresponsiveness characterizing asthma [22, 23].

We have studied the mechanisms of airway hyperresonsiveness and M_2 receptor dysfunction after viral infections, allergen inhalation and ozone inhalation. The mechanisms of M_2 receptor dysfunction differ among these models, and the differences, as well as the similarities and overlaps, may be instructive as potential mechanisms of asthma and of asthma attacks.

Allergen-induced airway hyperresponsiveness and M_2 receptor dysfunction – the role of eosinophils

Sensitization of guinea pigs to ovalbumin, given via intraperitoneal injections, followed by inhalation of an aerosol of ovalbumin, causes airway hyperresponsiveness [22, 23]. This is accompanied by, and due to, loss of function of airway M_2 receptor function, potentiating vagally mediated reflex bronchoconstriction.

Inhalation of allergens in humans and of ovalbumin in sensitized guinea pigs causes an influx of eosinophils into the airways. This has long been suspected to be of pathogenic significance in increasing airway responsiveness [24], although the mechanisms responsible were incompletely understood. The importance of eosinophils in allergen-induced hyperresponsiveness and M_2 receptor dysfunction in guinea pigs are highlighted by studies in which eosinophils were depleted by treatment with a monoclonal antibody to interleukin-5 (IL-5) [25]. This prevents allergen-induced airway eosinophilia, as well as the attendant hyperresponsiveness and M_2 receptor dysfunction. Likewise, prevention of the migration of eosinophils into the airways by treatment with a monoclonal antibody to the adhesion molecule VLA-4 (which normally facilitates migration of the eosinophil by binding to VCAM on the endothelial cell) also prevents allergen-induced hyperresponsiveness and M_2 receptor dysfunction [26].

Having established the role of the eosinophil in the airway response to allergen inhalation, the question arises of the mechanism by which the eosinophil causes M_2 receptor dysfunction. As the M_2 receptor is heavily sialated [27], giving it a net negative charge, and because some positively charged proteins are known to bind to M_2 receptors [28], we used a combination of in vitro and in vivo approaches to investigate the properties of positively charged eosinophil proteins on M_2 receptor function and airway responsiveness.

The eosinophil contains three strongly cationic proteins: major basic protein (MBP), eosinophil cationic protein and eosinophil peroxidase. In addition, the eosinophil granule contains a less charged protein called eosinophil-derived neurotoxin. We tested the effects of these proteins (purified from human eosinophils) on the binding of n-methylscopolamine (NMS) to M_2 muscarinic receptors in vitro

[29]. The most plentiful of the eosinophil proteins, MBP, was also the most potent M_2 receptor antagonist, displacing ^{3}HNMS from M_2 receptors with a K_i of 15 μM. We demonstrated that MBP is an allosteric antagonist, a property that is shared by eosinophil peroxidase, but not by eosinophil cationic protein. Thus other factors, in addition to the positive charge, are important in determining binding to M_2 receptors. The non-cationic eosinophil-derived neurotoxin had no effect on M_2 receptors.

We have taken several approaches to demonstrating the relevance of the *in vitro* findings to allergen-induced airway hyperresponsiveness and M_2 receptor dysfunction. Initially, we took advantage of the fact that heparin, an anionic polysaccharide, binds and neutralizes MBP. We were able to demonstrate that heparin, injected intravenously, has no effect on airway responsiveness in control guinea pigs. However, when injected into allergen-challenged guinea pigs, heparin causes a 50% decrease in vagally mediated bronchoconstriction over the course of 20 min [17]. We showed that this is due to restored function of the airway M_2 receptors. A similar effect can be demonstrated using poly-l-glutamate, which is also anionic and binds MBP. The potential therapeutic relevance of these findings is underscored by the fact that partially desulfated heparin, which is devoid of anticoagulant activity, has similar effects on airway responsiveness and M_2 receptor function [30].

That heparin was working by neutralizing MBP was subsequently demonstrated by our studies showing that an antibody to MBP prevented allergen-induced M_2 receptor dysfunction [31]. Thus allergen inhalation causes eosinophils to move into the airways and to become activated, releasing MBP which binds to vagal M_2 receptors, preventing negative feedback inhibition of acetylcholine release and potentiating vagally mediated reflex bronchoconstriction.

We have subsequently investigated the mechanisms of recruitment and activation of eosinophils. An important observation in this regard is that eosinophils appear to be selectively recruited to the airway nerves, appearing in and around airway nerve bundles in higher density than in other airway tissues (Fig. 2) [32]. This can be demonstrated in allergen-challenged guinea pigs and in the airways of patients that have died with severe asthma [32]. In allergen-challenged guinea pigs, the degree of M_2 receptor dysfunction correlates with the number of eosinophils in the immediate proximity of the airway nerves, further supporting the importance of this phenomenon.

As tachykinins are known to be chemotactic for eosinophils and to activate eosinophils, we investigated the role of tachykinins in the airway response to allergen challenge. Pretreating guinea pigs with NK2 receptor antagonists has no effect on either the influx of eosinophils or the subsequent airway hyperresponsiveness and M_2 receptor dysfunction. Pretreating with NK1 receptor antagonists has no effect on the number of eosinophils entering the airways, but did prevent allergen-induced hyperresponsiveness and M_2 receptor dysfunction [33].

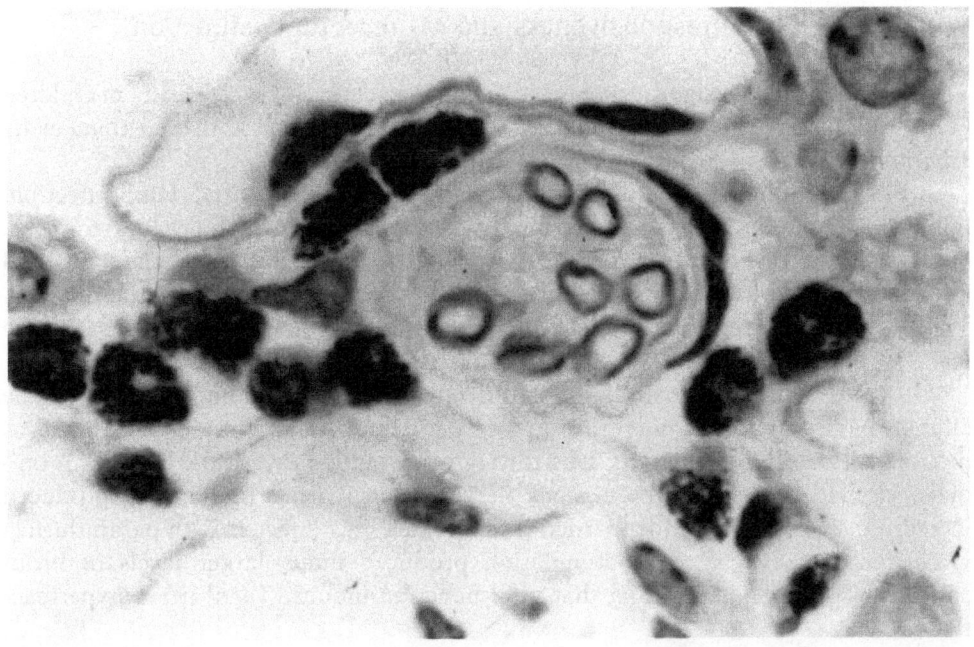

Figure 2
Histological section of a nerve bundle from the airway of an antigen-challenged guinea pig.
Note the clustering of eosinophils around the nerve.

Based on these studies, we hypothesized that airway tachykinins, while not responsible for recruiting eosinophils, were involved in the activation and release of MBP by eosinophils. To add credence to this possibility, we tested the effects of exogenous substance P on airway responsiveness and M_2 receptor function. Doses of substance P too small to cause bronchoconstriction caused airway hyperresponsiveness and M_2 receptor dysfunction [34]. This was not accompanied by a change in the number of eosinophils in the airways, but was clearly the result of eosinophil activation as it was blocked by an antibody to MBP and reversed by treatment with heparin.

Having established tachykinins as important mediators of eosinophil activation, but at the same time eliminating them as key mediators for eosinophil recruitment, the question remains as to the mechanisms of eosinophil attraction to the airway nerves. Our preliminary results suggest that the airway parasympathetic nerves can express the selective eosinophil chemoattractant eotaxin [35]. The role of neuronal eotaxin, as well as other cytokines, in eosinophil chemotaxis and activation remains to be determined.

Virus-induced hyperresponsiveness and M_2 receptor dysfunction

Viral infections play a significant role in asthma exacerbations, especially in children [1, 36]. Johnston [1] found that in 85% of children presenting with an asthma exacerbation, evidence of a viral infection could be found.

In 1976, Empey [37] found that during naturally acquired viral infection, nonasthmatic individuals developed airway hyperreactivity which lasted 4–6 weeks. He was able to reverse this hyperreactivity by giving atropine, suggesting that viral infection induced an abnormality in the parasympathetic control of the airways. This group speculated that epithelial damage exposed irritant nerve endings in the airways, potentiating the afferent limb of the reflex arc.

However, subsequent studies demonstrated substantial abnormalities in the efferent parasympathetic nerves. In guinea pigs, Buckner confirmed that this virus induced airway hyperreactivity was through the vagus nerve [20]. In control animals, electrical stimulation of vagus nerves at increasing frequencies produced increasing levels of bronchoconstriction. In guinea pigs infected with parainfluenza virus, the same vagal nerve stimulation produced much larger levels of bronchoconstriction, demonstrating that viral infection induced vagal nerve hyperreactivity.

We have shown that parainfluenza virus infection induces loss of M_2 muscarinic receptor function in guinea pigs [15]. As in Buckner's study, viral infection potentiated vagally induced bronchoconstriction non-infected control guinea pigs. This confirms that viral infection causes M_2 muscarinic receptor dysfunction and vagal hyperreactivity.

Multiple mechanisms contribute to virus-induced M_2 receptor dysfunction. Effects of multiple inflammatory cells and cytokines contribute, as well as effects of virus that appear to be independent of an inflammatory response. In guinea pigs pretreated with cyclophosphamide to deplete inflammatory cells non-selectively, the inflammatory cell response to viral infection was significantly diminished. Despite this, cyclophosphamide was only able to prevent virus induced M_2 receptor dysfunction and vagal hyperreactivity in approximately half of the animals [38]. When viral titers were measured in both groups of guinea pigs, the animals with dysfunctional M_2 receptors were found to have significantly higher viral loads. Thus in animal with severe viral infections, viral infection can directly induce M_2 muscarinic receptor dysfunction regardless of the inflammatory response.

Potential effects of virus independent of the inflammatory response

The M_2 receptor is N-glycosylated at three asparagine residues (Asn2, Asn3 and Asn6) in the extracellular portion of the receptor. Carbohydrates make up 26.5% of the molecular weight of the receptor [27]. Gies and Landry [39] showed that treatment of M$_2$ receptors with the enzyme neuraminidase, which cleaves the sialic acid residues from the receptor, decreases the affinity of muscarinic agonists for the M$_2$ receptor but not for other muscarinic receptor subtypes. Neuraminidase is contained in the coat of both influenza and parainfluenza viruses [40] and is expressed in large quantities in infected tissues [41]. We performed ligand-receptor binding studies of M$_2$ receptors incubated *in vitro* with parainfluenza virus or with neuraminidase [42]. These treatments did not affect total receptor number. However, agonist affinity for the M$_2$ receptor was decreased by an order of magnitude. The addition of a neuraminidase inhibitor prevented virus-induced loss of agonist affinity. This confirmed that parainfluenza virus could directly induce M$_2$ muscarinic receptor dysfunction by deglycosylating the receptor.

Viral infection can also directly inhibit M$_2$ muscarinic receptor gene expression. We have demonstrated that viral infection of primary cultures of airway parasympathetic nerve cells markedly potentiates the release of acetylcholine in response to electrical stimulation. In uninfected cells, blocking M$_2$ receptors using atropine potentiated acetylcholine release, while stimulating with methacholine suppressed acetylcholine release. In virus-infected cells, the release of acetylcholine was neither potentiated by atropine nor suppressed by methacholine, demonstrating loss of M$_2$ receptor function (Fig. 3) [43]. In order to determine whether the effects of viral infection were the result of a decrease in the expression of the gene encoding the M$_2$ receptor, we used a competitive reverse transcription-polymerase chain reaction assay to measure M$_2$ receptor mRNA in the cells. In virus-infected cells, M$_2$ muscarinic receptor mRNA was decreased to less than one-tenth of the control levels. Thus, virus induced M$_2$ muscarinic receptor dysfunction may also occur by decreased gene expression and loss of receptor synthesis.

The specific mechanism by which viruses induce M$_2$ muscarinic receptor indirectly through inflammation is not yet known. One potential mechanism is via the release of interferon-γ (IFNγ). When cultured airway parasympathetic neurons are treated with this cytokine, loss of M$_2$ receptor function and gene expression results [43]. It may be significant that dexamethasone, which increases expression of the M$_2$ receptor in these cells, can block the effects of IFNγ in this system [44].

Because eosinophils are responsible for the loss of M$_2$ receptor function in antigen sensitized and challenged guinea pigs, we studied their role in virus induced M$_2$ receptor dysfunction [45]. Unlike the allergen-challenged guinea pigs, depletion of eosinophils in nonsensitized virus infected animals by pretreatment with an antibody to IL-5 before viral infection did not prevent loss of M$_2$ muscarinic receptor function. Therefore inflammatory cells other than eosinophils must also play a role in virus induced M$_2$ receptor dysfunction. Determining which cells are responsible is a topic of current investigation.

Although viral infection causes airway hyperresponsiveness in both asthmatic and nonasthmatic individuals, the airway pathology and obstruction of asthma is

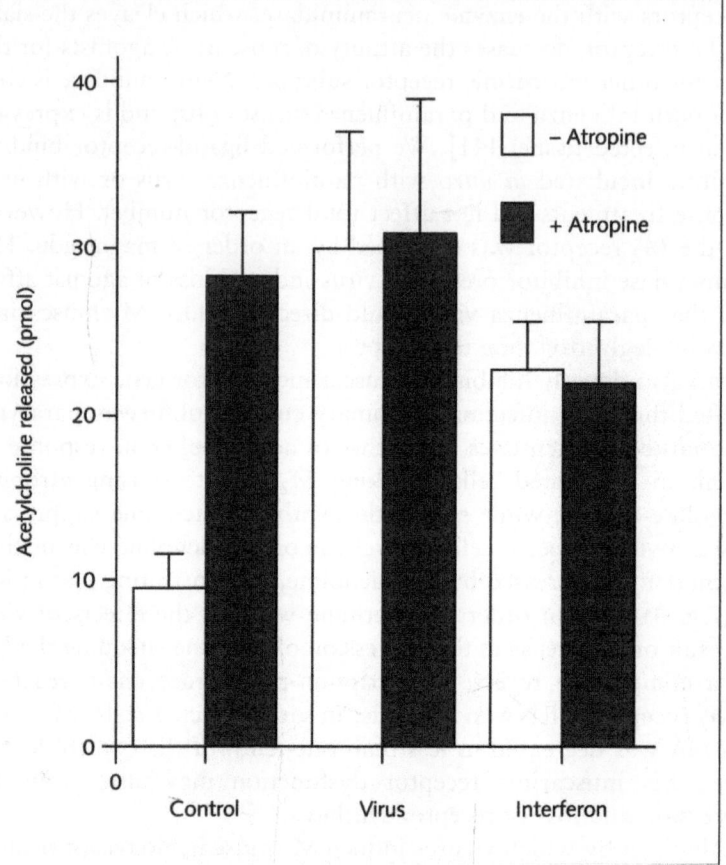

Figure 3
Release of acetylcholine from cultured airway parasympathetic nerve cells in response to electrical field stimulation. In control cells, blocking the M_2 receptors with atropine increases acetylcholine release. Both viral infection (parainfluenza virus type 1) and exposure to IFNγ (300 U/ml for 24 h) increase acetylcholine release. In both cases, there is no further potentiation after treatment with atropine, demonstrating loss of M_2 receptor function. Competitive RT-PCR analysis in these cells demonstrates loss of expression of the M_2 receptor gene (not shown) [43].

unique. While the inflammatory pathology of a viral infection in non-asthmatics consists of lymphocytes, mononuclear and polymorphonuclear cells [46], asthma pathology during exacerbations consists of the significant addition of eosinophils and their byproducts. [47]. Coyle [48] looked at the possibility that viral antigens could induce a different inflammatory reaction in an atopic milieu in mice. A strain of transgenic mice in which the CD8[+] T cells expressed the receptor for a glycoprotein of the lymphochoriomeningitis virus was challenged with aerosolized viral glycoprotein protein. CD8[+] T cells from the mice produced IL-2 and IFNγ as expected and lung lavage from the animals contained increased numbers of mononuclear cells and neutrophils. When this strain of mice was sensitized to a nonviral protein, ovalbumin (given by intraperitoneal injections), before being exposed to the viral protein aerosol, the inflammatory response was switched. After sensitization, the CD8[+] T cells produced IL-4 and IL-5 in addition to IL-2. The lung lavage of the sensitized mice had increased eosinophils in addition to the polymorphonuclear cells. Therefore, sensitization before viral infection switched the immune response to involve eosinophils as seen in asthmatics during asthma exacerbation.

Because viral infection in a sensitized or atopic host induced an eosinophilic immune response, we investigated the possibility that virus induced M_2 muscarinic receptor dysfunction in an atopic host might also be switched to involve eosinophils [45]. Guinea pigs were sensitized to ovalbumin via intraperitoneal injection three weeks before intranasal infection with parainfluenza virus. While sensitization alone had no effect on M_2 receptor function, all sensitized virus infected guinea pigs developed M_2 muscarinic receptor dysfunction and associated vagal hyperreactivity. However, in contrast to non-sensitized guinea pigs, in which eosinophils do not participate in virus-induced hyperresponsiveness and M_2 receptor dysfunction, in sensitized virus-infected guinea pigs both hyperresponsiveness and M_2 receptor dysfunction were mediated by eosinophils. In these sensitized virus-infected animals, both hyperresponsiveness and M_2 receptor dysfunction were blocked when eosinophils were depleted using an antibody to IL-5, or when MBP was neutralized with an antibody. Likewise, in these animals hyperresponsiveness and M_2 receptor dysfunction were reversed when MBP was neutralized using heparin. Thus sensitization to a non-viral antigen, without antigen challenge, changes the response to a subsequent viral infection to one which, in many ways, resembles the response to allergen challenge and is completely mediated by eosinophils and MBP. This suggests that a possible reason for the pathology seen in asthmatic individuals with simple viral infections may be a predisposition to eosinophilia in their airways on the basis of an atopic background.

Hyperresponsiveness and M_2 receptor dysfunction after ozone inhalation

Exposure of guinea pigs to ozone results in hyperreactivity and loss of neuronal M_2 muscarinic receptor function [18]. Immediately and one day after exposure to ozone, hyperreactivity is mediated by eosinophils and by loss of neuronal M_2 receptor function, since depletion of eosinophils or removal of MBP (with an antibody to MBP) protected neuronal M_2 receptor function and prevented the development of hyperreactivity [49]. Likewise, removal of MBP with heparin reversed hyperreactivity, while simultaneously restoring the function of the neuronal M_2 receptors [49].

Two days after exposure to ozone, the neuronal M_2 receptors are again functional, demonstrating that the effects of MBP on the M_2 receptors last less than two days (Fryer, unpublished data). In conclusion, the mechanism of M_2 receptor dysfunction following exposure to ozone involves eosinophils. This may be relevant to hyperreactivity in humans with asthma since exposure of human asthmatics to ozone also results in loss of M_2 receptor function and in hyperreactivity (Fryer, unpublished data).

Conclusions

Loss of function of M_2 muscarinic receptors on the airway parasympathetic nerves is seen after a variety of environmental insults. Multiple mechanisms may participate, including loss of M_2 receptor gene expression, deglycosylation of the receptor by neuraminidase and blockade of the receptor by the endogenous antagonist eosinophil MBP. Which of these mechanisms predominates may depend on the atopic status of the host. The eosinophil is most important in the M_2 receptor dysfunction that follows allergen inhalation. While the eosinophil does not participate in the response to most airway viruses under normal circumstances, in an atopic host the eosinophil is once again responsible for virus-induced loss of M_2 receptor function. Likewise degranulation of airway eosinophils mediates M_2 receptor dysfunction after inhalation of ozone.

The mechanisms of eosinophil recruitment and activation are at present incompletely understood. Eosinophils accumulate in the vicinity of the airway nerves, both in allergen-challenged guinea pigs and in humans with fatal asthma, suggesting that airway nerves may release an eosinophil chemoattractant, perhaps eotaxin. Airway tachykinins are important in eosinophil activation, stimulating the release of MBP.

Thus there are multiple points in the response to environmental stimuli at which the resulting M_2 receptor dysfunction and airway hyperresponsiveness might be interrupted.

Acknowledgments
This work was funded by grants from the Maryland Thoracic Society and from the Royal College of Physicians and Surgeons of Canada, by National Institute of Health grants HL-44727, HL-55543 and HL-54659, by the Center for Indoor Air Research and by the American Heart Association.

References

1   Johnston SL, Pattemore PK, Sanderson G, Smith S, Lampe F, Josephs L, Symington P, O'Toole S, Myint SH, Tyrrell D et al (1995) Community study of role of viral infections in exacerbations of asthma in 9–11 year old children. *Br Med J* 310: 1225–1229

2   Booij-Noord H, Orie NGM, deVries K (1971) Immediate and late bronchial obstructive reactions to inhalation of house dust and protective effect of disodium cromoglycate and prednisolone. *J Allergy Clin Immunol* 48: 344–53

3   Krzyzanowski M, Quachenboss JJ, Lebowitz MD (1992) Relation of peak expiratory flow rates and symptoms to ambient ozone. *Arch Env Health* 47: 107–115

4   Stern BR, Raizenne ME, Burnett RT, Jones L, Kearney J, Franklin CA (1994) Air pollution and childhood respiratory health: Exposure to sulfate and ozone in 10 Canadian rural communities. *Env Res* 66: 125–142

5   White MC, Etzel RA, Wilcox WD, Lloyd C (1994) Exacerbations of childhood asthma and ozone pollution in Atlanta. *Env Res* 65: 56–68

6   Fryer AD, Maclagan J (1984) Muscarinic inhibitory receptors in pulmonary parasympathetic nerves in the guinea-pig. *Br J Pharmacol* 83: 973–978

7   Ito Y, Yoshitomi T (1988) Autoregulation of acetylcholine release from vagus nerve terminals through activation of muscarinic receptors in the dog trachea. *Br J Pharmacol* 93: 636–646

8   Blaber LC, Fryer AD, Maclagan J (1985) Neuronal muscarinic receptors attenuate vagally induced contraction of feline bronchial smooth muscle. *Br J Pharmacol* 86: 723–728

9   Killingsworth CR, Mingfu Y, Robinson NE (1992) Evidence for the absence of a functional role for muscarinic M₂ inhibitory receptors in cat trachea *in vivo*; contrast with *in vitro* results. *Br J Pharmacol* 105: 263–270

10 Aas P, Maclagan J (1990) Evidence for prejunctional M₂ muscarinic receptors in pulmonary cholinergic nerves of the rat. *Br J Pharmacol* 101: 73–76

11 Minette P, Barnes PJ (1988) Prejunctional inhibitory muscarinic receptors on cholinergic nerves in human and guinea-pig airways. *J Appl Physiol* 64: 2532–2537

12 Ayala LE, Ahmed T (1989) Is there loss of a protective muscarinic receptor in asthma? *Chest* 96: 1285–1291

13 Minette PJ, Lammers JWJ, Dixon CMS, McCusker MT, Barnes PJ (1989) A muscarinic agonist inhibits reflex bronchoconstriction in normal but not asthmatic subjects. *J Appl Physiol* 67: 2461–2465

14 Okayama M, Shen T, Midorikawa J, Lin JT, Inoue H, Takishima T, Shirato K (1994) Effect of pilocarpine on propranolol-induced bronchoconstriction in asthma. *Am J Respir Crit Care Med* 149: 76–80

15 Fryer AD, Jacoby DB (1991) Parainfluenza virus infection damages inhibitory M₂ muscarinic receptors on pulmonary parasympathetic nerves in the guinea-pig. *Br J Pharmacol* 102: 267–271

16 Fryer AD, Wills-Karp M (1991) Dysfunction of M₂ muscarinic receptors in pulmonary

parasympathetic nerves after antigen challenge in guinea-pigs. *J Appl Physiol* 71: 2255–2261

17 Fryer AD, Jacoby DB (1992) Function of pulmonary M_2 muscarinic receptors in antigen challenged guinea-pigs is restored by heparin and poly-l-glutamate. *J Clin Invest* 90: 2292–2298

18 Schultheis A, Bassett D, Fryer A (1994) Ozone-induced airway hyperresponsiveness and loss of neuronal M2 muscarinic receptor function. *J Appl Physiol* 76: 1088–1097

19 Boushey H, Holtzman M (1985) Experimental airway inflammation and hyperreactivity; searching for cells and mediators. *Am Rev Respir Dis* 131: 312–313

20 Buckner CK, Songsiridej V, Dick EC, Busse WW (1985) *In vivo* and *in vitro* studies of the use of the guinea pig as a model for virus-provoked airway hyperreactivity. *Am Rev Respir Dis* 132: 305–310

21 McCaig DJ (1987) Comparison of autonomic responses in the trachea isolated from normal and albumin-sensitive guinea-pigs. *Br J Pharmacol* 92: 809–816

22 Santing RE, Pasman Y, Olymulder CG, Roffel AF, Meurs H, Zaagsma J (1995) Contribution of a cholinergic reflex mechanism to allergen-induced bronchial hyperreactivity in permanently instrumented, unrestrained guinea-pigs. *Br J Pharmacol* 114: 414–418

23 Costello RW, Evans CE, Yost BL, Belmonte KE, Gleich GJ, Jacoby DB, Fryer AD (1999) Antigen-induced hyperreactivty to histamine: role of the vagus nerve and eosinophils. *Am J Physiol* 276: L709–L714

24 Santing RE, Hoekstra Y, Pasman Y, Zaagsma J, Meurs H (1994) The importance of eosinophil activation for the development of allergen-induced bronchial hyperreactivity in conscious, unrestrained guinea pigs. *Clin Exp Allergy* 24: 1157–1163

25 Elbon CL, Jacoby DB, Fryer AD (1995) Pretreatment with an antibody to interleukin-5 prevents loss of pulmonary M_2 muscarinic receptor function in antigen-challenged guinea-pigs. *Am J Respir Cell Mol Biol* 12: 320–328

26 Fryer AD, Costello RW, Yost BL, Lobb RR, Tedder TF, Steeber DA, Bochner BS (1997) Antibody to VLA-4, but not to L-selectin, protects neuronal M_2 muscarinic receptors in antigen-challenged guinea pig airways. *J Clin Invest* 99: 2036–44

27 Peterson GL, Rosenbaum LC, Broderick DJ, Schimerlik MI (1986) Physical properties of the purified cardiac muscarinic acetylcholine receptor. *Biochemistry* 25: 3189–3202

28 Hu J, Wang S-Z, Forray C, El-Fakahany EE (1992) Complex allosteric modulation of cardiac muscarinic receptors by protamine: a potential model for putative endogenous ligands. *Mol Pharmacol* 42: 311–324

29 Jacoby DB, Gleich GJ, Fryer AD (1993) Human eosinophil major basic protein is an endogenous allosteric antagonist at the inhibitory muscarinic M_2 receptor. *J Clin Invest* 91: 1314–1318

30 Fryer A, Huang YC, Rao G, Jacoby D, Mancilla E, Whorton R, Piantadosi CA, Kennedy T, Hoidal J (1997) Selective O-desulfation produces nonanticoagulant heparin that retains pharmacological activity in the lung. *J Pharmacol Exp Ther* 282: 208–19

31 Evans CM, Jacoby DB, Gleich GJ, Fryer AD, Costello RW (1997) Antibody to

eosinophil major basic protein protects M$_2$ receptor function of antigen challenged guinea pigs *in vivo*. *J Clin Invest* 100: 2254–2262

32 Costello RW, Schofield BH, Kephart GM, Gleich GJ, Jacoby DB, Fryer AD (1997) Localization of eosinophils to airway nerves and effect on neuronal M$_2$ muscarinic receptor function. *Am J Physiol* 273: L93–103

33 Costello R, Fryer A, Belmonte K, Jacoby D (1998) Effects of tachykinin NK1 receptor antagonists on vagal hyperreactivity and neuronal M$_2$ muscarinic receptor function in antigen-challenged guinea-pigs. *Br J Pharmacol* 124: 267–276

34 Evans CM, Jacoby DB, Gleich GJ, Fryer AD (1999) Substance P-induced hyperreactivity is caused by eosinophil major basic protein and M$_2$ receptor dysfunction. *Am J Respir Crit Care Med* 159: A280

35 Evans CM, Fryer AD, Jacoby DB (1998) Eotaxin mRNA in primary cultures of parasympathetic nerves from guinea pig tracheas. *Am J Respir Crit Care Med* 157: A599

36 Welliver RC (1983) Upper respiratory infections in asthma. *J Allergy Clin Immunol* 72: 341–346

37 Empey DW, Laitinen LA, Jacobs L, Gold WM, Nadel JA (1976) Mechanisms of bronchial hyperreactivity in normal subjects following upper respiratory tract infection. *Am Rev Respir Dis* 113: 523–527

38 Fryer AD, Yarkony KA, Jacoby DB (1994) The effect of leukocyte depletion on pulmonary M2 muscarinic receptor function in parainfluenza virus-infected guinea-pigs. *Br J Pharmacol* 112: 588–594

39 Gies J-P, Landry Y (1988) Sialic acid is selectively involved in the interaction of agonists with M$_2$ muscarinic acetylcholine receptors. *Biochem Biophys Res Comm* 150: 673–680

40 Scheid A, Caliguiri LA, Compans RW, Choppin PW (1972) Isolation of paramyxovirus glycoproteins Association of both hemagglutinating and neuraminidase activities with the larger SV5 glycoprotein. *Virology* 50: 640–652

41 Boulan ER, Pendergast M (1980) Polarized distribution of viral envelope proteins in the plasma membrane of infected epithelial cells. *Cell* 20: 45–54

42 Fryer AD, El-Fakahany EE, Jacoby DB (1990) Parainfluenza virus type 1 reduces the affinity of agonists for muscarinic receptors in guinea-pig heart and lung. *Eur J Pharmacol* 181: 51–58

43 Jacoby DB, Xiao HQ, Lee NH, Chan-Li Y, Fryer AD (1998) Virus- and interferon-induced loss of inhibitory M$_2$ muscarinic receptor function and gene expression in guinea-pig airway parasympathetic neurons. *J Clin Invest* 102: 242–248

44 Jacoby DB, Chani-Li Y, Xiao HQ, Fryer AD (1998) Dexamethasone increases M$_2$ muscarinic receptor expression and decreases acetylcholine release in cultured airway parasympathetic neurons. *Am J Respir Crit Care Med* 157: A715

45 Adamko DJ, Yost BL, Gleich GJ, Fryer AD F, Jacoby DB (1999) Ovalbumin sensitization changes the inflammatory response to subsequent parainfluenza infection. Eosinophils mediate airway hyperresponsiveness, M2 muscarinic receptor dysfunction, and antiviral effects. *J Exp Med* 190: 1465–1478

46   Walsh JJ, Dietlein LF, Low FN, Burch GE, Mogabgab WJ (1960) Bronchotracheal response in human influenza. *Arch Int Med* 108: 376–388

47   Frigas E, Loegering DA, Solley GO, Farrow GM, Gleich GJ (1981) Elevated levels of the eosinophil granule MBP in the sputum of patients with bronchial asthma. *Mayo Clin Proc* 56: 345–353

48   Coyle AJ, Erard F, Bertrand C, Walti S, Pircher H, Le GG (1995) Virus-specific CD8[+] cells can switch to interleukin 5 production and induce airway eosinophilia. J *Exp Med* 181: 1229–33

49   Yost BL, Gleich GJ, Fryer AD (1999) Ozone-induced hyperresponsiveness and blockade of M_2 muscarinic receptors by eosinophil major basic protein. *J Appl Physiol* 87: 1272–1278

Muscarinic receptor-β-adrenoceptor cross-talk in airways smooth muscle

Herman Meurs[1], Ad F. Roffel[2], Carolina R.S. Elzinga[1] and Johan Zaagsma[1]

[1]Department of Molecular Pharmacology, University Centre for Pharmacy, Antonius Deusinglaan 1, 9713 AV Groningen, The Netherlands; [2]Department of Research Management, Pharma Bio-Research Group B.V., P.O. Box 200, 9470 AE Zuidlaren, The Netherlands

Introduction

Cholinergic airway smooth muscle contraction and its functional antagonism by relaxing stimuli, including catecholamines, are major determinants in the control of airway function under physiological and pathophysiological conditions such as bronchial asthma. In addition, an altered responsiveness of airway smooth muscle to contractile and/or relaxing agonists has been implicated in the pathogenesis of bronchial hyperreactivity that is commonly observed in these patients [1–3]. Consequently, considerable research effort has been focussed on elucidating the molecular mechanisms underlying muscarinic receptor-mediated contraction and β_2-adrenoceptor-induced relaxation of airway smooth muscle [4]. In addition, there is now substantial evidence that the signalling mechanisms of muscarinic receptors and β-adrenoceptors in smooth muscle are not independent pathways that converge at the level of the contractile apparatus, i.e., actin-myosin interaction, but that these mechanisms may interact at a number of different homeostatic levels, including the receptor transduction level [5–7], leading to a highly integrated, finely tuned system to adapt the responsiveness of the tissue to multiple incoming signals.

In this contribution, we will present evidence that cross-talk between muscarinic and β-adrenergic receptor transduction mechanisms may be involved in the functional antagonism between contractile and relaxing stimuli and that this process could lead to an altered airway smooth muscle responsiveness in asthma. In addition, the possible role of muscarinic receptors as well as muscarinic receptor-β-adrenoceptor cross-talk in airway smooth muscle proliferation, another important determinant of airway hyperreactivity in particularly chronic and severe asthma, will be discussed briefly.

Muscarinic Receptors in Airways Diseases, edited by Johan Zaagsma, Herman Meurs and Ad F. Roffel
© 2001 Birkhäuser Verlag Basel/Switzerland

Muscarinic receptor and β-adrenoceptor signalling in airway smooth muscle

Muscarinic M_3 receptor signalling

Both in animal [8–17] and in human [18–20] airway smooth muscle it has now been well established that phosphoinositide (PI) metabolism induced by various contractile stimuli, including muscarinic agonists, may be involved in the pharmacomechanical coupling of contraction. Thus, stimulation of muscarinic M_3 receptors in airway smooth muscle leads to activation of PI-specific phospholipase C (PLC) *via* the α-subunit of a G_q regulatory protein ($G_{q\alpha}$), causing hydrolysis of phosphatidylinositol 4,5-bisphosphate to generate inositol 1,4,5-trisphosphate (IP_3) and *sn*-1,2-diacylglycerol (DAG) (Fig. 1).

IP_3 mobilizes Ca^{2+} from intracellular stores [21], producing a rapid and transient rise in intracellular Ca^{2+} concentration ($[Ca^{2+}]_i$) [22–28] which initiates contraction *via* activation of a Ca^{2+}-calmodulin-dependent myosin light chain kinase (MLCK) and subsequent phosphorylation of myosin light chain (MLC) [29–31]. In addition, a sustained influx of extracellular Ca^{2+} through receptor-operated Ca^{2+} channels as well as DAG-induced activation of protein kinase C (PKC) has been implicated in the tonic phase of contraction [17, 23, 24, 26, 28, 32–35].

In various smooth muscles, including airway smooth muscle, it has been shown that PKC may cause phosphorylation of several proteins which are involved in the regulation of actin-myosin interaction; however, the physiological importance of many of these phosphorylation reactions in agonist-induced contraction has not yet been established [35, 36]. Several lines of evidence suggest that PKC may be involved in contractile agonist-induced increased myofilament sensitivity to Ca^{2+} (Ca^{2+}-sensitization) [23, 32, 35], which could explain the maintenance of the contractile response after the fall in agonist-induced Ca^{2+} intracellular mobilization and in MLCK phosphorylation [29–31]. It has been proposed that PKC may cause Ca^{2+}-sensitization by inhibition of myosin light chain phosphatase (PP-1M) [35, 37]. Inhibition of PP-1M increases MLC phosphorylation, thereby facilitating Ca^{2+}-induced contraction [35]. However, although PKC activation by phorbol 12,13-dibutyrate increased Ca^{2+} sensitivity in permeabilized airway smooth muscle [38], specific PKC inhibitors did not affect acetylcholine-induced Ca^{2+} sensitization in this tissue [39].

In vascular smooth muscle, evidence has been obtained for a role of PKC in direct or indirect phosphorylation of the thin-filament-associated, actin-binding proteins caldesmon and calponin, thus relieving the inhibitory action of these regulatory proteins on actin-myosin interaction (see [35]). A role for muscarinic agonist- and PKC-induced calponin phosphorylation was recently also indicated in canine tracheal smooth muscle [40]. In addition, it was demonstrated that muscarinic agonist-induced activation of ERK-2 MAPK and p38 MAPK may be involved in caldesmon phosphorylation in airway smooth muscle [41, 42]; however, the role of PKC in these processes has thus far not been established.

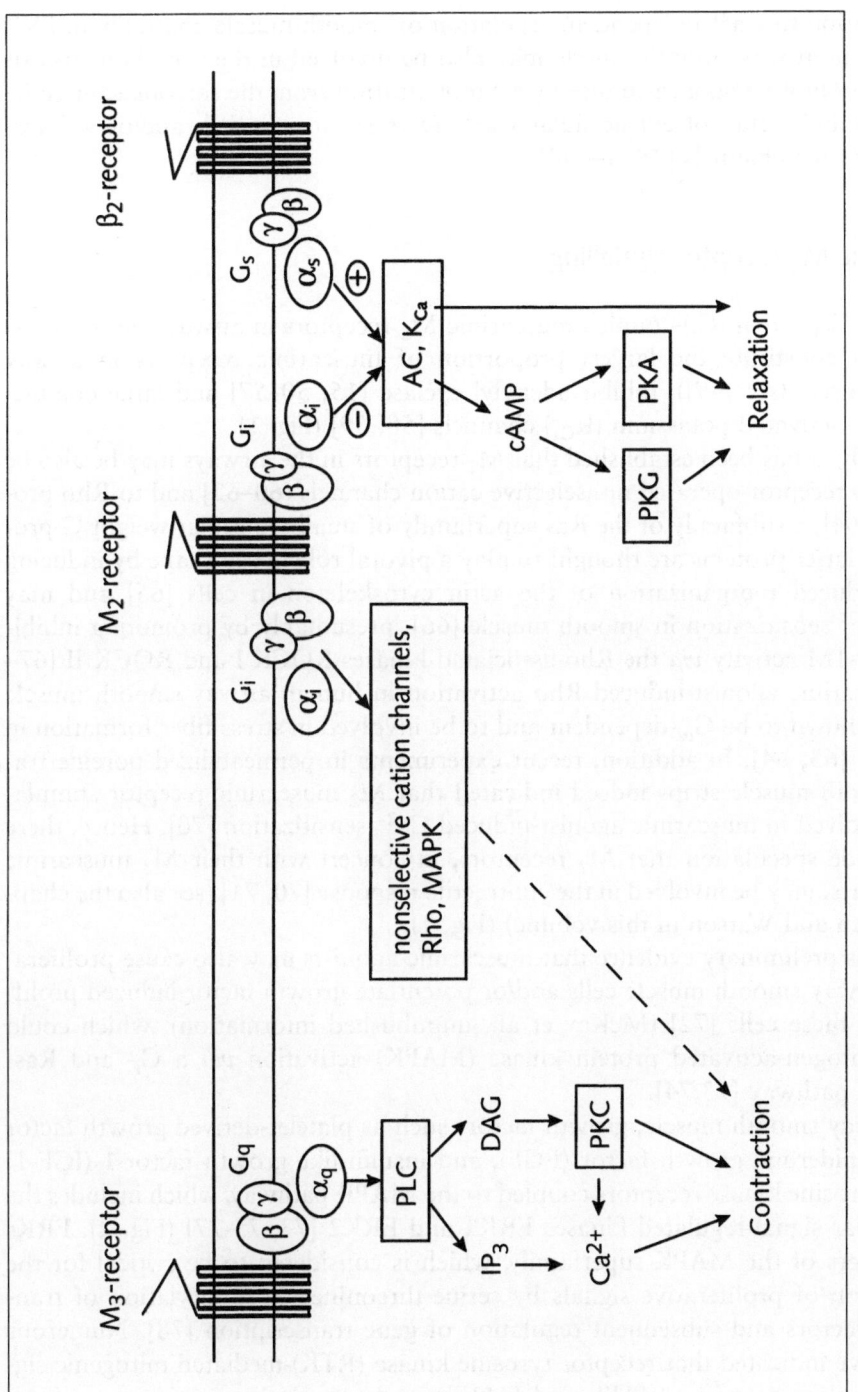

Figure 1

Muscarinic M_2 and M_3 receptor- and β_2-adrenoceptor-mediated signalling pathways involved in airway smooth muscle contraction and relaxation. See text for abbreviations and full description.

In addition to Ca^{2+}-independent regulation of smooth muscle contraction, PKC activation in airway smooth muscle may also be involved in the Ca^{2+} homeostasis itself, by promoting agonist-induced Ca^{2+} mobilization from the sarcoplasmic reticulum [43] and influx of extracellular Ca^{2+} *via* receptor-operated- and/or voltage-dependent Ca^{2+}-channels [28, 44–48].

Muscarinic M_2 receptor signalling

Inhibitory G-protein (G_i)-coupled muscarinic M_2 receptors in airway smooth muscle, which constitute the largest proportion of muscarinic receptors in airway smooth muscle (see [49]), inhibit adenylyl cyclase [15, 50–57] and large conductance Ca^{2+}-activated potassium (K_{Ca}) channels [58, 59] (Fig. 1).

Recently, it has been established that M_2 receptors in the airways may be also be coupled to receptor-operated nonselective cation channels [60–62] and to Rho proteins [63, 64], a subfamily of the Ras superfamily of small molecular weight G-proteins. The latter proteins are thought to play a pivotal role in cell shape by inducing agonist-induced reorganization of the actin cytoskeleton in cells [65] and may induce Ca^{2+} sensitization in smooth muscle [66], presumably by promoting inhibition of PP-1M activity *via* the Rho-associated kinases ROCK I and ROCK II [67–69]. Muscarinic agonist-induced Rho activation in human airway smooth muscle cells was shown to be $G_{\alpha i}$-dependent and to be involved in stress fiber formation in these cells [63, 64]. In addition, recent experiments in permeabilized porcine tracheal smooth muscle strips indeed indicated that M_2 muscarinic receptor stimulation is involved in muscarinic agonist-induced Ca^{2+}-sensitization [70]. Hence, there is now wide speculation that M_2 receptors, in concert with their M_3 muscarinic counterparts, may be involved in the contractile response [70, 71] (see also the chapter by Eglen and Watson in this volume) (Fig. 1).

There is preliminary evidence that muscarinic agonists may also cause proliferation of airway smooth muscle cells and/or potentiate growth factor-induced proliferation of these cells [72] (McKay et al., unpublished information), which could involve mitogen-activated protein kinase (MAPK) activation *via* a G_i- and Ras-dependent pathway [73,74].

In airway smooth muscle, growth factors such as platelet-derived growth factor (PDGF), epidermal growth factor (EGF), and insulin-like growth factor-1 (IGF-1) activate tyrosine kinase receptors coupled to the MAPK pathway, which includes the extracellular signal-regulated kinases ERK1 and ERK2 [73, 75-77] (Fig. 2). ERKs are members of the MAPK superfamily, which is considered to be critical for the transduction of proliferative signals by serine-threonine phosphorylation of transcription factors and subsequent regulation of gene transcription [78]. Numerous studies have indicated that receptor tyrosine kinase (RTK)-mediated mitogenic signalling involves a series of SH2- and SH3-dependent protein-protein interactions

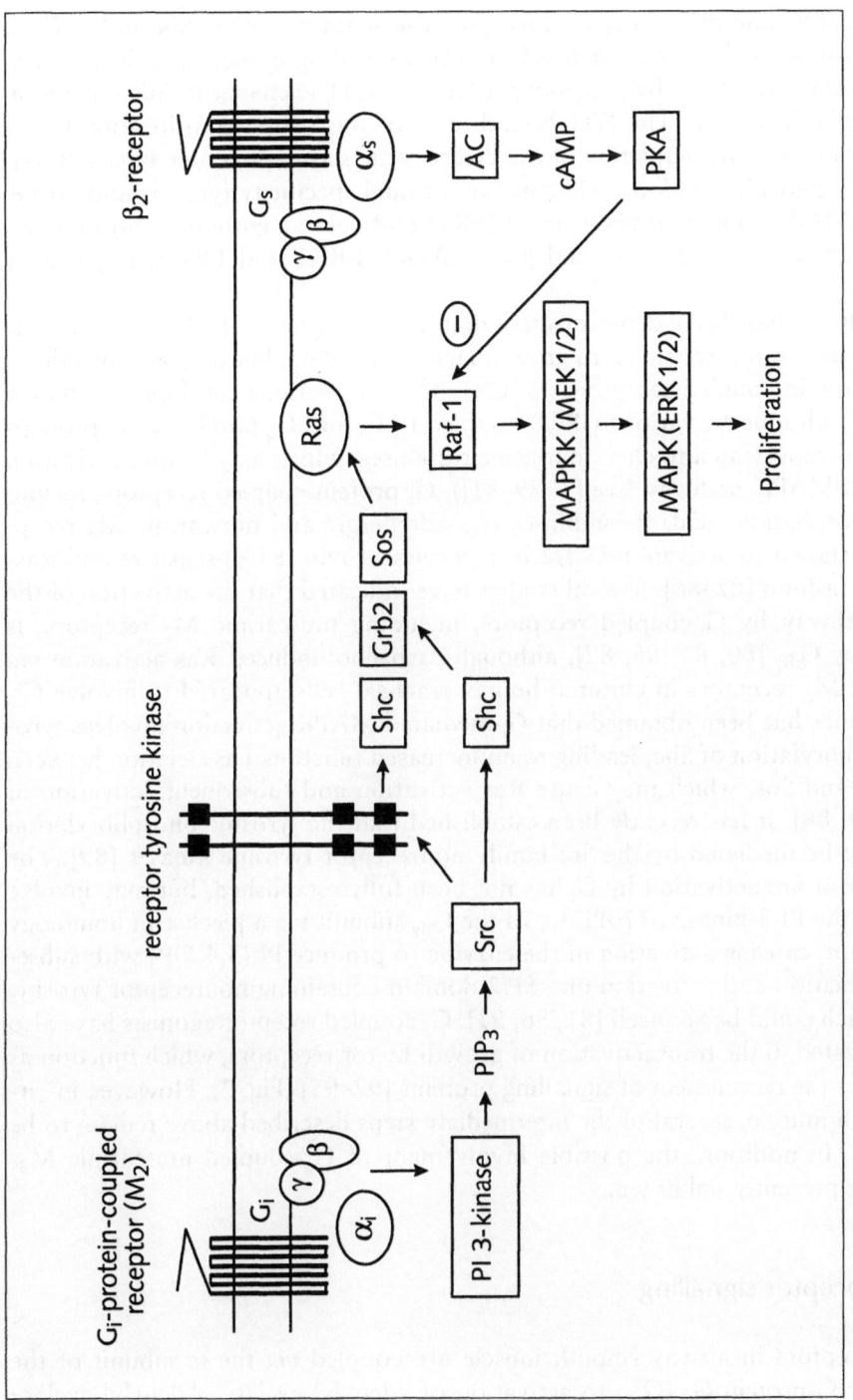

Figure 2

(Putative) signalling pathways involved in muscarinic M$_2$ receptor-, β$_2$-adrenoceptor- and receptor tyrosine kinase-induced modulation of airway smooth muscle proliferation. See text for abbreviations and full description.

between the tyrosine-phosphorylated receptor, the adapter proteins Shc and/or Grb2 and the guanine nucleotide exchange factor Sos, resulting in the activation of plasma membrane-bound Ras by promoting a GDP → GTP exchange in this small molecular weight G-protein. The GTP-bound form of Ras causes translocation to the plasma membrane and activation of the cytoplasmic serine-threonine kinase Raf-1, followed by phosphorylation and activation of dual specificity tyrosine and serine-threonine MAPK kinases (MEK1 and MEK2) and subsequently the phosphorylation and activation of the p44 and p42 MAPKs, ERK1 and ERK2, respectively (Fig. 2).

Recently, it has been demonstrated that proliferation and activation of the MAPK cascade is not restricted to growth factor signalling, but may also be initiated by G-protein-coupled receptors (GPCR). Thus, in various cell types, including airway smooth muscle, various GPCR coupled to G_i and G_q families of G-proteins have been demonstrated to elicit mitogenic responses, which may involve activation of the ERK/MAPK pathway (see [3, 79–81]). G_i-protein-coupled receptors, including lysophosphatidic acid, thrombin-α, α_{2A}-adrenergic and muscarinic M_2 receptors, were shown to activate ERK1/2 in a pertussis toxin (PTX)-sensitive and Ras-dependent fashion [82–86]. Several studies have indicated that the activation of the MAPK pathway by G_i-coupled receptors, including muscarinic M_2 receptors, is mediated by $G_{i\beta\gamma}$ [80, 81, 86, 87], although carbachol-induced Ras activation *via* muscarinic M_2-receptors in cultured human tracheal cells appeared to involve $G_{i\alpha}$ [74]. Evidence has been obtained that G_i-mediated MAPK activation involves tyrosine phosphorylation of Shc, leading to an increased functional association between Shc, Grb2 and Sos, which may cause Ras activation and subsequent activation of MAPK [80, 88]. It has recently been established that the tyrosine phosphorylation of Shc may be mediated by the Src family nonreceptor tyrosine kinases [89]. The mechanism of Src activation by G_i has not been fully established, but may involve binding of the PI 3-kinase p110PI3Kγ to the $G_{\beta\gamma}$-subunit *via* a pleckstrin homology (PH) domain, causing activation of the enzyme to produce PI $(3,4,5)P_3$ with subsequent association and activation of a SH2 domain-containing nonreceptor tyrosine kinase, which could be Src itself [81, 90, 91]. G_i-coupled receptor agonists have also been implicated in the transactivation of growth factor receptors, which function as scaffolds for the recruitment of signalling proteins [92–95] (Fig. 2). However, in airway smooth muscle, several of the intermediate steps described above remain to be established. In addition, the possible involvement of G_q-coupled muscarinic M_3-receptors is presently unknown.

β_2-Adrenoceptor signalling

β_2-Adrenoceptors in airway smooth muscle are coupled *via* the α subunit of the stimulatory G-protein G_s ($G_{s\alpha}$) to activation of adenylyl cyclase. Adenylyl cyclase activation leads to enhanced production of cyclic adenosine 3',5'-monophosphate (cAMP) and subsequent activation of cAMP-dependent protein kinase A (PKA) [96, 97] (Fig. 1). *Via* phosphorylation of serine and threonine residues on specific target proteins, this leads to various biochemical responses that induce airway smooth muscle relaxation [4, 97–99]. These mechanisms are based on reduction of $[Ca^{2+}]_i$ [22, 23, 28] and diminishing Ca^{2+} sensitivity of the contractile elements (Ca^{2+} desensitization) [31, 100].

Reduction of $[Ca^{2+}]_i$ may be caused by stimulating Ca^{2+} extrusion or sequestration to intracellular stores [101, 102], inhibition of IP_3-mediated Ca^{2+} release from the intracellular stores [103] and hyperpolarization by opening of K_{Ca} channels, which may proceed *via* both cAMP-dependent and cAMP-independent (directly *via* $G_{s\alpha}$) mechanisms [59, 100, 104, 105]. β-Adrenoceptor stimulation, as well as forskolin and cAMP analogues, has also been reported to inhibit histamine-induced PI metabolism; however, muscarinic agonist-induced PLC responses are largely resistant to such inhibition [16, 106–108] (see further below).

Ca^{2+} desensitization has long been considered to result from PKA-induced phosphorylation of MLCK, which reduces the sensitivity of MLCK to activation by Ca^{2+}-calmodulin [31]. However, it has recently been suggested that inhibition of PP-1M could be more important [100].

Furthermore, there is evidence that cAMP may activate protein kinase G (PKG) as well as PKA, and that this dual action may contribute to the relaxant activity of cAMP [97]. The mechanisms of PKG-induced relaxation have not yet fully been established, but may involve reduction of $[Ca^{2+}]_i$, activation of K_{Ca} channels and inhibition of Ca^{2+} sensitivity of the contractile apparatus [97, 109].

In HEK293 cells, it has recently been established that β-adrenoceptors can also couple to G_i proteins and subsequently activate MAPK [110]. Activation of this pathway required phosphorylation of the receptor by PKA, since inhibition of PKA blocked the response and a mutant receptor lacking the normal phosphorylation sites could activate adenylyl cyclase, but not MAPK. In addition, concurrent coupling of the $β_2$-adrenoceptor to G_s and G_i has recently been demonstrated in murine cardiac myocytes, in which coupling of the receptor to G_i appears to inhibit the $β_2$-adrenoceptor-mediated inotropic response [111]. Coupling of $β_2$-adrenoceptors to G_i and subsequent activation of MAPK in airway smooth muscle cells has thus far not been reported. In contrast, it has been established that β-adrenoceptor stimulation, as well as other stimuli that enhance activation of PKA by cAMP, may cause inhibition of growth factor- and GPCR-induced proliferation of airway smooth muscle cells [112–121], which, as in other cells [122, 123], may involve inhibition of the MAPK signalling pathway by inhibition of Raf-1 [115, 120] (Fig. 2). In addition, it was recently demonstrated that cAMP may cause inhibition of growth factor-induced mitogenesis of airway smooth muscle by inhibition of growth factor-induced PI-3K activity [124] and cAMP-driven gene expression [121].

Cross-talk between muscarinic M_3 receptor stimulation and β-adrenoceptor function in airway smooth muscle

Role of PI metabolism in the functional antagonism of muscarinic agonist-induced airway smooth muscle contraction by β-adrenoceptor stimulation

It is well known that exaggerated cholinergic stimulation of airway smooth muscle causes a reduced relaxability of the muscle by β-adrenoceptor agonists. In isolated preparations of human [19, 125] and animal [17, 126–131] airway smooth muscle, it has been observed that the potency (pD_2 = –log EC_{50}) and degree (E_{max}) of relaxation induced by β-adrenoceptor agonists are gradually reduced in the presence of increasing concentrations of various contractile stimuli, including muscarinic agonists. Functional antagonism between β-adrenoceptor and different contractile agonists has also been demonstrated *in vivo* [132] and might explain why $β_2$-adrenoceptor agonists become less effective when asthmatic episodes become more severe, whereas the efficacy is unchanged in patients with mild and asymptomatic asthma [133, 134].

In vitro studies in human airway smooth muscle have demonstrated that in the presence of cholinergic tone, the efficacy of partial β-agonists is reduced more than that of full β-agonists [135, 136]. This may be explained by the need of a full receptor occupancy by partial agonists for their maximal efficacy, while full agonists usually have receptor reserve and reach their maximal response at fractional receptor occupancy. Thus, at cholinergic tone, partial β-agonists are more sensitive to limitation of the functional receptor number in obtaining a maximal response than full agonists. Based on the pharmacological principle described above, partial agonists may also exert antagonistic behaviour in the presence of an agonist with higher efficacy [137]. In human airways, this was recently demonstrated by the antagonistic effect of the long-acting partial $β_2$-agonist salmeterol, but not of the long-acting full $β_2$-agonist formoterol, on various short-acting β-adrenoceptor agonists [136]. In clinical terms, long-lasting receptor occupancy by a long-acting partial β-agonist, having reduced intrinsic activity as a result of airway constriction, could possibly result in antagonism of the short-acting β-adrenoceptor agonists used for asthma rescue in severe asthmatics. However, the interaction between long- and short-acting β-agonists in patients with acute severe asthma remains to be established.

Several studies have also indicated that the β-adrenergic responsiveness in the presence of contractile agonists is not determined by the contraction level *per se*, but rather by the agonist under investigation, especially at higher and supra-maximal concentrations [127, 130–132]. For example, in guinea pig tracheal preparations we have found that the decrease of isoprenaline-induced relaxation in terms of pD_2 and E_{max} at increasing concentrations of the contractile stimulus was relatively large for methacholine, intermediate for oxotremorine and histamine, and only small for the partial muscarinic agonist McN-A-343 [131]. In bovine tracheal smooth muscle (BTSM) slices it was noticed that these agonists all have a different capacity to induce total inositol phosphates (IP) accumulation as an index of PLC activation [12, 131]. Analysis of the dose-response curves for contraction and IP accumulation with the muscarinic agonists demonstrated a direct relationship between the two parameters, with a considerable reserve of IP production for the full contractile agonists methacholine and oxotremorine, and no reserve for the partial agonist McN-A-343 [12]. The efficacy of the contractile agonists to induce IP production (methacholine > oxotremorine > histamine > McN-A-343) showed a striking correlation with their capacity to diminish pD_2 and E_{max} values of isoprenaline-induced relaxation of guinea pig trachea, indicating that the (reserve of) PI metabolism may be involved in the functional antagonism of the contraction by β-adrenoceptor agonists [131]. Similarly, in bovine [7] and human [19] bronchial smooth muscle a significant correlation between IP production induced by various concentrations of methacholine and histamine and the reduction of pD_2 and E_{max} values of isoprenaline was found, although in the human tissue both contractile agonists were equipotent with respect to IP production and reduction of β-adrenergic relaxation [19].

These findings clearly suggested a functional relationship between the levels of PI metabolism and β-adrenoceptor function in both human and animal airway smooth muscle. Two possible explanations could account for this relationship: (1) competition between the two transduction mechanisms for the functional response, and (2) direct interference of the PI metabolism with the β-adrenoceptor-mediated activation of adenylyl cyclase *via* PKC-mediated phosphorylation of either the β-adrenoceptor and/or G_s. Thus, in various cells or tissues, it was demonstrated that activation of PKC *via* agonist-induced PI metabolism or phorbol esters may lead to uncoupling of the β-adrenoceptor, presumably *via* phosphorylation of the β-adrenoceptor and/or G_s [5, 138, 139]. Since PKC activation, *via* receptor-mediated stimulation of the PI metabolism, is involved in the activation of various cells involved in the allergic asthmatic response, including airway smooth muscle and inflammatory cells, we hypothesised that PKC-induced (heterologous) desensitization could cause a reduced β-adrenergic responsiveness of these cells and thus contribute to enhanced airway reactivity in patients with active and severe asthma [1, 140].

Evidence for this hypothesis was first found in lymphocytes from asthmatic patients. It was demonstrated that allergen challenge of the patients caused β-adrenoceptor uncoupling and non-specific adenylyl cyclase desensitization in these cells, presumably *via* a change in G_s [140–142]. These changes were closely mimicked by treatment of normal lymphocytes with the selective PKC activator phorbol 12-myristate 13-acetate (PMA) [140, 143], indicating that PKC activation during the allergic response by an as yet unknown stimulus could play a role in the observed changes. This suggestion was further supported by observations that lymphocytes are activated after allergen challenge [144] and that this activation may indeed be associated with enhanced PI metabolism in these cells [145]. Moreover, it was recently demonstrated that the reduced β-adrenoceptor function in T cells from asthmatics after allergen challenge may cause a reduced control of interferon-γ (IFNγ) and interleukin-5 (IL-5) mRNA expression, which coincides with a selective priming of the IL-5 mRNA production [146].

Biochemical evidence that cross-talk between PI metabolism and adenylyl cyclase could similarly play a role in the regulation of β-adrenergic responsiveness of airway smooth muscle by contractile agonists was first obtained by Grandordy et al. [147], who found β-adrenoceptor uncoupling and a reduced β-agonist-induced cAMP accumulation in bovine tracheal smooth muscle slices after 40 min *in vitro* incubation of the tissue with PMA and carbachol. In addition, it was shown that pre-treatment (10–30 min) of guinea pig tracheal smooth muscle slices with methacholine caused a reduced β-agonist-induced adenylyl cyclase activation in isolated membranes, which involved a functional inactivation of G_s [52]. Surprisingly, it was also reported that chronic (18 h), but not acute (30 min, 2 h), pretreatment of bovine tracheal smooth muscle slices with carbachol caused reduced basal adenylyl cyclase activity as well as reduced adenylyl cyclase responses to isoprenaline, prostaglandin E_1, GTP and forskolin in this tissue [148]. The reduced adenylyl cyclase activity did not involve β-adrenoceptor down-regulation and changes in G_i expression and was inhibited by the (non-specific) PKC inhibitor staurosporin, which may suggest the involvement of PKC in this long-term effect.

Using the more specific PKC inhibitor GF109203X [149], we recently found direct evidence for the involvement of PKC in the (acute) functional antagonism of muscarinic agonist-induced bovine airway smooth muscle contraction by β-adrenoceptor agonists. Thus, in the presence of the PKC inhibitor, a significant increase in the sensitivity and maximal response to isoprenaline was observed for methacholine-induced contractions at fixed concentrations of the contractile agonist (Fig. 3). Although methacholine-induced contractions were reduced by GF109203X to some extent as expected, the changes in sensitivity to the β-agonist could only partially be explained by the reduced contractile tone. Thus, isoprenaline pD_2 values were higher in the presence of GF109203X compared to control at equal levels of contractile tone (Fig. 4). In addition, significantly enhanced pD_2 values, but not E_{max} values, were obtained in the presence of GF109203X, when reduced contractile tones were compensated for by additional agonist administration. In contrast to isoprenaline, forskolin- and 8-Br-cAMP-induced relaxation of methacholine-induced contraction were not affected by GF109203X, indicating that uncoupling of the β-adrenoceptor from the effector system by PKC may indeed be involved [2].

In line with these results, in isolated bovine tracheal smooth muscle cells we also demonstrated that the inhibitory effect of isoprenaline on both methacholine-induced intracellular Ca^{2+} mobilization and extracellular Ca^{2+} influx was markedly

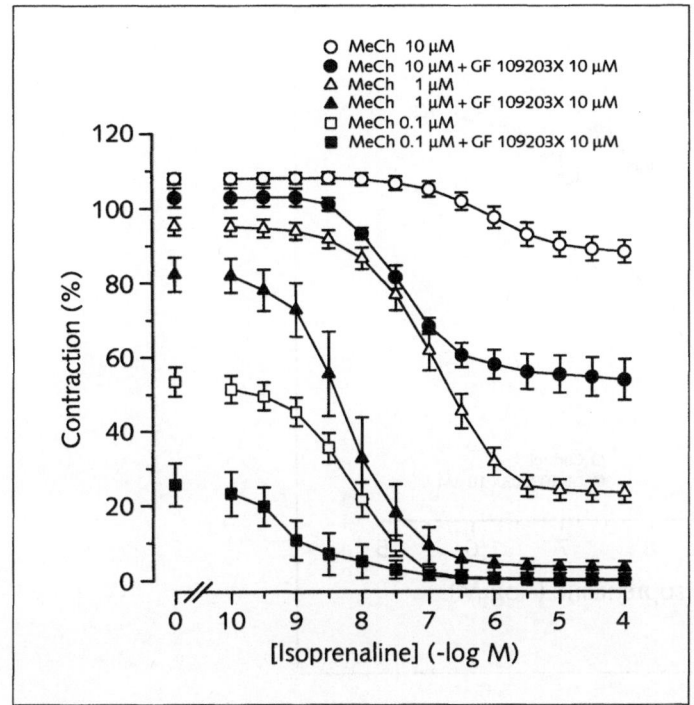

Figure 3
Isoprenaline-induced relaxation of bovine tracheal smooth muscle preparations following contraction by various concentrations of methacholine in the absence and presence of the specific protein kinase C inhibitor GF 109203X (10 μM). Results are means ± s.e.m. of 4–9 experiments.

potentiated by GF109203X (Meurs et al., unpublished information). This indicates that the reduced β-adrenoceptor function induced by muscarinic agonist-induced PKC activation may contribute to the functional antagonism between muscarinic and β-adrenoceptor agonists by reduced inhibition of the contractile agonist-induced Ca^{2+} response.

There is evidence that the described mechanism of cross-talk between PI metabolism and the β-adrenoceptor system may also apply to asthmatic airways. Thus, it has been observed that the relaxant response to isoprenaline is reduced in isolated airway preparations from patients with fatal [150–152] or non-fatal [153, 154] active and severe asthma, whereas a normal response has been found in patients with mild to moderate asthma [155, 156] (for review see [1]). Interestingly, the reduced β-adrenergic relaxation described above was found irrespective of prior

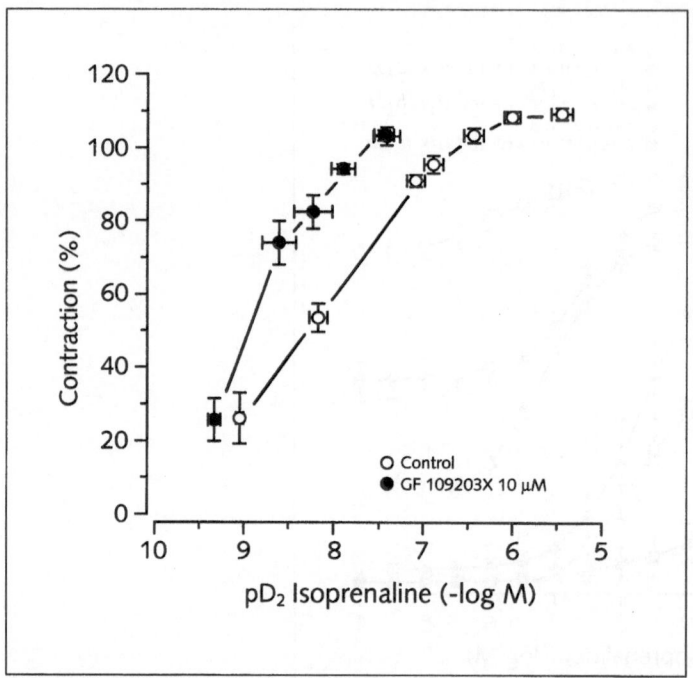

Figure 4

Relationships between isoprenaline relaxation potencies (pD$_2$) and contractile tone induced by increasing concentrations (0.03–100 μM, half log$_{10}$ concentration steps) of methacholine, in the absence (Control) and presence of the specific protein kinase C inhibitor GF 109203X (10 μM). Results are means ± s.e.m. of 4–9 experiments.

treatment with β-agonists, which may have a confounding effect (see further below). Autoradiographic studies in lungs from patients with fatal and non-fatal asthma demonstrated normal [157] or even enhanced [154, 158] β-adrenoceptor densities in the airway smooth muscle, while concomitantly the β-adrenergic relaxation was reduced in some of these studies [154, 158], indicating that the β-adrenoceptor is uncoupled from the effector system in patients with severe asthma, which could involve contractile agonist-induced activation of PKC. Similar to lymphocytes (see above), a reduced β-adrenoceptor responsiveness of airway smooth muscle in patients with active and severe asthma could result from the allergic asthmatic reaction. Thus, using a guinea pig model of allergic asthma, we have recently shown that single or repeated allergen challenge may lead to a reduced isoprenaline-induced relaxation of methacholine- or histamine-contracted tracheal preparations from these animals [159].

Modulation of agonist-induced PI metabolism and Ca^{2+} signalling by cAMP-dependent mechanisms

Another mechanism of cross-talk between contractile receptor-mediated PI metabolism and β-adrenoceptor-mediated adenylyl cyclase activation that could hypothetically contribute to the functional antagonism between contractile and relaxing stimuli is cAMP-dependent inhibition of the IP response. Thus, both in canine [107] and bovine [106, 108, 160] tracheal smooth muscle slices, it has been demonstrated that β-adrenoceptor agonists and other cAMP elevating agents, such as forskolin and dibutyryl cAMP, cause a marked attenuation of histamine-induced IP formation. In bovine tracheal smooth muscle, the β-agonist did not inhibit the initial transient increase in IP_3 formation induced by histamine, but rather the sustained phase of PI metabolism that follows this transient increase and which is dependent on influx of extracellular Ca^{2+} [6]. This effect may be at least partially explained by membrane hyperpolarization due to β-agonist-induced opening of K_{Ca} channels, thus causing reduced influx of Ca^{2+} through dihydropyridine-sensitive voltage-dependent Ca^{2+}-channels, an effect that can be mimicked by various K^+ channel openers [16].

Very surprisingly, no inhibitory effects of β-agonists or other cAMP-elevating agents were observed when IP formation was induced by full muscarinic agonists such as methacholine and carbachol, even at low concentrations [106–108, 160]. Furthermore, some, but not all, partial muscarinic agonist-induced responses were also insusceptible to such inhibition [108]. A role for M_2 receptor-mediated inhibition of adenylyl cyclase in the apparent insensitivity of muscarinic agonist-induced responses is unlikely, since the methacholine-induced PI metabolism was also shown to be resistant to cAMP elevations induced by forskolin and stable analogues of cAMP [107]. Although a good explanation is presently not at hand, insensitivity of muscarinic agonist-induced PI metabolism to influx of Ca^{2+}, and thus to inhibition by β-agonists and K^+ channel openers, could be involved [6]. Remarkably, in bovine tracheal smooth muscle slices, it was recently shown that combined inhibition of type 3 and type 4 phosphodiesterases can inhibit methacholine-induced IP_3 responses [161].

Since airway smooth muscle contractions to muscarinic agonists are relatively resistant to relaxation by β-adrenoceptor agonists, compared with contractions induced by histamine [130–132], it has been suggested that this resistance is related to the lack of effect of β-agonists on muscarinic agonist-induced IP formation. However, we have demonstrated that cAMP-dependent inhibition of contractile agonist-induced Ca^{2+} mobilization, as well as influx in isolated bovine tracheal smooth muscle cells, is not primarily caused by attenuation of IP production and that the relative resistance of muscarinic agonist-induced contraction to β-adrenoceptor agonists is largely determined by its higher potency in inducing PI metabolism and subsequent intracellular Ca^{2+} changes [28]. This was supported by the observation of a dissociation between salbutamol and salmeterol-induced cAMP accumulation and inhibition of histamine-induced PI metabolism in bovine tracheal smooth muscle slices [162].

β-Agonist-induced homologous β-adrenoceptor desensitization and bronchoprotection against bronchoconstrictive stimuli

Inhaled β_2-agonists are by far the most effective bronchodilators available and in recommended doses there are few, if any, adverse effects [4, 163]. However, some reports have raised the possibility that regular use of relatively high doses of β-adrenergic bronchodilator drugs by inhalation may increase the morbidity and the mortality of asthma (e.g., see [164]). Support has been given to this suggestion by epidemiological surveys which have detected an association between the intensity of inhaled β_2-agonist drug therapy and asthma death or near death [165–167].

Although it is virtually impossible to discriminate retrospectively between the possibility that increased use of β_2-agonists is simply a marker for more severe asthma and the possibility that β_2-agonists are directly responsible for the increased morbidity and mortality, there are a number of prospective observations that adverse effects of long-term β_2-agonist treatment may indeed occur. Thus, regular administration of inhaled fenoterol caused deterioration in asthma control compared to on demand treatment with the β-agonist [168]. During the regular treatment with this agonist, there was also an increase in bronchial hyperreactivity to methacholine [168, 169]. Accordingly, regular inhalation of salbutamol over 1 year by patients with asthma and COPD caused a small but significantly increased bronchial responsiveness to histamine [170], while during a 2-year treatment-period a more rapid annual decline of lung function was found compared to symptomatic use of this drug [171]. A rebound increase in bronchial reactivity after cessation of β_2-agonist treatment was first noticed by Koëter et al. [172], who demonstrated that treatment of allergic asthmatics with oral terbutaline for 2 weeks caused a significant increase in the responsiveness of the airways to the β-blocker propranolol. Evidence for rebound hyperreactivity to histamine or methacholine after long-term inhaled β_2-agonist treatment has subsequently been found by several other investigators [173–175]. In addition, it has been reported that the bronchial response to allergens may be increased after cessation of treatment with a β_2-agonist [176–178].

The mechanisms by which long-term treatment with β_2-agonists may increase the symptoms of asthma remain to be defined. Of the possible mechanisms, the development of desensitization to or tolerance of β_2-agonists has been most widely investigated. Impaired bronchodilator responses have been observed after regular treatment of asthmatic patients with both short- and long-acting β_2-agonists [179–185], although the results of β-agonist treatment have differed [169, 186–192].

Remarkably, there may be a dissociation between changes in the efficacy of β_2-agonists to induce bronchodilation in (usually mild) asthmatics after prolonged treatment with these agonists, and changes in their capacity to protect against bronchoconstrictor stimuli. Thus, regular use of terbutaline [189], salbutamol [178], salmeterol [190, 191] and formoterol [192] caused a significantly reduced protection against methacholine- [178, 189–191] and/or adenosine monophosphate [189, 192] -induced bronchoconstriction, with no apparent effect on the bronchodilator effect of the β_2-agonists. One possible explanation for this apparent dissociation could be that the investigations described above have been mainly performed in patients with mild to moderate asthma still having a good reversibility of airways obstruction, while agonist-induced (more severe) bronchoconstriction in these patients is necessary to reveal reduced functional antagonism upon β-adrenoceptor desensitization. Reduced bronchoprotection by β-agonists against bronchoconstrictor stimuli, including methacholine, was also observed in several other studies [185, 193–196].

Very remarkably, loss of protection was significantly more pronounced for adenosine monophosphate than for methacholine as the provocation stimulus [189], which could indicate that tolerance of the mast cell-stabilizing effects of β-agonists may be important for the deterioration of asthma symptoms and bronchial hyperreactivity after long-term treatment. These results are in line with the initial *in vitro* observation of Van der Heijden et al. [197], who found in sensitized guinea pigs that β-adrenoceptors mediating inhibition of allergen-induced mediator release are more susceptible to desensitization than β-adrenoceptors mediating airway smooth muscle relaxation, which could at least partially explain the reduced protection by β-agonists against allergic stimuli after long-term treatment [177, 178]. Accordingly, β_2-adrenoceptors on mast cells and MNL appear to be more sensitive to agonist-induced desensitization than β_2-adrenoceptors on airway smooth muscle [176, 198], and β-adrenoceptor desensitization in these cells after β-agonist treatment could be important for enhanced inflammation and subsequent development of airway hyperreactivity to both allergic and non-allergic stimuli [1, 4, 199, 200].

β-Agonist-induced homologous β-adrenoceptor desensitization and its heterologous modulation by PKC

Biochemical processes that have been shown to be associated with the development of β-adrenoceptor desensitization are: rapid (seconds to minutes) uncoupling and/or sequestration of the receptor, and a more gradual (hours) receptor down-regulation (e.g., see [201–203]).

Rapid uncoupling of the receptor from G_s is mediated by phosphorylation of cytoplasmic residues of the receptor by either PKA and G-protein coupled receptor kinases (GRKs), including β-adrenergic receptor kinase 1 (βARK1, also known as GRK2) and βARK2 (also known as GRK3) [203–205]. βARK1 and βARK2 have been characterized as cytosolic enzymes that specifically phosphorylate the agonist-occupied form of the receptor, after translocation to the plasma membrane by binding to free $G_{\beta\gamma}$-units made available by receptor activation. This phosphorylation facilitates the binding of a second cytosolic protein, β-arrestin-1 or β-arrestin-2, to the phosphorylated receptor, thus preventing coupling of the receptor to G proteins. GRKs mediate receptor-specific uncoupling in response to relatively high concentrations of agonist, which is independent of cAMP levels. On the other hand, a more generalized form of receptor uncoupling is catalyzed by the cAMP-dependent PKA, which requires only small increases in cAMP and which can be activated by a variety of agents, including (low concentrations of) β-agonists. PKA-mediated desensitization proceeds by phosphorylation of an intracellular receptor domain which is directly involved in the interaction with the α-subunit of G_s [201].

GRK-induced receptor phosphorylation and β-arrestin binding also appear to be involved in sequestration of the β-adrenoceptor by endocytosis. Thus, after binding to the GRK-phosphorylated receptor, β-arrestins also bind with high affinity to clathrin, which facilitates internalization of the receptors into clathrin-coated vesicles [203, 206, 207]. In these vesicles, receptors are dephosphorylated by a membrane-associated receptor phosphatase upon acidification, which precedes recycling of the receptors to the plasma membrane [203, 208–210].

Down-regulation is defined as the loss of total receptor binding sites in the cell with accompanying loss in effector stimulation and occurs at more prolonged agonist exposure. Proteolytic degradation of the β-adrenoceptor is thought to contribute to down-regulation since full recovery of binding sites is slow and requires new protein synthesis. In addition, down-regulation may be associated with reduced steady-state levels of mRNA of the receptor. This decreased level of receptor mRNA presumably contributes to the overall reduction in receptor number and responsiveness and is most likely maintained until recovery from down-regulation begins [202]. Furthermore, in recombinant cell studies and in cultured human bronchial smooth muscle, it has been demonstrated that allelic polymorphisms of the β_2-adrenoceptor at codons 16 (Arg or Gly) and 27 (Glu or Gln) in the N-terminal region of the receptor [211, 212] and different polymorphisms in the promoter region of the human β_2-adrenoceptor gene, including a 19 aminoacid 5' leader peptide [213, 214] which regulates the β_2-adrenoceptor expression by inhibition of mRNA translation [215], may influence the degree of agonist-induced down-regulation. However, the biochemical processes involved in agonist-induced down-regulation are not very well understood.

As indicated above, GRKs are importantly involved in agonist-induced homologous desensitization of receptor-mediated responses. Several studies have now indicated that the extent of such involvement may vary with the level of expression or activity of the kinases [204]. In human MNL it has been shown that the activity of βARK1 (GRK2), and as a consequence the level of β-adrenoceptor desensitization, can be subject to heterologous regulation by PKC. Thus, βARK enzymatic activity in MNL was enhanced by calcium ionophores in a PKC-dependent manner (i.e., the effect was blocked by inhibitors of PKC), and also by the PKC activator PMA. The biochemical mechanism underlying these observations was shown to be that purified βARK1 is directly phosphorylated by purified PKC, leading to increased phosphorylating activity of βARK1. Enhanced phosphorylation of βARK1 was also found after PMA treatment of intact MNL, and this was accompanied by increased agonist-induced desensitization of the β_2-adrenoceptor. Since this increase in desensitization was sensitive to heparin, it was attributed to βARK1 [216]. Increased βARK1 activity was also observed after stimulating T cells with the mitogen phytohaemagglutinin (PHA), presumably *via* activation of PKC [217]. In this study, increased enzymatic activity was accompanied by increases in both βARK1 mRNA and protein, suggesting that regulation is not only by phosphorylation of pre-existent protein but also at the level of gene expression. As in the former study, functional consequences of increased βARK1 activity were observed; a decreased formation of cAMP in response to β-adrenoceptor stimulation was reported.

Similar observations of PKC-mediated increase in βARK1 activity and of PKC-mediated phosphorylation of the enzyme have been reported in transfected CHO cells; it was suggested that the mechanism involves enhanced translocation of the enzyme to the plasma membrane [218].

Recent preliminary observations at our laboratory have indicated that activation of PKC by PMA may also potentiate fenoterol-induced desensitization of bovine tracheal smooth muscle relaxation (Fig. 5). Hence both at the level of MNL and at the level of airway smooth muscle, this novel form of cross-talk between PKC activity induced by cell-activating stimuli and βARK could be of considerable importance for the development of a reduced β-adrenoceptor function in patients with active and severe asthma, who are frequently using high doses of β-adrenergic bronchodilator drugs to relieve bronchoconstriction. However, in contrast to patients with mild or moderate asthma as discussed above, the clinical significance of β-agonist-induced desensitization in patients with severe asthma remains to be established.

Cross-talk between muscarinic M$_2$ receptor stimulation and β-adrenoceptor function in airway smooth muscle

Role of muscarinic M$_2$ receptor stimulation in the functional antagonism of airway smooth muscle contraction by β-adrenoceptor agonists

Studies on the functional role of M_2 receptors in airway smooth muscle have been guided by the observation that these receptors inhibit adenylyl cyclase activity and may thus attenuate β-adrenoceptor-mediated relaxation. Thus, muscarinic M_2 receptor-mediated inhibition of adenylyl cyclase (as well as potassium channel clo-

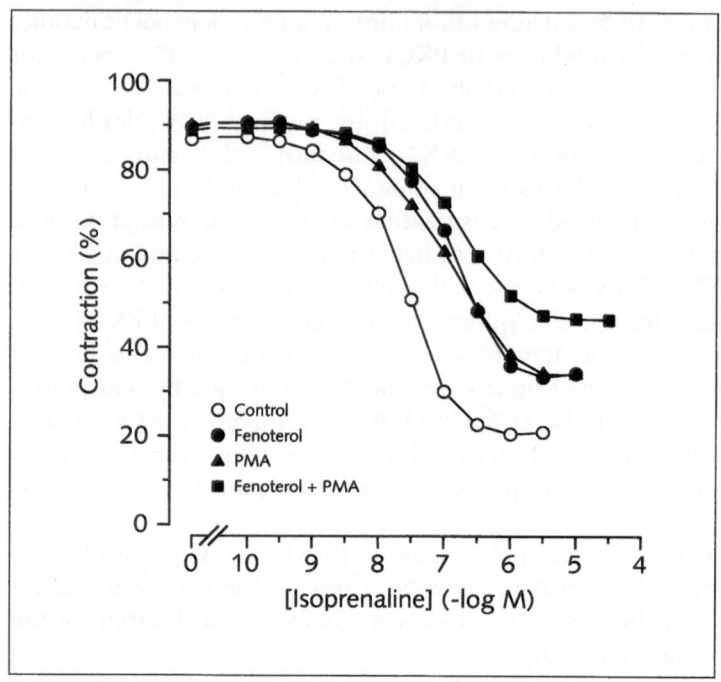

Figure 5
Relaxation of 1 µM methacholine-induced contraction of bovine tracheal smooth muscle by isoprenaline after preincubation (30 min, 37°C) with vehicle (Control), 1 µM fenoterol, 10 µM PMA, or 1 µM fenoterol plus 10 µM PMA, followed by extensive washout during 45 min. Results are representative of two similar experiments.

sure) could contribute to the phenomenon that isoprenaline-induced relaxation of cholinergic tone is hampered compared to similar contraction levels of histamine, and that muscarinic agonists promote a larger shift of isoprenaline potency than histamine at a given level of IP accumulation [131]. This subject is also covered in the chapter by Eglen and Watson in this volume and will be discussed below only briefly.

In canine, rabbit and guinea pig trachea precontracted with muscarinic agonists, selective blockade of M_2 receptors or inactivation of G_i enhanced the relaxant potency of isoprenaline (reviewed in [49, 219]). However, the effects were relatively small compared to the large changes in isoprenaline relaxant potency that can be obtained in these tissues as a result of varying cholinergic contraction levels [49]. Moreover, in guinea pig trachea such a role for muscarinic M_2 receptors was found to be completely absent when gallamine was used as the M_2-selective antagonist [220].

In BTSM, we observed significant correlations between the decrease of both pD_2 and E_{max} values of isoprenaline-induced relaxation and inhibition of adenylyl cyclase at increasing concentrations of methacholine, suggesting a possible role for muscarinic M_2 receptor-mediated inhibition of adenylyl cyclase in the decrease of β-adrenergic relaxability. However, quantitatively similar relationships of the pD_2 and E_{max} values with much lower levels of IP production were observed, indicating a more prominent role of M_3-receptor-mediated PI metabolism than of G_i-mediated adenylyl cyclase inhibition in the functional antagonism [7, 55]. This was supported by the observation that isoprenaline-induced relaxation of cholinergic tone was enhanced by the M_2-selective antagonists gallamine and AF-DX 116, and by the M_1 and M_3 receptor-selective antagonists pirenzepine and hexahydrosiladifenidol, which could be entirely explained by small changes in initial contractile tone in the presence of these antagonists [221]. In BTSM it was also recently demonstrated that M_2 receptors inhibit the relaxant effects of forskolin, but not of isoprenaline, presumably because part of the relaxant response to isoprenaline is mediated through a non-cAMP-dependent mechanism, which is largely unopposed by the M_2 receptor [57].

As in BTSM, a role for M_2 receptors in functional antagonism in human bronchial smooth muscle appears to be absent, as indicated by identical relationships between IP production and isoprenaline-induced relaxation for both methacholine- and histamine-induced tones [19] and the observation that β-adrenergic relaxation of carbachol-induced contraction is not potentiated by methoctramine [222].

Although, depending on the species under investigation, cross-talk between muscarinic M_2 receptor and β-adrenoceptor functions appears to play only a minor, if any, role in the functional antagonism of cholinergic contraction by β-agonists, such a role may become apparent in pathophysiological conditions, like allergic asthma. Thus, very recently it has been demonstrated that passive sensitization with human atopic serum as well as incubation with inflammatory cytokines, such as IL-1β and TNFα, may cause attenuated β-adrenoceptor-mediated rabbit tracheal smooth muscle relaxation due to enhanced muscarinic M_2 receptor-coupled activation of G_i, presumably by enhanced expression of the $G_{i\alpha}$ subunits [223–225]. In addition, enhanced expression of $G_{i\alpha}$ was also observed in a mouse strain with airway hyperreactivity [226].

Muscarinic receptor-β-adrenoceptor cross-talk and airway smooth muscle proliferation

As already indicated above, β-adrenoceptor stimulation of airway smooth muscle causes inhibition of growth factor- and GPCR-induced airway smooth muscle proliferation by enhanced cAMP production [112–114, 118] (Fig. 2).

Inhibition of muscarinic agonist-induced airway smooth muscle proliferation by β-adrenoceptor agonists or cAMP-elevating agents has thus far not been established. On the other hand, it can be envisaged that cross-talk between G_i-coupled muscarinic M_2 receptors (as well as G_q-coupled M_3 receptors *via* activation of PKC) and $β_2$-adrenoceptors causes reduced effectiveness of endogenous catecholamines, as well as exogenous β-agonist bronchodilator drugs, to inhibit proliferation in response to cholinergic agonists as well as other proliferative stimuli. Such a mechanism could possibly be involved in the ineffectiveness of β-agonists to inhibit airway smooth muscle proliferation observed after long-term repeated allergen challenges in a guinea pig model of asthma [227]; it may also be important in patients with chronic and severe asthma, who frequently show airway remodelling characterized by an increase in smooth muscle mass [3, 228–233].

Concluding remarks

In recent years, evidence has been obtained that cross-talk between muscarinic M_3 receptor-induced PI metabolism, *via* activation of PKC, and β-adrenoceptor function is important in the functional antagonism of cholinergic airway smooth muscle contraction by β-agonists and that this cross-talk may contribute importantly to the reduced bronchodilating potency of β-adrenoceptor agonists in patients with active and severe asthma. In addition, recent studies have indicated that activation of PKC may potentiate β-agonist-induced desensitization through activation and induction of βARK, which may similarly be harmful for severe asthmatics who frequently use high doses of β-agonists to relieve bronchoconstriction. Cross-talk between M_2-muscarinic receptor- and G_i-protein-mediated inhibition of adenylyl cyclase and β-adrenoceptor- and G_s-protein-mediated activation of adenylyl cyclase seems to play only a minor role in the functional antagonism of muscarinic (M_3-) receptor-mediated contraction by β-agonists, although its contribution may be enhanced in inflammatory conditions such as asthma, *via* the induction of G_i.

As indicated, PKC plays a central role in airway smooth muscle contraction as well as in the modulation of airway smooth muscle relaxation by β-agonists. It has been established that PKC is not a single protein kinase but rather a family of multiple isoenzymes with different regulatory requirements, substrate specificities and biochemical effects [234]. The different isoenzymes, which vary in their tissue distribution, have been classified into a group of conventional PKCs, dependent upon Ca^{2+}, DAG and phospholipid for their activity, including PKC $α$, $β_I$, $β_{II}$ and $γ$; novel PKCs, which are Ca^{2+}- independent, including the $δ$, $ε$, $η$ and $θ$ isoforms; and atypical PKCs, including PKC $ξ$, $μ$, $ι$ and $λ$, which require only phospholipid for activation. The isoenzymes present in canine, porcine, bovine and human airway smooth muscle have recently been identified [235–237]. The isoform pattern shows some variation between the different species as well as different airway regions and appears to depend on the proliferative condition of the cells [238]. Specific, but rather isoenzyme-unselective PKC inhibitors that are presently available have indicated the importance of PKC isoenzymes in airway smooth muscle contraction as well as β-adrenoceptor-mediated relaxation. The specific isoenzymes involved in these processes have thus far not been identified. Identification of these isoenzymes and development of selective PKC inhibitors may provide a valuable novel therapeutic approach in the treatment of airways obstruction in patients with asthma and COPD.

Recent observations that GPCR, including muscarinic receptors, may be coupled to the MAPK signalling pathway and cell proliferation, have opened an intriguing new area of asthma research on the role of neurotransmitters and mediators in airway smooth muscle proliferation and airway remodelling as occur in chronic disease. The role of muscarinic M_2 and M_3 receptors and other contractile GPCR in asthmatic airway smooth muscle proliferation, as well as the role of cross–talk of these receptors with β-adrenoceptors and growth factor receptors in this process, is unknown at present and warrants further investigation.

Acknowledgements
The authors wish to thank Dr. B.H. Hoiting, Drs. R. Kuipers, Drs. D. Wagemaker, Drs. P.B. Eppens, Drs. M. Boterman and Drs. S. McKay for their valuable contributions to some of the investigations described in this chapter. The Netherlands Asthma Foundation and the Groningen Graduate School for Behavioural and Cognitive Neuroscience are gratefully acknowledged for funding several of these studies.

References

1   Meurs H, Zaagsma J (1991) Pharmacological and biochemical changes in airway smooth muscle in relation to bronchial hyperresponsiveness. In: DK Agrawal, RG Townley (eds): *Inflammatory cells and mediators in bronchial asthma*. CRC Press, Inc., Boca Raton, 1–38

2   Zaagsma J, Roffel AF, Meurs H (1997) Muscarinic control of airway function. *Life Sci* 60: 1061–1068

3   Barnes PJ (1998) Pharmacology of airway smooth muscle. *Am J Respir Crit Care Med* 158: S123–S132

4   Barnes PJ (1995) Beta-adrenergic receptors and their regulation. State of the art. *Am J Respir Crit Care Med* 152: 838–860

5   Abdel-Latif AA (1991) Biochemical and functional interactions between the inositol 1,4,5-trisphosphate-Ca^{2+} and cyclic AMP signalling systems in smooth muscle. *Cell Signalling* 3: 371–85

6   Challiss RAJ, Boyle JP (1994) Modulation of agonist-stimulated phosphoinositide turnover in airways smooth muscle by cyclic nucleotide-dependent and independent mechanisms. In: D Raeburn, MA Giembycz (eds): *Airways smooth muscle: a reference source*. Birkhäuser Verlag, Basel, 309–327

7   Meurs H, Hoiting BH, Roffel AF, Elzinga CRS, Zaagsma J (1994) Cross-talk between receptor transduction mechanisms in relation to airways obstruction. In: DS Postma, J Gerritsen (eds): *Bronchitis V*. Van Gorcum, Assen, 135–148

8   Baron CB, Cunningham M, Strauss JF, Coburn RF (1984) Pharmacomechanical coupling in smooth muscle may involve phosphatidylinositol metabolism. *Proc Natl Acad Sci USA* 81: 6899–6903

9   Takuwa Y, Takuwa N, Rasmussen H (1986) Carbachol induces a rapid and sustained hydrolysis of polyphosphoinositide in bovine tracheal smooth muscle: measurements of the mass of polyphosphoinositides, 1,2-diacylglycerol and phosphatidic acid. *J Biol Chem* 261: 14670–14675

10   Grandordy BM, Cuss FM, Sampson AS, Palmer JB, Barnes PJ (1986) Phosphatidylinositol response to cholinergic agonists in airway smooth muscle: Relationship to contraction and muscarinic receptor occupancy. *J Pharmacol Exp Ther* 238: 273–279

11   Duncan RA, Krzanowski JJ Jr, Davis JS, Polson JB, Coffey RG, Shimoda T, Szentiranyi A (1987) Polyphosphoinositide metabolism in canine tracheal smooth muscle (CTSM) in response to a cholinergic stimulus. *Biochem Pharmacol* 36: 307–310

12   Meurs H, Roffel AF, Postema HB, Timmermans A, Elzinga CRS, Kauffman HF, Zaagsma J (1988) Evidence for a direct relationship between phosphoinositide metabolism and airway smooth muscle contraction induced by muscarinic agonists. *Eur J Pharmacol* 156: 271–274

13   Chilvers ER, Challiss RAJ, Barnes PJ, Nahorski SR (1989) Mass changes of inositol (1,4,5) trisphosphate in trachealis muscle following agonist stimulation. *Eur J Pharmacol* 164: 587–590

14   Roffel AF, Meurs H, Elzinga CRS, Zaagsma J (1990) Characterization of the muscarinic receptor subtype involved in phosphoinositide metabolisms in bovine tracheal smooth muscle. *Br J Pharmacol* 99: 293–296

15   Yang CM, Chou S-P, Sung T-C (1991) Muscarinic receptor subtypes coupled to generation of different second messengers in isolated trachea smooth muscle cells. *Br J Pharmacol* 104: 613–618

16   Challiss RAJ, Patel N, Arch JRS (1992) Comparative effects of BRL 38227, nitrendipine and isoprenaline on carbachol- and histamine-stimulated phosphoinositide metabolism in airway smooth muscle. *Br J Pharmacol* 105: 997–1003

17   Roux F, Mavoungou E, Naline E, Lacroix H, Tordet C, Advenier C, Grandordy BM (1995) Role of 1,2-sn diacylglycerol in airway smooth muscle stimulated by carbachol. *Am J Respir Crit Care Med* 151: 1745–1751

18   Meurs H, Timmermans A, Van Amsterdam RGM, Brouwer F, Kauffman HF, Zaagsma J (1989) Muscarinic receptors in human airway smooth muscle are coupled to phosphoinositide metabolism. *Eur J Pharmacol* 164: 369–371

19 Van Amsterdam RGM, Meurs H, Ten Berge REJ, Veninga CCM, Brouwer F, Zaagsma J (1990) Role of phosphoinositide metabolism in human bronchial smooth muscle contraction and in functional antagonism by beta-adrenoceptor agonists. *Am Rev Respir Dis* 142: 1124–1128

20 Marmy N, Durand-Arczynska W, Durand J (1992) Agonist-induced production of inositol phosphates in human airway smooth muscle cells in culture. *J Physiol (Paris)* 86: 185–194

21 Hashimoto T, Hirata M, Ito Y (1985) A role for inositol 1,4,5 trisphosphate in the initiation of agonist-induced contractions of dog tracheal smooth muscle. *Br J Pharmacol* 86: 191–199

22 Takuwa Y, Takuwa N, Rasmussen H (1988) The effects of isoproterenol on intracellular calcium concentration. *J Biol Chem* 263: 762–768

23 Ozaki H, Kwon S-C, Tajimi M, Karaki H (1990) Changes in cytosolic Ca^{2+} and contraction induced by various stimulants and relaxants in canine tracheal smooth muscle. *Pflügers Arch* 416: 351–359

24 Murray RK, Kotlikoff MI (1991) Receptor-activated calcium influx in human airway smooth muscle cells. *J Physiol* 435: 123–44

25 Gunst SJ, Gerthoffer WT, Al-Hassani MH (1992) Ca^{2+} sensitivity of contractile activation during muscarinic stimulation of tracheal muscle. *Am J Physiol* 263: C1258–C1265

26 Al-Hassani MH, Garcia JGN, Gunst SJ (1993) Differences in Ca^{2+} mobilization by muscarinic agonists in tracheal smooth muscle. *Am J Physiol* 264: L53–59

27 Yang CM, Yo Y-L, Wang Y-Y (1993) Intracellular calcium in canine cultured tracheal smooth muscle cells is regulated by M_3 muscarinic receptors. *Br J Pharmacol* 110: 983–988

28 Hoiting BH, Meurs H, Schuiling M, Kuipers R, Elzinga CRS, Zaagsma J (1996) Modulation of agonist-induced phosphoinositide metabolism, Ca^{2+} signalling and contraction of airway smooth muscle by cyclic AMP-dependent mechanisms. *Br J Pharmacol* 117: 419–426

29 Kamm KE, Stull JT (1985) The function of myosin and myosin light chain kinase phosphorylation in smooth muscle. *Ann Rev Pharmacol Toxicol* 25: 593–620

30 Gerthoffer WT (1991) Regulation of the contractile element of airway smooth muscle. *Am J Physiol* 261: L15–L28

31 De Lanerolle P, Paul RJ (1991) Myosin phosphorylation/dephosphorylation and regulation of airway smooth muscle. *Am J Physiol* 291: L1–L14

32 Rasmussen H, Takuwa Y, Park S (1987) Protein kinase C in the regulation of airway smooth muscle contraction. *FASEB J* 1: 177–185

33 Langlands JM, Diamond J (1992) Translocation of protein kinase C in bovine tracheal smooth muscle strips: the effect of methacholine and isoprenaline. *Eur J Pharmacol* 227: 131–138

34 Yang KXF, Black JL (1995) The involvement of protein kinase C in the contraction of human airway smooth muscle. *Eur J Pharmacol* 275: 283–289

35 Horowitz A, Menice CB, Laporte R, Morgan KG (1996) Mechanisms of smooth muscle contraction. *Physiol Rev* 76: 967–1002

36 Park S, Rasmussen H (1986) Carbachol-induced protein phosphorylation changes in bovine tracheal smooth muscle. *J Biol Chem* 261: 15734–15739

37 Masuo M, Reardon S, Ikebe M, Kitazawa T (1994) A novel mechanism for the Ca^{2+}-sensitizing effect of protein kinase C on vascular smooth muscle: inhibition of myosin light chain phosphatase. *J Gen Physiol* 104: 265–286

38 Bremerich DH, Warner DO, Lorenz RR, Shumway R, Jones KA (1997) Role of protein kinase C in calcium sensitization during muscarinic stimulation in airway smooth muscle. *Am J Physiol* 273: L775–L781

39 Bremerich DH, Hirasaki A, Jones KA, Warner DO (1997) Halothane attenuation of calcium sensitivity in airway smooth muscle. *Anesthesiology* 87: 94–101

40 Pohl J, Winder SJ, Allen BG, Walsh, MP, Sellers JR, Gerthoffer WT (1997) Phosphorylation of calponin in airway smooth muscle. *Am J Physiol* 272: L115–L123

41 Gerthoffer WT, Yamboliev IA, Pohl J, Haynes R, Dang S, McHugh J (1997) Activation of MAP kinases in airway smooth muscle. *Am J Physiol* 272: L244–L252

42 Hedges JC, Yamboliev IA, Ngo M, Horowitz B, Adam LP, Gerthoffer WT (1998) p38 Mitogen-activated protein kinase expression and activation in smooth muscle. *Am J Physiol* 275: C527–C534

43 Hoiting BH, Kuipers R, Elzinga CRS, Zaagsma J, Meurs H (1995) Feedforward control of agonist-induced Ca^{2+} signalling by protein kinase C in airway smooth muscle cells. *Eur J Pharmacol* 290: R5–R7

44 Souhrada M, Souhrada J (1989) Sodium and calcium influx induced by phorbol esters in airway muscle cells. *Am Rev Respir Dis* 13: 927–932

45 Schramm CM, Grunstein MM (1989) Mechanisms of protein kinase C regulation of airway contractility. *J Appl Physiol* 66: 1935–1941

46 Knox A, Boldwin D, Cragoa E, Ajao P (1993) The effect of sodium transport and calcium channel inhibitors on phorbol ester-induced contraction of bovine airway smooth muscle. *Pulm Pharmacol* 6: 241–246

47 Rossetti M, Savineau J-P, Crevel H, Marthan R (1995) Role of protein kinase C in non-sensitized and passively sensitized human isolated bronchial smooth muscle. *Am J Physiol* 268: L966–L971

48 Yang KXF, Black JL (1996) Protein kinase C induced changes in human airway smooth muscle tone: the effects of Ca^{2+} and Na^+ transport. *Eur J Pharmacol* 315: 65–71

49 Roffel AF, Zaagsma J (1995) Acetylcholine. In: D Raeburn, MA Giembycz (eds): *Airways smooth muscle: Neurotransmitters, amines, lipid mediators and signal transduction*. Birkhäuser Verlag, Basel, 81–130

50 Jones CA, Madison JM, Tom-Moy M, Brown JK (1987) Muscarinic cholinergic inhibition of adenylate cyclase in airway smooth muscle. *Am J Physiol* 253: C97–C104

51 Sankary RM, Jones CA, Madison JM, Brown JK (1988) Muscarinic cholinergic inhibition of cyclic AMP accumulation in airway smooth muscle. Role of a pertussis toxin-sensitive protein. *Am Rev Respir Dis* 138: 145–150

52 Pyne NJ, Grady MW, Shehnaz D, Stevens PA, Pyne S, Rodger IW (1992) Muscarinic blockade of β-adrenoceptor-stimulated adenylyl cyclase: the role of stimulatory and inhibitory guanine-nucleotide binding regulatory proteins (G_s and G_i). *Br J Pharmacol* 107: 881–887

53 Challiss RAJ, Adams D, Mistry R, Boyle JP (1993) Second messenger and ionic modulation of agonist-stimulated phophoinositide turnover in airway smooth muscle. *Biochem Soc Trans* 21: 1138–1145

54 Widdop S, Daykin K, Hall IP (1993) Expression of muscarinic M_2 receptors in cultured human airway smooth muscle cells. *Am J Respir Cell Mol Biol* 9: 541–546

55 Meurs H, Roffel AF, Elzinga CRS, De Boer REP, Zaagsma J (1993) Muscarinic receptor-mediated inhibition of adenylyl cyclase and its role in the functional antagonism of isoprenaline in airway smooth muscle. *Br J Pharmacol* 108: 208P

56 Shaefer OP, Ethier MF, Madison JM (1995) Muscarinic regulation of cyclic AMP in bovine trachealis cells. *Am J Respir Cell Mol Biol* 13: 217–226

57 Ostrom RS, Ehlert FJ (1998) M_2 muscarinic receptors inhibit forskolin- but not isoproterenol-mediated relaxation in bovine tracheal smooth muscle. *J Pharmacol Exp Ther* 286: 234–242

58 Kume H, Kotlikoff MI (1991) Muscarinic inhibition of single K_{Ca} channels in smooth muscle cells by a pertussis-sensitive G protein. *Am J Physiol* 216: C1204–C1209

59 Kume H, Graziano MP, Kotlikoff MI (1992) Stimulatory and inhibitory regulation of calcium-activated potassium channels by guanine nucleotide binding proteins. *Proc Natl Acad Sci USA* 89: 11051–11055

60 Jansen LJ, Sims SM (1992) Acetylcholine activates non-selective cation and chloride conductances in canine and guinea pig tracheal myocytes. *J Physiol* 453: 197–218

61 Wang Y-X, Fleishmann BK, Kotlikoff MI (1997) M_2 receptor activation of nonselective cation channels in smooth muscle cells: calcium and G_i/G_o requirements. *Am J Physiol* 273: C500–C508

62 Kotlikoff MI, Wang Y-X (1998) Calcium release and calcium-activated chloride channels in airway smooth muscle cells. *Am J Respir Crit Care Med* 158: S109–S114

63 Hirshman CA, Togashi H, Shao D, Emala CW (1998) $G\alpha_{i-2}$ is required for carbachol-induced stress fiber formation in human airway smooth muscle cells. *Am J Physiol* 275: L911–L916

64 Togashi H, Emala CW, Hall IP, Hirshman CA (1998) Carbachol-induced actin reorganization involves G_i activation of Rho in human airway smooth muscle cells. *Am J Physiol* 274: L803–L809

65 Hall A (1994) Small GTP-binding proteins and the regulation of the actin cytoskeleton. *Annu Rev Cell Biol* 10: 31–54.

66 Hirata K, Kikuchi A, Sasaki T, Kuroda S, Kaibuchi K, Matsuura Y, Seki H, Saida K, Takai Y (1992) Involvement of rho p21 in the GTP-enhanced calcium ion sensitivity of smooth muscle contraction. *J Biol Chem* 267: 8719–8722

67 Kimura K, Ito M, Amano M, Chihara K, Fukata Y, Nakafuku M, Yamamori B, Feng J,

Nakano T, Okawa K et al (1996) Regulation of myosin phosphatase by rho and rho-associated kinase (rho-kinase). *Science* 273: 245–248

68   Uehata M, Ishizaki T, Satoh H, Ono T, Kawahara T, Morishita T, Tamakawa H, Yamagami K, Inui J, Maekawa M et al (1997) Calcium sensitization of smooth muscle mediated by a Rho-associated protein kinase in hypertension. *Nature* 389: 990–994

69   Yoshii A, Iizuka K, Dobashi K, Horie T, Harada T, Nakazawa T, Mori M (1999) Relaxation of contracted rabbit tracheal and human bronchial smooth muscle by Y-27632 through inhibition of Ca^{2+} sensitization. *Am J Respir Cell Mol Biol* 20: 1190–1200

70   Hirshman CA, Lande B, Croxton TL (1999) Role of M_2 muscarinic receptors in airway smooth muscle contraction. *Life Sci* 64: 443–448

71   Kotlikoff MI, Dhulipala P, Wang Y-X (1999) M_2 signalling in smooth muscle cells. *Life Sci* 64: 437–442

72   Billington CK, Hall IP (1998) The effects of carbachol and PDGF-BB on DNA synthesis and the proliferation of human cultured airway smooth muscle cells. *Br J Pharmacol* 123: 34P

73   Gerthoffer WT, Yamboliev IA, Pohl J, Haynes R, Dang S, McHugh J (1997) Activation of MAP kinases in airway smooth muscle. *Am J Physiol* 272: L244–L252

74   Emala CW, Liu F, Hirshman CA (1999) $G_i\alpha$ but not $G_q\alpha$ is linked to activation of $p21^{ras}$ in human airway smooth muscle cells. *Am J Physiol* 276: L564–L570

75   Kelleher MD, Abe MK, Chao T-SO, Jain M, Green JM, Solway J, Rosner MR, Hershenson MB (1995) Role of MAP kinase activation in bovine tracheal smooth muscle mitogenesis. *Am J Physiol* 268: L894–L901

76   Karpova AY, Abe MK, Li J, Liu PT, Rhee JM, Kuo W-L, Hershenson MB (1997) MEK1 is required for PDGF-induced ERK activation and DNA synthesis in tracheal myocytes. *Am J Physiol* 272: L558–L565

77   Cohen MD, Ciocca V, Panettieri RA (1997) TGF-β1 modulates human airway smooth muscle cell proliferation induced by mitogens. *Am J Resp Cell Mol Biol* 16: 85–90

78   Seger R, Krebs EG (1995) The MAPK signaling cascade. *FASEB J* 9: 726–735

79   Panettieri RA (1998) Cellular and molecular mechanisms regulating airway smooth muscle proliferation and cell adhesion molecule expression. *Am J Respir Crit Care Med* 158: S133–S140

80   Van Biesen T, Luttrell LM, Hawes BE, Lefkowitz RJ (1996) Mitogenic signaling *via* G protein-coupled receptors. *Endocr Rev* 17: 698–714

81   Lopez-Ilasaca M (1998) Signaling from G-protein-coupled receptors to mitogen-activated protein (MAP)-kinase cascades. *Biochem Pharmacol* 56: 269–277

82   Van Corven EJ, Hordijk PL, Medema RH, Bos JL, Moolenaar WH (1993) Pertussis toxin-sensitive activation of p21ras by G protein-coupled receptor agonists in fibroblasts. *Proc Natl Acad Sci USA* 90: 1257–1261

83   Howe LR, Marshall CJ (1993) Lysophosphatidic acid stimulates mitogen-activated protein kinase activation *via* a G-protein-coupled pathway requiring p21ras and p74raf-1. *J Biol Chem* 268: 20717–20720

84   Alblas J, Van Corven EJ, Hordijk PL, Milligan G, Moolenaar WH (1993) G_i-mediated

activation of the p21ras-mitogen-activated protein kinase pathway by alpha 2-adrenergic receptors expressed in fibroblasts. *J Biol Chem* 268: 22235–22238

85 Winitz S, Russell M, Qian N-X, Gardner A, Dwyer L, Johnson GL (1993) Involvement of Ras and Raf in the G_i-coupled acetylcholine muscarinic M_2 receptor activation of mitogen-activated (MAP) kinase kinase and MAP kinase. *J Biol Chem* 268: 19196–19199

86 Gutkind JS, Crespo P, Xu N, Teramoto H, Coso OA (1997) The pathway connecting M_2 receptors to the nucleus involves small GTP-binding proteins acting on divergent MAP kinase cascades. *Life Sci* 60: 999–1006

87 Crespo P, Xu N, Simonds WF, Gutkind JS (1994) Ras-dependent activation of MAP kinase pathway mediated by G-protein βγ subunits. *Nature* 369: 418–420

88 Van Biesen T, Hawes BE, Luttrell DK, Krueger KM, Touhara K, Porfiri E, Sakaue M, Luttrell LM, Lefkowitz RJ (1995) Receptor-tyrosine-kinase- and $G_{βγ}$-mediated MAP kinase activation by a common signalling pathway. *Nature* 376: 781–784

89 Luttrell LM, Hawes BE, Van Biesen T, Luttrell DK, Lansing TJ, Lefkowitz RJ (1996) Role of c-Src tyrosine kinase in G protein-coupled receptor- and Gβγ subunit-mediated activation of mitogen-activated protein kinases. *J Biol Chem* 271: 19443–19450

90 Cheng G, Ye ZS, Baltimore D (1994) Binding of Bruton's tyrosine kinase to Fyn, Lyn, or Hck through a Src homology 3 domain-mediated interaction. *Proc Natl Acad Sci USA* 91: 8152–8155

91 Lopez-Ilasaca M, Crespo P, Pellici PG, Gutkind JS, Wetzker R (1997) Linkage of G protein-coupled receptors to the MAPK signaling pathway through PI 3-kinase γ. *Science* 275: 394–397

92 Linseman DA, Benjamin CW, Jones DA (1995) Convergence of angiotensin II and platelet-derived growth factor receptor signaling cascades in vascular smooth muscle cells. *J Biol Chem* 270: 12563–12568

93 Rao GN, Delafontaine P, Runge MS (1995) Thrombin stimulates phosphorylation of insulin-like growth factor-1 receptor, insulin receptor substrate-1, and phospholipase C-gamma 1 in rat aortic smooth muscle cells. *J Biol Chem* 270: 27871–27875

94 Daub H, Weiss FU, Wallasch C, Ullrich (1996) A role of transactivation of the EGF receptor in signalling by G-protein-coupled receptors. *Nature* 379: 557–560

95 Selbie LA, Hill SJ (1998) G protein-coupled-receptor cross-talk: the fine-tuning of multiple receptor-signalling pathways. *Trends Pharmacol Sci* 19: 87–93

96 Torphy TJ, Freese WB, Rinard GA, Brunton LL, Mayer SE (1982) Cyclic nucleotide-dependent protein kinases in airway smooth muscle. *J Biol Chem* 257: 11609–11616

97 Torphy TJ (1994) β-Adrenoceptors, cAMP and airway smooth muscle relaxation: challenges to the dogma. *Trends Pharmacol Sci* 15: 370–374

98 Giembycz MA, Raeburn D (1991) Putative substrates for cyclic nucleotide-dependent protein kinases and the control of airway smooth muscle tone. *J Auton Pharmacol* 166: 365–368

99 Nijkamp FP, Engels F, Henricks PAJ, Van Oosterhout AJM (1992) Mechanisms of β-

adrenergic receptor regulation in lungs and its implications for physiological responses. *Physiol Rev* 72: 323–367

100 Kotlikoff MI, Kamm KE (1996) Molecular mechanisms of β-adrenergic relaxation of airway smooth muscle. *Ann Rev Physiol* 58: 115–141

101 Furukawa K-I, Tawada Y, Shigekawa M (1988) Regulation of the plasma membrane Ca^{2+} pump by cyclic nucleotides in cultured vascular smooth muscle cells. *J Biol Chem* 263: 8058–8065

102 Twort CHC, Van Breemen C (1989) Human airway smooth muscle in cell culture: control of the intracellular calcium store. *Pulm Pharmacol* 2: 45–53

103 Schramm CM, Chuang ST, Grunstein MM (1995) cAMP generation inhibits inositol 1,4,5-trisphosphate binding in rabbit tracheal smooth muscle. *Am J Physiol* 269: L715–L719

104 Kume H, Takai A, Tokuno H, Tomita T (1989) Regulation of Ca^{2+}-dependent K^+-channel activity in tracheal myocytes by phosphorylation. *Nature* 341: 152–154

105 Kume H, Hall IP, Washabau RJ, Takagi K, Kotlikoff MI (1994) β-Adrenergic agonists regulate K_{Ca} channels in airway smooth muscle by cAMP-dependent and -independent mechanisms. *J Clin Invest* 94: 371–379

106 Hall IP, Hill SJ (1988) β-Adrenoceptor stimulation inhibits histamine-stimulated inositol phospholipid hydrolysis in bovine tracheal smooth muscle. *Br J Pharmacol* 95: 1204–1212

107 Madison JM, Brown JK (1988) Differential inhibitory effects of forskolin, isoproterenol, and dibutyryl cyclic adenosine monophosphate on phosphoinositide hydrolysis in canine tracheal smooth muscle. *J Clin Invest* 82: 1462–1465

108 Offer GJ, Chilvers ER, Nahorski SR (1991) β-Adrenoceptor-induced inhibition of muscarinic receptor-stimulated phosphoinositide metabolism is agonist-specific in bovine tracheal smooth muscle. *Eur J Pharmacol* 207: 243–248

109 Jones KA, Wong GY, Jankowski CJ, Akao M, Warner DO (1999) cGMP modulation of Ca^{2+} sensitivity in airway smooth muscle. *Am J Physiol* 276: L35–L40

110 Daaka Y, Luttrell LM, Lefkowitz RJ (1997) Switching of the coupling of the $β_2$-adrenergic receptor to different G proteins by protein kinase A. *Nature* 390: 88–91

111 Xiao R-P, Avdonin P, Zhou Y-Y, Cheng H, Akhter SA, Eschenhagen T, Lefkowitz RJ, Koch WJ, Lakatta EG (1999) Coupling of $β_2$-adrenoceptor to G_i proteins and its physiological relevance in murine cardiac myocytes. *Circ Res* 84: 43–52

112 Noveral JP, Grunstein MM (1994) Adrenergic receptor-mediated regulation of cultured rabbit airway smooth muscle cell proliferation. *Am J Physiol* 267: L291–L299.

113 Tomlinson PR, Wilson JW, Stewart AG (1995) Salbutamol inhibits the proliferation of human airway smooth muscle cells grown in culture: relationship to elevated cAMP levels. *Biochem Pharmacol* 49: 1809–1819

114 Young PG, Skinner SJ, Black PN (1995) Effects of glucocorticoids and beta-adrenoceptor agonists on the proliferation of airway smooth muscle. *Eur J Pharmacol* 273: 137–143

115 Hershenson MB, Chao TS, Abe MK, Gomes I, Kelleher MD, Solway J, Rosner MR

(1995) Histamine antagonizes serotonin and growth factor-induced mitogen-activated protein kinase activation in bovine tracheal smooth muscle cells. *J Biol Chem* 270: 19908–19913

116 Florio C, Martin JG, Styhler A, Heisler S (1994) Antiproliferative effect of prostaglandin E_2 in cultured guinea pig tracheal smooth muscle cells. *Am J Physiol* 266: L131–L137

117 Maruno K, Absood A, Said SI (1995) VIP inhibits basal and histamine-stimulated proliferation of human airway smooth muscle cells. *Am J Physiol* 268: L1047–L1051

118 Schramm CM, Omlor GJ, Quinn LM, Noveral JP (1996) Methylprednisolone and isoproterenol inhibit airway smooth muscle proliferation by separate and additive mechanisms. *Life Sci* 59: PL9–14

119 Stewart AG, Tomlinson PR, Wilson JW (1997) Beta 2-adrenoceptor agonist-mediated inhibition of human airway smooth muscle cell proliferation: importance of the duration of beta 2-adrenoceptor stimulation. *Br J Pharmacol* 121: 361–368

120 Pyne NJ, Pyne S (1998) PDGF-stimulated cyclic AMP formation in airway smooth muscle: assessment of the roles of MAP kinase, cytosolic phospholipase A_2, and arachidonate metabolites. *Cell Signalling* 10: 363–369

121 Billington CK, Joseph SK, Swan C, Scott MGH, Jobson TM, Hall IP (1999) Modulation of human airway smooth muscle proliferaton by type 3 phosphodiesterase inhibition. *Am J Physiol* 276: L412–L419

122 Cook SJ, McCormick F (1993) Inhibition by cAMP of ras-dependent activation of raf. *Science* 262: 1069–1072

123 Wu J, Dent P, Jelinek T, Wolfman A, Weber MJ, Sturgill TW (1993) Inhibition of the EGF-activated MAP kinase signaling pathway by adenosine 3',5'-monophosphate. *Science* 262: 1065–1069

124 Scott PH, Belham CM, Al-Hafidh J, Chilvers ER, Peacock AJ, Gould GW, Plevin R (1996) A regulatory role for cAMP in phosphatidylinositol 3-kinase/p70 ribosomal S6 kinase-mediated DNA synthesis in platelet-derived-growth-factor-stimulated bovine airway smooth muscle cells. *Biochem J* 318: 965–971

125 Raffestin B, Cerrina J, Boullet C, Labat C, Benveniste J, Brink C (1985) Response and sensitizing of isolated human pulmonary muscle preparation to pharmacological agents. *J Pharmacol Exp Ther* 233: 186–194

126 Van den Brink FG (1973) The model of functional interaction. II. Experimental verification of a new model: the antagonism of β-adrenoceptor stimulants and other agonists. *Eur J Pharmacol* 22: 279–286

127 Torphy TJ (1984) Differential relaxant effects of isoproterenol on methacholine versus leukotriene D_4-induced contraction in the guinea-pig trachea. *Eur J Pharmacol* 102: 549–553

128 Torphy TJ, Rinard GA, Rietow MG, Mayer SE (1983) Functional antagonism in canine tracheal smooth muscle: Inhibition by methacholine of the mechanical and biochemical responses to isoproterenol. *J Pharmacol Exp Ther* 227: 694–699

129 Torphy TJ, Zheng C, Peterson SM, Fiscus RR, Rinard GA, Mayer SE (1985) Inhibitory effect of methacholine on drug-induced relaxation, cyclic AMP-dependent protein

kinase activation in canine tracheal smooth muscle. *J Pharmacol Exp Ther* 233: 409–417

130 Russel JA (1984) Differential inhibitory effect of isoproterenol on contractions of canine airways. *J Appl Physiol* 57: 801–807

131 Van Amsterdam RGM, Meurs H, Brouwer F, Postema JB, Timmermans A, Zaagsma J (1989) Role of phosphoinositide metabolism in functional antagonism of airway smooth muscle contraction by β-adrenoceptor agonists. *Eur J Pharmacol* 172: 175–183

132 Jenne JW, Shaughnessy TK, Druz WS, Manfredi CJ, Vestal RE (1987) *In vivo* functional antagonism between isoproterenol and bronchoconstrictants in the dog. *J Appl Physiol* 63: 812–819

133 Barnes PJ, Pride NB (1983) Dose-response curves to inhaled beta-adrenoceptor agonists in normal and asthmatic subjects. *Br J Clin Pharmacol* 15: 677–682

134 Tattersfield AE, Holgate ST, Harvey JE, Gribbin HR (1983) Is asthma due to partial beta-blockade of airways? *Agents Actions* 13: 265–271

135 Naline E, Zhang Y, Qian Y, Mairon N, Anderson GP, Grandordy B, Advenier C (1994) Relaxant effects and durations of action of formoterol and salmeterol on the isolated human bronchus. *Eur Respir J* 7: 914–920

136 Molimard M, Naline E, Zhang Y, Le Gros V, Begaud B, Advenier C (1998) Long- and short-acting β$_2$ adrenoceptor agonists: interactions in human contracted bronchi. *Eur Respir J* 11: 583–588

137 Ariëns EJ, Simonis AM, Van Rossum JM (1964) Drug-receptor interaction of one or more drugs with one receptor system. In: EJ Ariëns (ed): *Molecular pharmacology*. Academic Press, New York, 119–286

138 Sibley DR, Lefkowitz RJ (1985) Molecular mechanisms of receptor desensitization using the beta-adrenergic receptor-coupled adenylate cyclase system as a model. *Nature* 317: 124–129

139 Houslay MD (1991) "Crosstalk": a pivotal role for protein kinase C in modulating relationships between signal transduction pathways. *Eur J Biochem* 195: 9–27

140 Meurs H, Kauffman HF, Koëter GH, Timmermans A, De Vries K (1987) Regulation of the beta-receptor-adenylate cyclase system in lymphocytes of allergic asthmatic patients. Possible role for protein kinase C in allergen-induced non-specific refractoriness of adenylate cyclase. *J Allergy Clin Immunol* 80: 306–339

141 Meurs H, Koëter GH, De Vries K, Kauffman HF (1982) The beta-adrenergic system in allergic and bronchial asthma: changes in lymphocyte beta-adrenergic receptor number and adenylate cyclase activity after an allergen-induced asthmatic attack. *J Allergy Clin Immunol* 70: 272–280.

142 Dooper MWSM, Timmermans A, Aalbers R, De Monchy JGR, Kauffman HF (1993) Desensitization of the adenylyl cyclase system in peripheral blood mononuclear cells from patients with asthma three hours after allergen challenge. *J Allergy Clin Immunol* 92: 559–566

143 Meurs H, Kauffman HF, Timmermans A, Van Amsterdam FThM, Koëter GH, De Vries K (1986) Phorbol 12-myristate 13-acetate induces beta-adrenergic receptor uncoupling

and non-specific desensitization of adenylate cyclase in human mononuclear leukocytes. *Biochem Pharmacol* 35: 4217–4222

144 Gerblich AA, Campbell AE, Schuyler MR (1984) Changes in T lymphocyte subpopulations after antigenic bronchial provocation in asthmatics. *N Engl J Med* 310: 1349–1352

145 Meurs H, Timmermans A, De Monchy JGR, Zaagsma J, Kauffman HF (1993) Lack of coupling of muscarinic receptors to phosphoinositide metabolism and adenylyl cyclase in human lymphocytes and polymorphonuclear leukocytes: Studies in healthy subjects and allergic asthmatic patients. *Int Arch Allergy Immunol* 100: 19–27

146 Borger P, Jonker G, Vellenga E, Postma DS, De Monchy JG, Kauffman HF (1999) Allergen challenge primes for IL-5 mRNA production and abrogates beta-adrenergic function in peripheral blood T lymphocytes from asthmatics. *Clin Exp Allergy* 29: 933–940

147 Grandordy BM, Mak JCW, Barnes PJ (1993) Modulation of airway smooth muscle β-adrenoceptor function by a muscarinic agonist. *Life Sci* 54: 185–191

148 Schears G, Clancy J, Hirshman CA, Emala CW (1997) Chronic carbachol pretreatment decreases adenylyl cyclase activity in airway smooth muscle. *Am J Physiol* 273: L640–L647

149 Toullec D, Pianetti P, Coste H, Bellevergue P, Grande-Perret T, Ajakane M, Baudet V, Boissin P, Boursier E, Loriolle F et al (1991) The bisindolylmaleimide GF 109203X is a potent and selective inhibitor of protein kinase C. *J Biol Chem* 266: 15771–15781

150 Goldie RG, Spina S, Henry PJ, Lulich KM, Paterson JW (1986) *In vitro* responsiveness of human asthmatic bronchus to carbachol, histamine, β-adrenoceptor agonists and theophylline. *Br J Clin Pharmacol* 22: 669–676

151 Bai TR (1990) Abnormalities in airway smooth muscle in fatal asthma. *Am Rev Respir Dis* 141: 552–557

152 Bai TR (1991) Abnormalities in airway smooth muscle in fatal asthma: a comparison of tracheal and bronchial smooth muscle. *Am Rev Respir Dis* 143: 441–443

153 Cerrina J, Le Roy Ladurie M, Labat C, Raffestin B, Bayol A, Brink C (1986) Comparison of human bronchial muscle responses to histamine *in vivo* with histamine and isoproterenol agonists *in vitro*. *Am Rev Respir Dis* 134: 57–61

154 Spina D, Rigby PJ, Paterson JW, Goldie RG (1989) Autoradiographic localization of beta-adrenoceptors in human asthmatic lung. *Am Rev Respir Dis* 140: 1410–1415

155 De Jongste JC, Mons H, Bonta R, Kerrebijn KF (1987) Human asthmatic airway responses *in vitro* – A case report. *Eur J Respir Dis* 70: 23–29

156 Whicker SD, Armour CL, Black JL (1988) Responsiveness of bronchial smooth muscle from asthmatic patients to relaxant and contractile agonists. *Pulmon Pharmacol* 1: 25–31

157 Sharma RK, Jeffery PK (1990) Airway β-adrenoceptor number in cystic fibrosis and asthma. *Clin Sci* 78: 409–417

158 Bai TR, Mak JC, Barnes PJ (1992) A comparison of β-adrenergic receptors and *in vitro* relaxant responses to isoproterenol in asthmatic airway smooth muscle. *Am J Respir Cell Mol Biol* 6: 647–651

159 Santing RE, Schraa EO, Vos BG, Gores R-JJ, Olymulder CG, Meurs H, Zaagsma J (1994) Dissociation between bronchial hyperreactivity *in vivo* and reduced β-adrenoceptor sensitivity *in vitro* in allergen-challenged guinea pigs. *Eur J Pharmacol* 257: 145–152

160 Hall IP, Donaldson S, Hill SJ (1989) Inhibition of histamine-stimulated inositol phospholipid hydrolysis by agents which increase cyclic AMP levels in bovine tracheal smooth muscle. *Br J Pharmacol* 97: 603–613

161 Challiss RAJ, Adams D, Mistry R, Nicholson CD (1998) Modulation of spasmogen-stimulated Ins(1,4,5)P$_3$ generation and functional responses by selective inhibitors of types 3 and 4 phosphodiesterase in airways smooth muscle. *Br J Pharmacol* 124: 47–54

162 Chilvers ER, Lynch BJ, Challiss RAJ (1997) Dissociation between β-adrenoceptor-mediated cyclic AMP accumulation and inhibition of histamine-stimulated phosphoinositide metabolism in airways smooth muscle. *Biochem Pharmacol* 53: 1565–1568

163 Nelson HS (1995) Beta-adrenergic bronchodilators. *N Engl J Med* 333: 499–506

164 Barnes PJ, Chung KF (1992) Questions about inhaled β$_2$-adrenoceptor agonists in asthma. *Trends Pharmacol Sci* 13: 20–23

165 Speizer FE, Doll R, Heaf P, Strang LB (1968) Investigation into use of drugs preceding death from asthma. *Br Med J* 1: 339–343

166 Grainger J, Woodman K, Pearce N, Crane J, Burgess C, Culling C, Windom H, Beasley R (1991) Prescribed fenoterol and death from asthma in New Zealand, 1981–7: a further case-control study. *Thorax* 46: 105–111

167 Spitzer WO, Suissa S, Ernst P, Horwitz RI, Habbick B, Cockroft D, Boivin J-F, McNutt M, Buist AS, Rebuck AS (1992) Asthma death and near-fatal asthma in relation to beta-agonist use. *N Engl J Med* 326: 501–506

168 Sears MR, Taylor DR, Print CG, Lake DG, Li Q, Flannery EM, Yates DM, Lucas MK, Herbison GP (1990) Regular inhaled beta-agonist treatment in bronchial asthma. *Lancet* 336: 1391–1396

169 Taylor DR, Sears MR, Herbison GP, Flannery EM, Print CG, Lake DC, Yates DM, Lucas MK, Li Q (1993) Regular inhaled beta agonist in asthma: effects on exacerbations and lung function. *Thorax* 48: 134–8

170 Van Schayk CP, Graafsma SJ, Visch MB, Dompeling E, Van Weel C, Van Herwaarden CLA (1990) Increased bronchial hyperresponsiveness after inhaling salbutamol during 1 year is not caused by subsensitization to salbutamol. *J Allergy Clin Immunol* 86: 793–800

171 Van Schayk CP, Dompeling E, Van Herwaarden CLA, Folgering H, Verbeck ALM, Van der Hoogen JM, Van Weel C (1991) Bronchodilator treatment in moderate asthma or chronic bronchitis: continuous or on demand? A randomised control study. *Br Med J* 303: 1426–1431

172 Koëter GH, Meurs H, Kauffman HF, De Monchy JGR, Sluiter HJ, De Vries K (1983) Changes in the beta-adrenergic system in bronchial asthma induced by terbutaline. *Agents Actions* 13: 259–264

173 Kraan J, Koëter GH, Van der Mark ThW, Sluiter HJ, De Vries K (1985) Changes in

bronchial hyperreactivity induced by 4 weeks of treatment with antiasthmatic drugs in patients with allergic asthma: a comparison between budesonide and terbutaline. *J Allergy Clin Immunol* 76: 628–636

174 Kerrebijn KF, Van Essen-Zandvliet EEM, Neijens HJ (1987) Effects of long-term treatment with inhaled corticosteroids and beta-agonists on the bronchial responsiveness in children with asthma. *J Allergy Clin Immunol* 79: 653–659

175 Vathenen AS, Knox AJ, Higgins BG, Britton JR, Tattersfield AE (1988) Rebound increase in bronchial responsiveness after treatment with inhaled terbutaline. *Lancet* i: 554–558

176 Tashkin DP, Connoly ME, Deutsch RI, Hui KK, Littner M, Scarpace P (1982) Subsensitization of beta-adrenoceptors in airways and lymphocytes of healthy and asthmatic subjects. *Am Rev Respir Dis* 125: 185–193

177 Larsson K, Martinsson A, Hjemdahl P (1992) Influence of β-adrenergic receptor function during terbutaline treatment on allergen sensitivity and bronchodilator response to terbutaline in asthmatic subjects. *Chest* 101: 953–960

178 Cockroft D, McParland CP, Britto SA, Swystun VA, Rutherford C (1993) Regular inhaled salbutamol and airway responsiveness to allergen. *Lancet* 342: 833–837

179 Jenne JW, Chick TW, Strickland RD, Wall FJ (1977) Subsensitivity of beta responses during therapy with a long acting beta-2 preparation. *J Allergy Clin Immunol* 60: 346–356

180 Plummer AL (1978) The development of drug tolerance to beta$_2$-adrenergic agents. *Chest* 76: 949–956

181 Weber RW, Smith JA, Nelson HS (1982) Aerosolized terbutaline in asthmatics: development of subsensitivity with long-term administration. *J Allergy Clin Immunol* 70: 417–422

182 Newnham DM, McDevitt DG, Lipworth BJ (1994) Bronchodilator subsensitivity alter chronic dosing with eformoterol in patients with asthma. *Am J Med* 97: 29–37

183 Newnham DM, Grove A, McDervitt DG, Lipworth BJ (1995) Subsensitivity of bronchodilator and systemic beta-2 adrenoceptor responses after regular twice daily treatment with eformoterol dry powder in asthmatic patients. *Thorax* 50: 497–504

184 Grove A, Lipworth BJ (1995) Bronchodilator subsensitivity to salbutamol after twice daily salmeterol in asthmatic patients. *Lancet* 346: 201–206

185 Yates DH, Sussman H, Shaw MJ, Barnes PJ, Chung KF (1995) Regular formoterol treatment in mild asthma: effects on bronchial responsiveness during and after treatment. *Am J Respir Crit Care Med* 152: 1170–1174

186 Larsson S, Svedmyr N, Thiringer G (1977) Lack of bronchial beta-adrenoceptor resistance in asthmatics during long-term treatment with terbutaline. *J Allergy Clin Immunol* 59: 93–100

187 Peel ET, Gibson GJ (1980) Effects of long-term inhaled salbutamol therapy on the provocation of asthma by histamine. *Am Rev Respir Dis* 121: 973–978

188 Connolly ME, Jenne JW, Hui KK, Borst SE (1987) Beta-adrenergic tachyphylaxis

(desensitization) and functional antagonism. In: JW Jenne, S Murphy (eds): *Drug therapy for asthma: research and clinical practice*. Dekker Inc, New York, 259–296

189 O'Connor BJ, Aikman SL, Barnes PJ (1992) Tolerance to the nonbronchodilator effects of inhaled β_2-agonists in asthma. *N Engl J Med* 327: 1204–1208

190 Cheung D, Timmers MC, Zwinderman AH, Bel EH, Dijkman J, Sterk PJ (1992) Long-term effects of a long-acting β_2-adrenoceptor agonist, salmeterol, on airway hyperresponsiveness in patients with mild asthma. *N Engl J Med* 327: 1198–1203

191 Booth H, Bish R, Walters J, Whitehead F, Walters EH (1996) Salmeterol tachyphylaxis in steroid treated asthmatic subjects. *Thorax* 51: 1100–1104

192 Aziz I, Tan KS, Hall IP, Devlin MM, Lipworth BJ (1998) Subsensitivity to bronchoprotection against adenosine monophosphate challenge following regular once-daily formoterol. *Eur Respir J* 12: 580–584

193 Bhagat R, Kalra S,. Swystun VA, Cockcroft DW (1995) Rapid onset of tolerance to the bronchoprotective effect of salmeterol. *Chest* 108: 1235–1239

194 Yates DH, Kharitonov SA, Barnes PJ (1996) An inhaled glucocorticoid does not prevent tolerance to the bronchoprotective effect of a long-acting inhaled beta-2 agonist. *Am J Respir Crit Care Med* 154: 1603–1607

195 Kalra S, Swystun V, Bhagat R, Cockcroft DW (1996) Inhaled corticosteroids do not prevent the development of tolerance to the bronchoprotective effect of salmeterol. *Chest* 109: 953–956

196 Lipworth BJ, Hall IP, Aziz I, Tan KS, Wheatley A (1999) β_2-Adrenoceptor polymorphism and bronchoprotective sensitivity with regular short- and long-acting β_2-agonist therapy. *Clin Sci* 96: 253–259

197 Van der Heijden PJCM, Van Amsterdam JGC, Zaagsma J (1984) Desensitization of smooth muscle and mast cell β-adrenoceptors in the airways of the guinea pig. *Eur J Respir Dis* 65 (Suppl. 135): 128–134

198 McGraw DW, Liggett SB (1997) Heterogeneity in β-adrenergic receptor kinase expression in the lung accounts for cell-specific desensitization of the β_2-adrenergic receptor. *J Biol Chem* 272: 7338–7344

199 Manolitsas DN, Wang J, Devalia JL, Trigg CJ, McAulay AE, Davies RJ (1995) Regular albuterol, nedocromil sodium, and bronchial inflammation in asthma. *Am J Respir Crit Med* 151: 1925–1930

200 Gavreau GM, Jordana M, Watson RM, Cockcroft DW, O'Byrne PM (1997) Effect of regular inhaled albuterol on allergen-induced late responses and sputum eosinophils in asthmatic subjects. *Am J Respir Crit Care Med* 156: 1738–1745

201 Lefkowitz RJ, Cotecchia S, Kjelsberg MA, Pitcher J, Koch WJ, Inglese J, Caron MG (1993) Adrenergic receptors: recent insights into their mechanism of activation and desensitisation. In: BL Brown, PRM Dobson (eds): *Advances in second messenger and phosphoprotein research* Vol. 28, Raven Press Ltd, New York, 1–9

202 Collins S (1993) Recent perspectives on the molecular structure and regulation of the β_2-adrenoceptor. *Life Sci* 52: 2083–2091

203 Lefkowitz RJ (1998) G protein-coupled receptors: III. New roles for receptor kinases

and β-arrestins in receptor signalling and desensitization. *J Biol Chem* 273: 18677–18680

204 Chuang TT, Iacovelli L, Sallese M, DeBlasi A (1996) G protein-coupled receptors: heterologous regulation of homologous desensitization and its implications. *Trends Pharmacol Sci* 17: 416–421

205 Pitcher JA, Freedman NJ, Lefkowitz RJ (1998) G protein-coupled receptor kinases. *Annu Rev Biochem* 67: 653–692

206 Ferguson SSG, Downey WE III, Colapietro A-M, Barak LS, Ménard I, Caron MG (1996) Role of β-arrestin in mediating agonist-promoted G protein-coupled receptor internalization. *Science* 271: 363–366

207 Goodman OB Jr, Krupnick JG, Santini F, Gurevich VV, Penn RB, Gagnon AW, Keen JH, Benovic JL (1996) β-Arrestin acts as a clathrin adaptor in endocytosis of the β_2-adrenergic receptor. *Nature* 383: 447–450

208 Yu SS, Lefkowitz RJ, Hausdorff WP (1993) Beta-adrenergic receptor sequestration: A potential mechanism of receptor resensitization. *J Biol Chem* 268: 337–341

209 Pitcher JA, Payne ES, Csortos C, DePaoli-Roach AA, Lefkowitz RJ (1995) The G protein-coupled receptor phosphatase: A protein phosphatase type 2A with a distinct subcellular distribution and substrate specificity. *Proc Natl Acad Sci USA* 92: 8343–8347

210 Krueger KK, Daaka Y, Pitcher JA, Lefkowitz RJ (1997) The role of sequestration in G protein-coupled receptor resensitization: Regulation of β_2-adrenergic receptor dephosphorylation by vesicular acidification. *J Biol Chem* 272: 5–8

211 Green SA, Turki J, Innis M, Liggett SB (1994) Amino-terminal polymorphisms of the human beta2-adrenergic receptor impart distinct agonist-promoted regulatory properties. *Biochemistry* 33: 9414–9419

212 Green SA, Turki J, Bejarano P, Hall IP, Liggett SB (1995) Influence of the beta2-adrenergic receptor genotypes on signal transduction in human airway smooth muscle cells. *Am J Resp Cell Mol Biol* 13: 25–33

213 McGraw DW, Forbes SL, Kramer LA, Liggett SB (1998) Polymorphisms of the 5' leader cistron of the human β_2-adrenergic receptor regulate receptor expression. *J Clin Invest* 102: 1927–1932

214 Scott MGH, Swan C, Wheatley AP, Hall IP (1999) Identification of novel polymorphisms within the promotor region of the human β_2 adrenergic receptor gene. *Br J Pharmacol* 126: 841–844

215 Parola AL, Kobilka BK (1994) The peptide product of a 5' leader cistron in the beta 2 adrenergic receptor mRNA inhibits receptor synthesis. *J Biol Chem* 269: 4497–4505

216 Chuang TT, LeVine III H, DeBlasi A (1995) Phosphorylation and activation of β-adrenergic receptor kinase by protein kinase C. *J Biol Chem* 270: 18660–18665

217 DeBlasi A, Parruti G, Sallese M (1995) Regulation of G protein-coupled receptor kinase subtypes in activated T lymphocytes. *J Clin Invest* 95: 203–210

218 Winstel R, Freund S, Krasel C, Hoppe E, Lohse MJ (1996) Protein kinase cross-talk: membrane targeting of the β-adrenergic receptor kinase by protein kinase C. *Proc Natl Acad Sci USA* 93: 2105–2109

219 Eglen RM, Reddy H, Watson N, Challis RAJ (1994) Muscarinic acetylcholine receptor subtypes in smooth muscle. *Trends Pharmacol Sci* 15: 114–119

220 Roffel AF, Meurs H, Elzinga CRS, Zaagsma J (1993) Muscarinic M_2 receptors do not participate in the functional antagonism between methacholine and isoprenaline in guinea pig tracheal smooth muscle. *Eur J Pharmacol* 249: 235–238

221 Roffel AF, Meurs H, Elzinga CRS, Zaagsma J (1995) No evidence for a role of muscarinic M_2-receptors in functional antagonism in bovine trachea. *Br J Pharmacol* 115: 665–671

222 Watson N, Magnussen H, Rabe KF (1995) Antagonism of β-adrenoceptor-mediated relaxations of human bronchial smooth muscle by carbachol. *Eur J Pharmacol* 275: 307–310

223 Hakonarson H, Herrick DJ, Grunstein MM (1995) Mechanism of impaired β-adrenoceptors responsiveness in atopic sensitized airway smooth muscle. *Am J Physiol* 269: L645–L652

224 Hakonarson H, Herrick DJ, Gonzalez Serrano P, Grunstein MM (1996) Mechanism of cytokine-induced modulation of β-adrenoceptor responsiveness in airway smooth muscle. *J Clin Invest* 97: 2593–2600

225 Hakonarson H, Herrick DJ, Gonzalez Serrano P, Grunstein MM (1997) Autocrine role of IL-1β in altered responsiveness of atopic asthmatic sensitized airway smooth muscle. *J Clin Invest* 99: 117–124

226 Gavett SH, Wills-Karp M (1993) Elevated lung G protein levels and muscarinic receptor affinity in a mouse model of airway hyperreactivity. *Am J Physiol* 265: L493–L500

227 Wang Z-L, Walker BAM, Weir TD, Yarema MC, Roberts CR, Okazawa M, Paré PD, Bai TR (1995) Effect of chronic antigen and β_2 agonist exposure on airway remodeling in guinea pigs. *Am J Respir Crit Care Med* 152: 2097–2104

228 Dunnill MS, Massarella GR, Anderson JA (1969) A comparison of the quantitative anatomy of the bronchi in normal subjects, in status asthmaticus, in chronic bronchitis, and in emphysema. *Thorax* 24: 176–179

229 Ebina M, Yaegashi H, Chiba T, Takahashi T, Motomiya M, Tanemura M (1990) Hyperreactive site in the airway tree of asthmatic patients revealed by thickening of bronchial muscles. *Am Rev Respir Dis* 141: 1327–1332

230 Ebina M, Takahashi T, Chiba T, Motomiya M (1993) Cellular hypertrophy and hyperplasia of ASM underlying bronchial asthma. *Am Rev Respir Dis* 148: 720–726

231 Lambert RK, Wiggs BR, Kuwano K, Hogg JC, Paré PD (1993) Functional significance of increased airway smooth muscle in asthma and COPD. *J Appl Physiol* 74: 2771–2781

232 Stewart AG, Tomlinson PR, Wilson J (1993) Airway wall remodelling in asthma: A novel target for the development of anti-asthma drugs. *Trends Pharmacol Sci* 14: 275–279

233 Knox AJ (1994) Airway re-modelling in asthma: role of airway smooth muscle. *Clin Sci* 86: 647–652

234 Nishizuka Y (1992) Intracellular signaling by hydrolysis of phospholipids and activation of protein kinase C. *Science* 258: 607–614

235 Donnelly R, Yang KXF, Omary MB, Azhar S, Black JL (1995) Expression of multiple isoenzymes of protein kinase C in airway smooth muscle. *Am J Respir Cell Mol Biol* 13: 253–256

236 Togashi H, Hirshman CA, Emala CW (1997) Qualitative immunoblot analysis of PKC isoforms expressed in airway smooth muscle. *Am J Physiol* 272: L603–L607

237 Webb BLJ, Lindsay MA, Barnes PJ, Giembycz MA (1997) Protein kinase C isoenzymes in airway smooth muscle. *Biochem J* 324: 167–175

238 Carlin S, Yang KXF, Donnely R, Black JL (1999) Protein kinase C isoforms in human airway smooth muscle cells: activation of PKC-ζ during proliferation. *Am J Physiol* 276: L506–L512

Gene regulation of muscarinic receptor subtypes

Peter J. Barnes

Department of Thoracic Medicine, National Heart and Lung Institute, Imperial College, Dovehouse St, London SW3 6LY, UK

Introduction

Of the five cloned muscarinic receptor subtypes, human airways express only three receptor subtypes (M_1, M_2 and M_3) and these subtypes are differentially distributed in the airways [1]. M_1-receptors are localised to alveolar walls, parasympathetic ganglia and submucosal glands, whereas M_3-receptors are predominant in airway smooth muscle and submucosal glands. In contrast, M_2-receptors are localised to airway smooth muscle of small airways, although there is little evidence that these receptors play an important functional role in the regulation of airway smooth muscle tone, at least in human airways. Functional studies suggest that M_2-receptors play an important role in regulating the release of acetylcholine from parasympathetic nerves in the airways. M_2-receptors which inhibit cholinergic neural contraction have been demonstrated in functional studies of human airways *in vitro* and pre-junctional M_2-receptors inhibit the release of acetylcholine in animal and human airways [2, 3]. There is considerable evidence that M_2-receptors may be dysfunctional in patients with asthma and this might contribute to exaggerated cholinergic reflex bronchoconstriction in asthma [4]. There are several possible mechanisms that might result in dysfunction of M_2-receptors, including viral infections, eosinophil basic proteins and oxidants. The inflammatory process may also result in impaired function of M_2-receptors.

Regulation of receptor expression

Inflammatory diseases may affect the expression of G-protein-coupled receptors in a number of ways. The cloning and expression of multiple G-protein-coupled receptors have made it possible to explore their molecular regulation in disease. Site-directed mutagenesis has made it possible to explore which parts of the receptor molecule are involved in regulation of function and in desensitisation. More recently, sequencing of the promoter region of receptors has made it possible to explore

Muscarinic Receptors in Airways Diseases, edited by Johan Zaagsma, Herman Meurs and Ad F. Roffel
© 2001 Birkhäuser Verlag Basel/Switzerland

159

which transcription factors are involved in the expression of receptors at a gene level. Receptor function may be regulated at several sites: at the level of gene expression through the activation of transcription factors, by alterations in messenger RNA stability, by affecting the translation of mRNA to protein, and by phosphorylation of the receptor protein *via* various kinases and phosphatases (Fig. 1). Thus there are several sites where receptor function may be affected in disease to alter their function and change tissue responsiveness. In inflammatory diseases, various inflammatory mediators have the capacity to alter gene expression *via* activation of transcription factors and to alter function by increasing the activity of kinases or phosphatases which affect the phosphorylation of the receptor protein itself, of G-proteins or of effector mechanisms. These effects may be manifested by altered expression of the receptor (up-regulation or down-regulation) or by altered capacity to desensitise in response to endogenous or exogenous agonists.

While β-adrenergic receptors have been extensively characterised in terms of gene regulation [5, 6], there is relatively little information about the molecular regulation of muscarinic receptors. This is partly because the promoter sequence of muscarinic receptors has proved difficult to sequence. However, the recent sequencing of the upstream promoter regions for several muscarinic receptor genes has revealed the presence of several consensus sites for the binding of transcription factors, indicating that several factors may regulate the expression of these genes. The rat M_4 promoter has several potential binding sites for transcription factor binding, including activator protein-2 (AP-2) and multiple Sp-1 sites [7]. Interestingly, a silencer element prevents the expression of the m4 gene in non-neuronal cells, but when removed the gene can be expressed in non-neuronal cells. This silencer region has now been identified as neurone-specific silencer/repressor element-1, which is localised 835 base pairs upstream from the transcription start site [8]. The rat M_1 receptor promoter also contains several consensus sequences for transcription factors, including AP-1, nuclear factor-κB (NF-κB) and Sp-1 [9]. The chick M_2 receptor promoter has several sites that may interact with transcription factors, including GATA transcription factors [10, 11].

In human lung there is no difference in muscarinic receptor density between normal and asthmatic subjects [12]. Since in human lung and airways most of the receptors detected by radioligand binding are of the M_1 and M_3 receptor subtypes [13, 14], this suggests that there is unlikely to be a major abnormality in expression of these receptor subtypes in asthma. However, it is possible that there is an abnormality in expression of M_2 receptors, in view of the functional studies indicating that there may be a defective function of pre-junctional M_2 receptors, resulting in increased release of acetylcholine (ACh) on nerve activation. There is evidence for abnormal functioning of pre-junctional M_2 receptors in experimental asthma in guinea pigs [4, 15] and in asthmatic patients [16].

Most attention has therefore focused on the gene regulation of M_2 receptors and there is increasing evidence that several factors relevant to inflammatory diseases

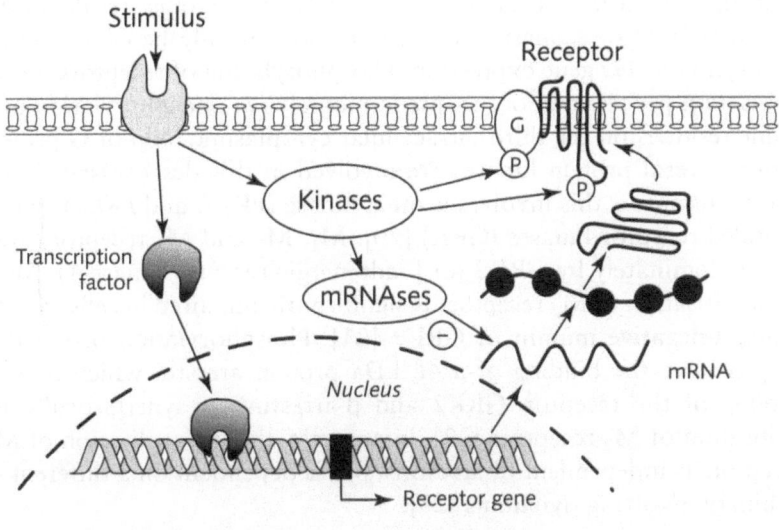

Figure 1

Regulation of receptor expression. Receptors may be regulated by extracellular stimuli which may activate transcription factors that control increased or decreased expression of the receptor gene, or through activation of kinases that phosphorylate the receptor or G-proteins, resulting in uncoupling and down-regulation of receptors, or by increasing the breakdown of messenger RNA, thus reducing receptor synthesis.

may regulate their expression [17, 18]. Because it is not possible to examine molecular mechanisms in parasympathetic nerves, studies have been conducted using a human cell line, human embryonic lung fibroblasts (HEL 299 cells), that expresses M_2 receptors in the absence of M_1 and M_3 receptors. These studies have revealed that several factors regulate the expression of human M_2 receptors.

Homologous regulation

Receptor desensitisation

The molecular mechanisms involved in desensitisation of G-protein-coupled receptors have now been explored in some detail, particularly in the case of β_2-adrenergic receptors [19]. Desensitisation may occur within seconds of exposure to an agonist and involves phosphorylation of the receptor and internalisation of the receptor from the cell surface. Prolonged exposure to the agonist results in degradation of the internalised or sequestered receptor and can only be overcome by *de novo* receptor synthesis *via* gene expression. Phosphorylation of receptors is critical in the desensitisation process and occurs predominately *via* phosphorylation of serine and threonine residues on the third intracellular cytoplasmic loop of G-protein-coupled receptors. Several protein kinases are involved in this desensitisation process. For muscarinic receptors this involves protein kinase (PK) A and PKC as well as G-protein-coupled receptor kinases (GRK) [20]. M_1, M_2 and M_3 receptors are phosphorylated predominately by GRK2 (or β-adrenergic receptor kinase-1). Agonist-mediated desensitisation of M_2 receptors is significantly impaired in cells transfected with a dominant-negative mutant of GRK2 [21]. Phosphorylation of the receptors by GRKs promotes the binding of a 48 kDa protein arrestin which is necessary for uncoupling of the receptor. GRK2 and β-arrestin act synergistically to enhance desensitisation of M_2 receptors [22]. In contrast, the internalisation of M_1, M_3 and M_4 receptors is independent of arrestins, but is dependent on a different endocytotic machinery involving dynamins [23].

Effect of cholinergic agonists

Using the non-selective muscarinic receptor antagonist radioligands [^{3}H]N-methylscopolamine ([^{3}H]NMS) and [^{3}H]quinuclidinyl benzilate ([^{3}H]QNB), the binding to muscarinic receptors in HEL 299 cells is best described by interaction with a single population of high-affinity binding sites. The cholinergic agonist carbachol (1 mM) induces a time-dependent decrease in the number of muscarinic receptors measured with the hydrophilic ([^{3}H]NMS) and the lipophilic ([^{3}H]QNB) ligands, without any change in the affinity of the remaining binding sites [24]. This suggests that the detected receptor down-regulation is due to a decrease in receptor density and is not a result of the presence of residual agonist in the binding assay. The loss of the lipophilic [^{3}H]QNB binding sites during the first 2 h of carbachol treatment occurs at slower rate than does the loss of [^{3}H]NMS binding sites as a result of receptor sequestration. Within 12 h, the process approaches steady state with 40 to 60% loss of receptors. The down-regulation seen after long-term carbachol treatment is probably due to receptor sequestration and subsequent degradation triggered by prolonged carbachol occupancy. This down-regulation is accompanied by uncoupling of the M_2-receptors after 24 h carbachol treatment. These results suggest that homologous sequestration, desensitisation and down-regulation of muscarinic M_2-receptors do not involve transcriptional or post-transcriptional modification of m2 receptor mRNAs. There is no reduction in either m2-receptor mRNA stability, or in m2-receptor gene transcription measured by nuclear run-on assay [24].

Agonist-induced down-regulation of M_4-receptors also occurs without changes in steady-state receptor mRNA in a neuroblastoma cell line [25]. However, in chick heart cells, down-regulation of M_2 and M_4 receptors occurs at the level of gene transcription, indicating possible differences between species and cell type in receptor regulation [26, 27]. Agonists result in a reduction in steady state mRNA for M_1 receptors in CHO cells, which appears to be due to destabilisation of m1 receptor mRNA and involves a critical region in the 3'-untranslated part of the receptor [28].

Effect of cholinergic antagonists

Exposure of HEL 299 cells to the non-selective muscarinic antagonists atropine and NMS results in up-regulation of M_2 receptors, with an increase in mRNA, suggesting an increase in gene transcription [24]. Similar findings have been reported for cerebellar M_2 and M_3 receptors [26] and cortical M_1 and M_3 receptors [29] and is due to increased gene transcription. These results are unexpected as they suggest that there is a constitutive activation of muscarinic receptors in the absence of endogenous acetylcholine and that atropine behaves like an inverse agonist. Constitutive activation of muscarinic receptors is regulated by the amount of G_q expression, thus allowing atropine to display inverse agonist properties [30]. In rabbits *in vivo* chronic administration of atropine results in up-regulation of pulmonary M_2 and M_3 receptors and increased methacholine-induced contraction of airway smooth muscle [31]. This may underlie the transient increase in cholinergic responsiveness seen in patients treated with long-term anticholinergic therapy [32].

Heterologous regulation

β_2-Adrenergic agonists

HEL 299 cells express functional β_2-adrenergic receptors; the $_b$-adrenergic antagonist [^{125}I]iodopindolol identifies a single population of binding sites on these cells with a K_D value and binding capacity of 20.9 ± 1.7 pM and 49.6 ± 6.6 fmol/mg protein respectively. These receptors display characteristic properties of the β_2-adrenoceptor subtype with regard to ligand binding, functional response (cyclic AMP accumulation) and mRNA expression.

The presence of functional M_2 and β_2-adrenoceptors in HEL 299 cells makes it possible to investigate cross-talk between these receptors. Carbachol (1 mM) treatment has no effect on the density or the affinity of β_2-adrenoceptors in these cells. However, short-term incubation with the β_2-adrenergic receptor agonist, procaterol

(5 μM), induces sequestration of the muscarinic receptors followed by their recycling to the cell surface before subsequent down-regulation [33]. Down-regulation is accompanied by functional uncoupling of the M_2 receptors, as inhibition of cAMP accumulation by carbachol is lost after 24 h treatment with procaterol. Down-regulation in receptor number is not a consequence of changes in m2 gene expression, however, as 24 h procaterol treatment fails to alter m2 mRNA levels, although shorter treatments resulted in a modest (25%), but significant, increase in m2 mRNA between 0.5 and 2 h. The loss in receptor density appears to be cyclic AMP-dependent, as it is mimicked by forskolin and by the stable cyclic AMP analogue, 8-bromo-cyclic AMP. The cellular kinases, PKA and PKC, are also implicated in the down-regulation and desensitisation process in response to β_2-agonists, as selective inhibitors of both PKA (H-8) and PKC (GF 109203X) fully and partially attenuate down-regulation and desensitisation respectively [33]. In chick heart cells, however, isoprenaline results in an increase in M_2 receptors and mRNA, but not in M_4 receptors, through a cyclic AMP-dependent mechanism [34].

Protein kinase C

Direct stimulation of PKC with the phorbol ester phorbol 13,14-dibutyrate (PDBu: 100 nM) results in a time-dependent decrease in [^{3}H]NMS binding and the steady-state levels of m2-muscarinic receptor mRNA and leads to a functional uncoupling of M_2 receptors in HEL 299 cells [35]. The loss of m2-receptor mRNA and protein after exposure to PDBu appears to be a PKC-mediated effect, since pre-treatment with the PKC inhibitor GF 109203X completely inhibits the PDBu-induced reduction in m2-receptor mRNA and significantly inhibits the reduction in M_2-receptors. Incubation with the inactive 4α-PDBu (100 nM) confirms a PKC-mediated effect as 24 h treatment had no effect on [^{3}H]NMS binding or m2-receptor mRNA levels. Potential PKC desensitisation following long-term treatment with PDBu was not observed as the calcium ionophore A23187, which is thought to potentiate the effect of PKC stimulation, in combination with PDBu does not produce any further down-regulation of M_2 muscarinic receptor protein or mRNA. This result indicates a relative insensitivity of PKC present in HEL 299 to calcium, which may relate to a particular PKC isoform in these cells. Elevation of intracellular Ca^{2+} by the ionophore A23187 has no effect on the [$_3$H]NMS binding capacity or on the level of muscarinic m2 mRNA. This argues against the involvement of the Ca^{2+}-calmodulin-dependent protein kinase in the down-regulation of M_2-receptors.

The reduction in muscarinic m2-receptor mRNA is not due to post-transcriptional modification of the mRNA, but rather was mediated through a reduction in the rate of transcription of the m2-receptor gene, as measured by nuclear run-on assays. Furthermore, this down-regulation requires protein synthesis, as the translation inhibitor cyclohexamide (10 µg/ml) protects against receptor down-regulation. Thus, synthesis of at least one other protein factor is required after PKC stimulation to alter m2-receptor mRNA levels. The nature of the protein(s) induced by PKC activation is not yet known. However, PKC is known to phosphorylate and induce DNA binding activity of a number of proteins, including transcription factors such as NF-κB and AP-1, which may in turn alter the expression of other genes. In contrast, endothelin increases the expression of M_2 and M_3 receptors in cerebellar cells; this effect is mediated *via* PKC and the activation of AP-1 [36].

MAP kinase

Several cascades of mitogen-activated protein (MAP) kinase pathways have now been identified and these play an important role in the long term regulation of cell function and the response to stress through the co-ordinated activation of transcription factors [37]. Several relatively selective inhibitors of MAP kinase pathways have recently been introduced. There is increasing evidence that G-protein-coupled receptors may activate MAP kinase cascades and that these can feed back to influence receptor function [38]. One pathway that is activated by growth factors leads to activation of Raf-1 and extracellular-signal regulated kinases (ERK1, ERK-2). ERK activation is selectively blocked by the inhibitor PD 098059, which inhibits MAPKK (MEK). This inhibitor partially blocks phorbol ester and platelet-derived growth factor (PDGF)-induced reduction in m2-receptor mRNA, indicating that some of the effects of PKC activation are mediated *via* MAP kinase and ERK activation [39]. Another MAP kinase pathway, the N-terminal c-Jun kinase (JNK) pathway, may also contribute to the down-regulation of m2 receptors mediated *via* pro-inflammatory cytokines in HEL 299 cells [40].

Effects of cytokines

Multiple cytokines orchestrate and perpetuate the chronic inflammation in asthma and COPD [41, 42]. There is increasing evidence that cytokines may influence the expression of muscarinic receptors.

Transforming growth factor-β

Transforming growth factor β (TGFβ) occurs as a group of disulphide-linked proteins comprising 12.5 kDa homodimers which are synthesised and secreted by most cell types as latent high molecular weight complexes. They exert their action by binding to specific cell surface serine/threonine kinase receptors. TGFβ1 has important physiological roles in the regulation of embryogenesis, tissue repair, inflammation or cell adhesion, growth and differentiation. In HEL 299 cells, TGFβ1 induces a time-dependent down-regulation of M_2 muscarinic receptor binding sites as measured by [^{3}H]NMS. This down-regulation is slow, with 58% loss of total receptors after 24 h of treatment. The affinity of [^{3}H]NMS for the remaining sites is unaltered by TGFβ1. Northern blot analyses show a 72% decrease in the steady-state levels of m2 muscarinic receptor mRNA following TGFβ1 treatment for 24 h [43]. The loss of [^{3}H]NMS binding sites occurs slowly, which reflects a fall in the steady-state levels of m2-receptor mRNA, rather than internalisation of the receptors through phosphorylation. The delay between protein loss and the fall in mRNA levels may be indicative of the rate of receptor turnover within the cell. The TGFβ1 effect was long-lasting ($t^1/_2 \sim 8$ h), as at least 12 h were required for m2-receptor mRNA to return to basal levels after TGFβ1 washout. Previous results obtained in the same cell line have shown that the recovery of M_2-receptor protein after receptor alkylation with propylbenzilylcholine mustard is mainly through the synthetic pathway, with an estimated half-life of receptor synthesis around 12 h [44]. The loss in [^{3}H]NMS binding is accompanied by a reduction in adenylyl cyclase activity and functional desensitisation of M_2 muscarinic receptors. There is no effect of TGFβ1 on the muscarinic m2-receptor mRNA half-life measured in the presence of actinomycin D, but the rate of m2-receptor gene transcription measured with nuclear run-on assay is reduced by 50%, indicating reduced gene transcription [43]. In chick embroyonic heart cells, TGFβ1 results in down-regulation of M_2 receptors and m2 mRNA, with a similar effect on M_4 receptors [45].

There is a requirement for *de novo* protein synthesis for receptor down-regulation. The nature of the protein(s) induced by TGFβ1 activation is not known. However, TGFβ1 is known to induce DNA binding activity of a number of proteins, including AP-1. Electrophoretic DNA mobility shift assays show a rapid and concomitant increase in AP-1 and NF-κB but not Oct-1 DNA binding activity to nuclear extracts from cells treated with TGFβ1 stimulation [43]. This increase peaks at 15–30 min after treatment and declines to control levels thereafter. The anti-oxidant pyrrolidine dithiocarbamate significantly represses the induction of NF-κB, but not AP-1, by TGFβ1. The same treatment provides significant protection against TGFβ1-induced down-regulation of M_2-receptor protein and gene expression. These results suggest the involvement, at least in part, of NF-κB in the down-regulation process. However, these results do not rule out the involvement of other transcription factors. Indeed, the kinetics of AP-1 and NF-κB induction by TGFβ1 are very rapid, suggesting activation of DNA binding of pre-existing molecules, rather than the occurrence of *de novo* protein synthesis. On the other hand, the cyclohexamide data suggest that there is a requirement for protein synthesis for receptor down-regulation. Direct interactions of these transcription factors with the m2-receptor gene promoter cannot be measured directly as no sequence data are available to date.

Proinflammatory cytokines

Cytokines released by immune and inflammatory cells infiltrating the airways are well recognised as key mediators in the orchestration and perpetuation of the chronic inflammation in asthma. The expression of the proinflammatory cytokines tumour necrosis factor α (TNFα) and interleukin-1β (IL-1β) is increased in asthma [42]. Stimulation of HEL 299 cells with TNFα or IL-1β has no effect on M_2 muscarinic receptor expression. However, the combination of these two cytokines markedly down-regulate muscarinic M_2 receptor protein and uncouple M_2 receptors from adenylyl cyclase [40]. Similarly, down-regulation of m2 receptor mRNA is absent with either TNFα or IL-1β alone, but there is a marked and sustained decrease in m2 mRNA when the two cytokines are administered in combination. The m2 muscarinic receptor mRNA slowly decreases, becoming apparent first after 4 h of stimulation, reaching a plateau of 89% control at 14 h and remaining stable for up to 24 h. TNFα and IL-1β have no effect on m2 muscarinic receptor mRNA stability; nuclear run-on assays show a reduced m2 receptor gene transcription. Sequential cytokine addition suggests that the synergy involves post-receptor events. The intracellular signalling pathways leading to receptor down-regulation have been investigated using kinase inhibitors. While the PKA inhibitor H-8 provides significant protection against M_2 receptor down-regulation, the PKC inhibitor GF109203X has no effect. Beside the classical cyclic AMP and PKC pathways, a third phosphorylation pathway known to be activated by these cytokines is represented by the lipid second messenger ceramide [46]. IL-1β and TNFα rapidly increase the cellular content of ceramide produced following the hydrolysis of sphingomyelin by two types of sphingomyelinase, a membrane-associated neutral and an endosomal acidic sphingomyelinase. Treatment of HEL 299 cells with N-acetyl-sphingosine (or C2-ceramide), a cell-permeable analogue of natural ceramide, did not affect the steady-state levels of m2 muscarinic receptor mRNA over the time-course investigated in an analogous manner to TNFα and IL-1β alone. However, the combination of C2-ceramide either with TNFα or IL-1β markedly down-regulated m2 receptor mRNA expression after 24 h of treatment to a comparable extent to that produced by the combination of the two cytokines. These results are consistent with a role for a ceramide pathway in m2 receptor down-regulation induced by the combination of TNFα and IL-1β treatment.

A further downstream signalling event triggered by TNFα and IL-1β is activation of the ERK and JNK MAP kinase cascades. Using an in gel kinase assay we have shown that TNFα and/or IL-1β activate JNK-1 and JNK-2 and to a lesser extent p42 and p44 MAP kinase isoforms [40]. This suggests that JNK pathway is preferentially activated by these cytokines. These results are in agreement with previous observations showing that the ERK cascade is primarily activated by mitogenic stimuli, whereas JNKs are mainly activated by ceramide, cellular stress such as UV irradiation and by cytokines such as TNFα and IL-1β. However, the absence of synergy between IL-1β and TNFα at the level of ERK or JNK activation suggests that activation of MAP kinases is necessary but not sufficient to cause m2 receptor down-regulation. These results suggest that TNFα and IL-1β synergise to induce transcriptional down-regulation of the M_2 muscarinic receptor, which seems to be mediated through activation of both ceramide and PKA pathways. Furthermore, these results demonstrate that M_2 receptor expression is under the control of a cytokine network.

Platelet-derived growth factor

PDGF also down-regulates M_2-receptors in HEL 299 cells and this appears to be secondary to a fall in m2 mRNA and reduced gene transcription [47]. As for the other cytokines, the reduction in gene transcription is blocked by cyclohexamide and is therefore dependent on the synthesis of some unidentified protein. The PDGF-induced reduction in m2-receptor mRNA was not accompanied by uncoupling of the remaining receptors, unlike the findings with other cytokines, suggesting that PDGF does not result in activation of kinases that phosphorylate muscarinic receptors. PDGF activates several signal transduction pathways with the involvement of multiple kinases. The down-regulation of m2 receptor mRNA is not inhibited by the PKC inhibitor GF 109203X, by the PKA inhibitor H-8, or by the PI-3 kinase inhibitor wortmannin. However PD 098059 completely blocked the down-regulation, indicating that MAPKK and ERKs are involved in the mechanism of down-regulation [39].

Nerve growth factors

Neural trophic factors, including nerve growth factor (NGF), ciliary neurotrophic factor (CNTF) and leukaemia inhibitory factor (LIF), also modulate expression of muscarinic receptors. CNTF and LIF markedly increase the gene expression of m2 receptors in a murine neuroblastoma cell line [10]. NGF increases the expression of M_4 receptors and m4 mRNA in PC12 cells by increasing mRNA stability [48]. This occurs through a MAP kinase-mediated effect and is blocked by the ERK inhibitor PD 098059.

Clinical implications

Several factors appear to influence the expression of M_2-receptors and this may have functional significance in inflammatory diseases (Fig. 2, Table 1). Thus PKC activation, which may occur in inflammation in response to inflammatory mediators, such

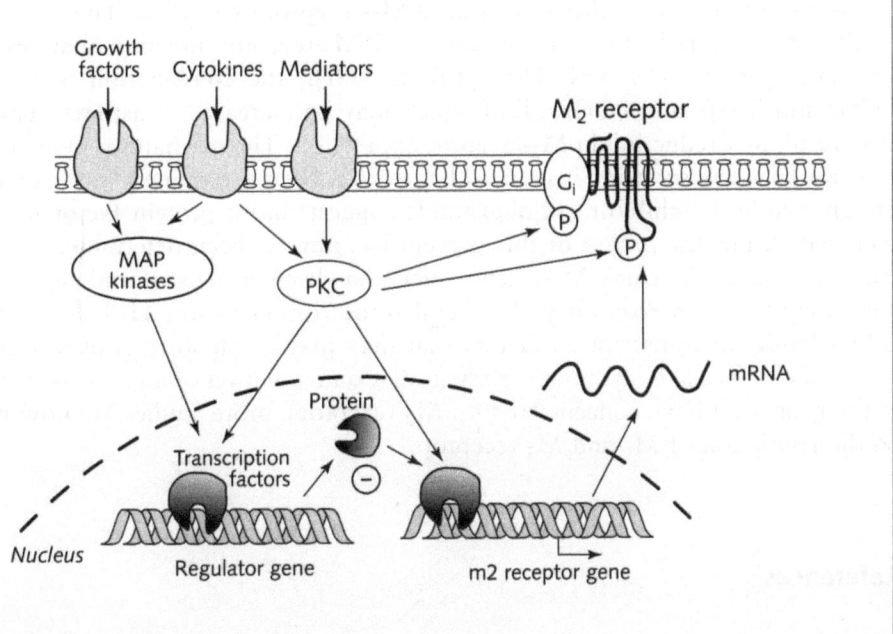

Figure 2
Regulation of M₂ receptor expression. Several heterologous stimuli influence the expression
of M₂ receptors, including inflammatory mediators, cytokines and growth factors, which
activate mitogen-activated protein (MAP) kinases or protein kinase C (PKC). This may
induce activation of transcription factors and the synthesis of an intermediary protein from
a so far unknown regulator gene.

Table 1 - Effect of various agents on M₂-receptor expression in HEL 299 cells

	Receptor density	mRNA level	Gene transcription	Coupling
Carbachol	↓	–	–	↓
β₂-agonists	↓	–	–	↓
Forskolin	↓	–	–	↓
PKC activation	↓	↓	↓	↓
TGFβ1	↓	↓	↓	↓
IL-1β + TNFα	↓	↓	↓	↓
PDGF	↓	↓	↓	–

as bradykinin, endothelin-1 and platelet-activating factor, may inhibit the transcription of m2-receptors, resulting in reduced M_2-receptor expression. This may underlie the apparent reduction in pre-junctional M_2-receptor function that occurs in asthmatic patients [16, 49]. The cytokine TGFβ, the combination of cytokines TNFα and IL-1β, and PDGF, all of which may be increased in asthma, may similarly result in a reduction in M_2-receptor expression. The mechanism by which M_2-receptor gene transcription is reduced has not yet been determined. Studies with the protein synthesis inhibitor cyclohexamide suggest that a protein factor has to be synthesised, but the nature of this protein has not yet been determined. The promoter region of the chick M_2-receptor gene has been sequenced and suggests that there may be an interaction with several transcription factors [10], including the GATA family of transcription factors that may play a role in responses to inflammatory cytokines [11]. Although most studies on regulation of muscarinic receptors at the gene level have concentrated on M_2 receptors, more studies are now needed on the regulation of M_1 and M_3 receptors.

References

1 Barnes PJ (1993) Muscarinic receptor subtypes in airways. *Life Sci* 52: 521–528
2 Barnes PJ (1992) Modulation of neurotransmission in airways. *Physiol Rev* 72: 699–729
3 Patel HJ, Barnes PJ, Takahashi T, Tadjkarimi S, Yacoub MH, Belvisi MG (1995) Characterization of prejunctional muscarinic autoreceptors in human and guinea-pig trachea *in vitro*. *Am J Resp Crit Care Med* 152: 872–878
4 Fryer AD, Jacoby DB (1993) Effect of inflammatory cell mediators on M_2 muscarinic receptors in the lungs. *Life Sci* 52: 529–536
5 Lefkowitz RJ, Pitcher J, Krueger K, Daaka Y (1998) Mechanisms of beta-adrenergic receptor desensitization and resensitization. *Adv Pharmacol* 42: 416–420
6 Barnes PJ (1995) Beta-adrenergic receptors and their regulation. *Am J Respir Crit Care Med* 152: 838–860
7 Wood IC, Roopra A, Harrington C, Buckley NJ (1995) Structure of the m4 cholinergic muscarinic receptor gene and its promoter. *J Biol Chem* 270: 30933–30940
8 Mieda M, Haga T, Saffen DW (1997) Expression of the rat m4 muscarinic acetylcholine receptor gene is regulated by the neuron-restrictive silencer element/repressor element 1. *J Biol Chem* 272: 5854–5860
9 Pepitoni S, Wood IC, Buckley NJ (1997) Structure of the m1 muscarinic acetylcholine receptor gene and its promoter. *J Biol Chem* 272: 17112–17117
10 Rosoff ML, Wei J, Nathanson NM (1996) Isolation and characterization of the chicken m2 acetylcholine receptor promoter region: induction of gene transcription by leukemia

inhibitory factor and ciliary neurotrophic factor. *Proc Natl Acad Sci USA* 93: 14889–14894

11   Rosoff ML, Nathanson NM (1998) GATA factor-dependent regulation of cardiac m2 muscarinic acetylcholine gene transcription. *J Biol Chem* 273: 9124–9129

12   Haddad E-B, Mak JCW, Barnes PJ (1996) Expression of β-adrenergic and muscarinic receptors in human lung. *Am J Physiol* 270: L947–L953

13   Mak JCW, Barnes PJ (1990) Autoradiographic visualization of muscarinic receptor subtypes in human and guinea pig lung. *Am Rev Respir Dis* 141: 1559–1568

14   Mak JCW, Baraniuk JN, Barnes PJ (1992) Localization of muscarinic receptor subtype mRNAs in human lung. *Am J Respir Cell Mol Biol* 7: 344–348

15   ten Berge REJ, Santing RE, Hamstra TJ, Roffel AF, Zaagsma J (1995) Dysfunction of muscarinic M_2-receptors after the early allergen reaction: possible contribution to bronchial hyperresponsiveness in allergic guinea pigs. *Br J Pharmacol* 114: 881–887

16   Minette PAH, Lammers J, Dixon CMS, McCusker MT, Barnes PJ (1989) A muscarinic agonist inhibits reflex bronchoconstriction in normal but not in asthmatic subjects. *J Appl Physiol* 67: 2461–2465

17   Barnes PJ, Haddad E, Rousell JA (1997) Regulation of muscarinic M_2-receptors. *Life Sci* 60: 1015–1021

18   Haddad el-B, Rousell J (1998) Regulation of the expression and function of the M_2 muscarinic receptor. *Trends Pharmacol Sci* 19: 322–327

19   Lefkowitz RJ, Pitcher J, Krueger K, Daaka Y (1998) Mechanisms of beta-adrenergic receptor desensitization and resensitization. *Adv Pharmacol* 42: 416–420

20   Pitcher JA, Freedman NJ, Lefkowitz RJ (1998) G protein-coupled receptor kinases. *Ann Rev Biochem* 67: 653–692

21   Pals-Rylaarsdam R, Gurevich VV, Lee KB, Ptasienski JA, Benovic JL, Hosey MM (1997) Internalization of the m2 muscarinic acetylcholine receptor. Arrestin-independent and dependent pathways. *J Biol Chem* 272: 23682–23689

22   Schlador ML, Nathanson NM (1997) Synergistic regulation of m2 muscarinic acetylcholine receptor desensitization and sequestration by G protein-coupled receptor kinase-2 and beta-arrestin-1. *J Biol Chem* 272: 18882–18890

23   Lee KB, Pals-Rylaarsdam R, Benovic JL, Hosey MM (1998) Arrestin-independent internalization of the m1, m3, and m4 subtypes of muscarinic cholinergic receptors. *J Biol Chem* 273: 12967–12972

24   Haddad E-B, Rousell J, Mak JCW, Barnes PJ (1995) Long-term carbachol treatment-induced down-regulation of muscarinic M_2 receptors but not m2 receptor mRNA in a human lung cell line. *Br J Pharmacol* 116: 2027–2032

25   Lenz W, Petrusch C, Jakobs KH, van Koppen CJ (1994) Agonist-induced down-regulation of the m4 muscarinic acetylcholine receptor occurs without changes in receptor mRNA steady-state levels. *Naunyn Schmiedeberg's Arch Pharmacol* 350: 507–513

26   Fukamauchi F, Saunders PA, Hough C, Chuang D (1993) Agonist-induced down-regulation and antagonist-induced up regulation of m2- and m3-muscarinic acetylcholine

receptor mRNA and protein in cultured cerebellar granule cells. *Mol Pharmacol* 44: 940–949

27 Habecker BA, Nathanson NM (1992) Regulation of muscarinic acetylcholine receptor mRNA expression by activation of homologous and heterologous receptors. *Proc Natl Acad Sci USA* 89: 5035–5038

28 Lee NH, Earle-Hughes J, Fraser CM (1994) Agonist-activated destabilization of ml muscarinic receptor mRNA. Elements involved in mRNA stability are localized in the 3'-untranslated region. *J Biol Chem* 269: 4291–4298

29 Brusa R, Gamalero SR, Genazzani E, Eva C (1995) In primary neuronal cultures muscarinic m1 and m3 receptor mRNA levels are regulated by agonists, partial agonists and antagonists. *Eur J Pharmacol* 289: 9–16

30 Burstein ES, Spalding TA, Brann MR (1997) Pharmacology of muscarinic receptor subtypes constitutively activated by G proteins. *Mol Pharmacol* 51: 312–319

31 Witt-Enderby PA, Yamamura HI, Halonen M, Lai J, Palmer JD, Bloom JW (1995) Regulation of airway muscarinic cholinergic receptor subtypes by chronic anticholinergic treatment. *Mol Pharmacol* 47: 485–490

32 Newcomb R, Tashkin DP, Hui KK, Connolly ME, Lee E, Dauphinee B (1985) Rebound hyperresponsiveness to muscarinic stimulation after chronic therapy with an inhaled muscarinic antagonist. *Am Rev Respir Dis* 132: 12–15

33 Rousell J, Haddad E-B, Webb BLJ, Giembycz MA, Mak JCW, Barnes PJ (1996) β-Adrenoceptor-mediated down-regulation of M_2-muscarinic receptors: role of cAMP-dependent protein kinases and protein kinase C. *Mol Pharmacol* 49: 629–635

34 Jackson DA, Nathanson NM (1995) Subtype-specific regulation of muscarinic receptor expression and function by heterologous receptor activation. *J Biol Chem* 270: 22374–22377

35 Rousell J, Haddad E, Mak JCW, Barnes PJ (1995) Transcriptional down-regulation of m2 muscarinic receptor gene expression in human embroyonic lung (HEL 299) cells by protein kinase C. *J Biol Chem* 270: 7213–7218

36 Fukamauchi F, Chuang DM (1994) Endothelin-1 increases the levels of mRNA and protein of muscarinic acetylcholine receptors and c-fos mRNA in cerebellar granule cells. *FEBS Lett* 348: 263–267

37 Karin M (1998) Mitogen-activated protein kinase cascades as regulators of stress responses. *Ann NY Acad Sci* 851: 139–146

38 Lopez-Ilasaca M (1998) Signalling from G-protein-coupled receptors to mitogen-activated protein (MAP)-kinase cascades. *Biochem Pharmacol* 56: 269–277

39 Haddad E-B, Rousell J (1998) Regulation of the expression and function of the M_2 muscarinic receptor. *Trends Pharmacol Sci* 19: 322–327

40 Haddad E-B, Rousell J, Lindsay MA, Barnes PJ (1996) Synergy between TNF-α and IL-1β in inducing down-regulation of muscarinic M_2 receptor gene expression. *J Biol Chem* 271: 32586–32592

41 Barnes PJ, Chung KF, Page CP (1998) Inflammatory mediators of asthma: an update. *Pharmacol Rev* 50: 515–596

42 Chung KF, Barnes PJ (1999) Cytokines in asthma. *Thorax* 54: 825–857
43 Haddad E-B, Rousell J, Mak JCW, Adcock IM, Barnes PJ (1996) Transforming growth factor-β1 induces transcriptional down-regulation of the m2 muscarinic receptor gene. *Mol Pharmacol* 49: 781–787
44 Haddad E-B, Rousell J, Barnes PJ (1995) Muscarinic M_2 receptor synthesis: study of receptor turnover with propylbenzilylcholine mustard. *Eur J Pharmacol Mol Pharmacol Sect* 283: 255–258
45 Jackson DA, Nathanson NM (1997) Regulation of expression and function of m2 and m4 muscarinic receptors in cultured embryonic chick heart cells by transforming growth factor- beta 1. *Biochem Pharmacol* 54: 525–527
46 Kolesnick R, Golde DW (1994) The sphingomyelin pathway in tumor necrosis factor and interleukin-1 signalling. *Cell* 77: 325–328
47 Rousell J, Haddad el-B, Lindsay MA, Barnes PJ (1997) Regulation of m2 muscarinic receptor gene expression by platelet-derived growth factor: involvement of extracellular signal-regulated protein kinases in the down-regulation process. *Mol Pharmacol* 52: 966–973
48 Lee NH, Malek RL (1998) Nerve growth factor regulation of m4 muscarinic receptor mRNA stability but not gene transcription requires mitogen-activated protein kinase activity. *J Biol Chem* 273: 22317–22325
49 Ayala LE, Ahmed T (1991) Is there a loss of a protective muscarinic receptor mechanism in asthma? *Chest* 96: 1285–1291

Muscarinic control of airway mucus secretion

Duncan F. Rogers

Thoracic Medicine, National Heart & Lung Institute (Imperial College), Dovehouse Street, London SW3 6LY, UK

Introduction

Secretory elements in the airways have a parasympathetic (cholinergic) supply, and activation of cholinergic nerves or administration of cholinomimetic drugs induces marked increases in secretion. In health, secreted high molecular weight glycoproteins (mucins) comprise a small proportion (~ 1%) of the thin layer of aqueous liquid (often termed "mucus") which overlies the surface epithelium of the airways. The liquid also contains electrolytes, immunoglobulins, plasma-derived proteins (for example albumin), lipids, enzymes, anti-enzymes, antibacterials, cell products and mediators. The relative contribution of each component to the normal functioning of the liquid is comparatively unknown, and the relevance of the liquid to airway homeostasis is, for the most part, only theoretical [1]. Mucins possess viscoelastic properties which confer upon the liquid the ability to be transported on the tips of beating cilia. Airborne particles are trapped in the liquid and are removed from the airways by mucociliary clearance. In disease, changes in the amount of liquid, or its viscoelasticity, or in the relative proportion of its components, convert a normally protective mechanism into a pathophysiological one. The altered mucus is difficult to clear and accumulates in the airways. The resulting airflow limitation contributes to morbidity and mortality in bronchial diseases such as asthma, chronic bronchitis and cystic fibrosis (CF).

Three neural pathways are currently recognised in the airways: sympathetic (adrenergic), parasympathetic (cholinergic) and a third system which remains active after adrenoceptor and cholinoceptor blockade and has consequently been termed non-adrenergic, non-cholinergic (NANC) [2] (Fig. 1). Cholinergic pathways represent the dominant control system in all species studied, including humans [3]. Sympathetic control of airway secretion can be demonstrated in laboratory animals, but does not appear to be of any significance in human bronchi *in vitro* [3]. The NANC system may be divided into excitatory (e-NANC) and inhibitory (i-NANC) neural systems. The e-NANC system comprises a discrete network of C-fibres (afferents) which subserve a motor function. The neurotransmitters of these nerves are small molecular weight peptides (collectively termed sensory neuropeptides), including

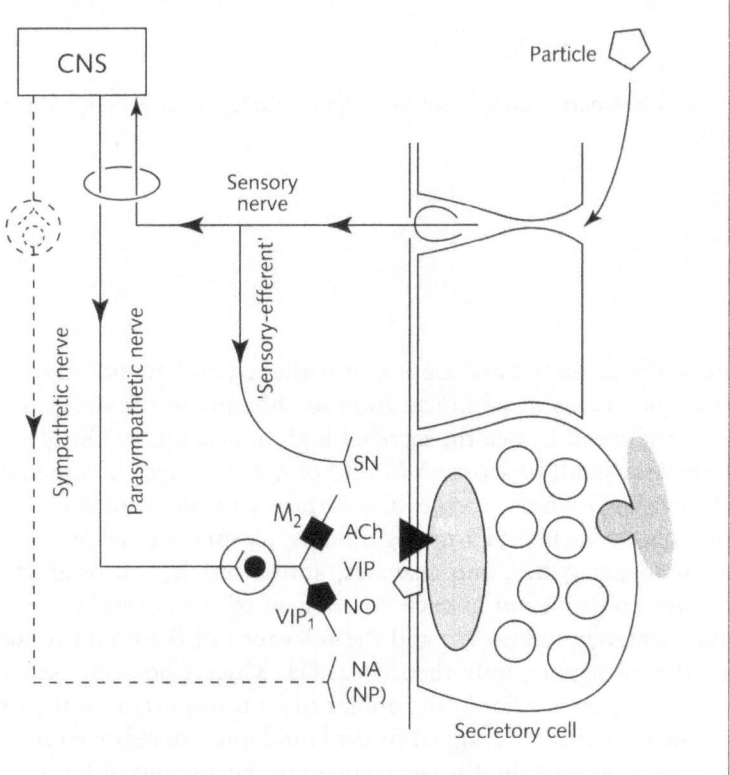

Figure 1
Cholinergic innervation of airway mucus secreting cells. Parasympathetic (cholinergic) nerves are the dominant neural pathway mediating neurogenic airway secretion. In these nerves, acetylcholine (ACh) is colocalised with vasoactive intestinal peptide (VIP) (and other VIP-related neuropeptides, for example peptide histidine isoleucine), and nitric oxide synthase, which generates nitric oxide (NO). ACh induces mucus secretion via muscarinic M_3 receptors on the secretory cells. The magnitude of secretion is regulated via prejunctional M_2 and VIP receptors, and via NO (see text and Figs. 2 and 5). Parasympathetic (adrenergic) neural control of airway secretion is species-specific and absent from human airways (NA, noradrenaline; NPY, neuropeptide tyrosine). Sensory nerve terminals in the epithelium detect airway irritation, relay impulses via afferent pathways to central systems (CNS) and initiate reflex cholinergic secretion. Local, or axonal, motor neurotransmission via collateral "sensory-efferent" pathways is species specific and is difficult to demonstrate in human airways (SN, sensory neuropeptides, for example substance P). The figure is simplified in that it does not illustrate overlap between nerves in their neuropeptide content, only shows the major neuropeptides involved and gives no indication of interaction between different components of the neural network.

calcitonin gene-related peptide (CGRP) and the tachykinins substance P and neurokinin A. Tachykininergic control of airway secretion can be readily demonstrated in laboratory animals [4], but is either insignificant in human bronchi [3] or can only be "uncovered" by removing inhibitory influences [5]. Inhibitory-NANC neurotransmitters are co-localised with "classical" neurotransmitters in adrenergic and cholinergic nerves where they regulate the activity of noradrenaline and acetylcholine. Vasoactive intestinal peptide (VIP), VIP-related peptides (for example peptide histidine isoleucine, PHI, or PH-methionine, PHM in humans) and nitric oxide (NO) are considered to be i-NANC neurotransmitters in cholinergic nerves. However, despite the convenience of dividing neural systems into defined categories, the close association between different nerve types, the co-localisation of different neurotransmitters within the same nerve, and the interaction between neurotransmitters make it unlikely that there are separate neural systems with each being responsible for a defined aspect of airway secretion. The final secretory product in response to cholinergic activity is the result of the interlinking activities of a variety of neural systems. An alteration in one part of the system, as part of a pathophysiological process, may lead to an inappropriate emphasis on another aspect which in turn may contribute to disease. In addition, mucus secretion is only one process contributing to the volume and physico-chemical properties of the airway surface liquid. Cholinergic activity not only increases mucus secretion but may also influence the transepithelial movement of electrolytes across the epithelial membrane, into and out of the airway lumen [6]. Water flow across the epithelium is linked to ion transport [7], which in turn contributes to the control of hydration of the airway liquid. Similarly, electrolyte (and water) secretion from submucosal glands is under cholinergic control [8]. Thus, the net effect of cholinergic activity is to increase the volume of water, ions and macromolecules in the airway surface liquid (Tab. 1). The increased liquid may be required to produce greater protection of the airway, by increasing the thickness of the barrier to inhaled insult, by facilitating mucociliary clearance, or by inducing cough. In addition, cholinergic activity will affect the degree of airway tone, essentially causing constriction and reducing airway patency, an effect which may be exacerbated by increased luminal mucus.

The last few years has witnessed considerable advances in understanding of the cholinergic control of airways mucus secretion. The present article focuses upon these advances in the nasal mucosa and the lower airways. Many studies are of necessity in animals other than man, but information in humans will be highlighted where it is available.

Sources of airway mucus

Mucus-secreting tissue is found in both the surface epithelium and submucosa. Of the 22 airway cell types identified, the epithelial goblet cell and serous cell, and the

Table 1 – Effect of cholinergic stimulation on production of airway "mucus"

Stimulus	Secretory cell type	Response
Nerve stimulation	goblet cell	mucus secretion
	submucosal gland	mucus secretion
	epithelium	–
Cholinoceptor agonists	goblet cell	mucus secretion hyperplasia*
	submucosal gland	mucus secretion hypertrophy*
	epithelium	electrolyte/water flux albumin flux

–, not published; *, after subacute repeated injection

mucous and serous cells in the acini of the submucosal glands, contain intracellular membrane-bound granules whose contents stain for mucus (Fig. 2). In addition, although lacking distinct secretory vesicles, the epithelial ciliated cell may produce a mucus layer termed the glycocalyx. The Clara cell may also be secretory, but the nature of its secretory product and its relevance to the formation of airway surface liquid are unclear. In the present article, cholinergic control of secretion from the goblet cell and sero-mucous acini of the submucosal glands will be discussed in detail. Descriptions of these cells have been published previously [9, 10]; only essential elements are presented below.

Goblet cells

The predominant secretory cell in the epithelium is the goblet cell (Fig. 2), named originally according to its characteristic shape after chemical fixation [11]. Its distribution varies with airway region, animal species and disease status. In general, airways of large animals or larger diameter airways in small animals have more goblet cells than the airways of small animals or smaller diameter airways in large animals. Exceptions to this generalisation are the ferret which has very few goblet cells in the trachea [12–14] and the specific pathogen-free (SPF) rat which has virtually no airway goblet cells [15]. In contrast, the epithelial serous cell is characteristic of airways maintained in irritant-free environments, for example foetal lung and the SPF rat, but is virtually absent in at least nine adult animal species, including man [9]. Thus, under normal conditions, goblet cells are the major surface epithelial contributor to the production of airway mucus [11].

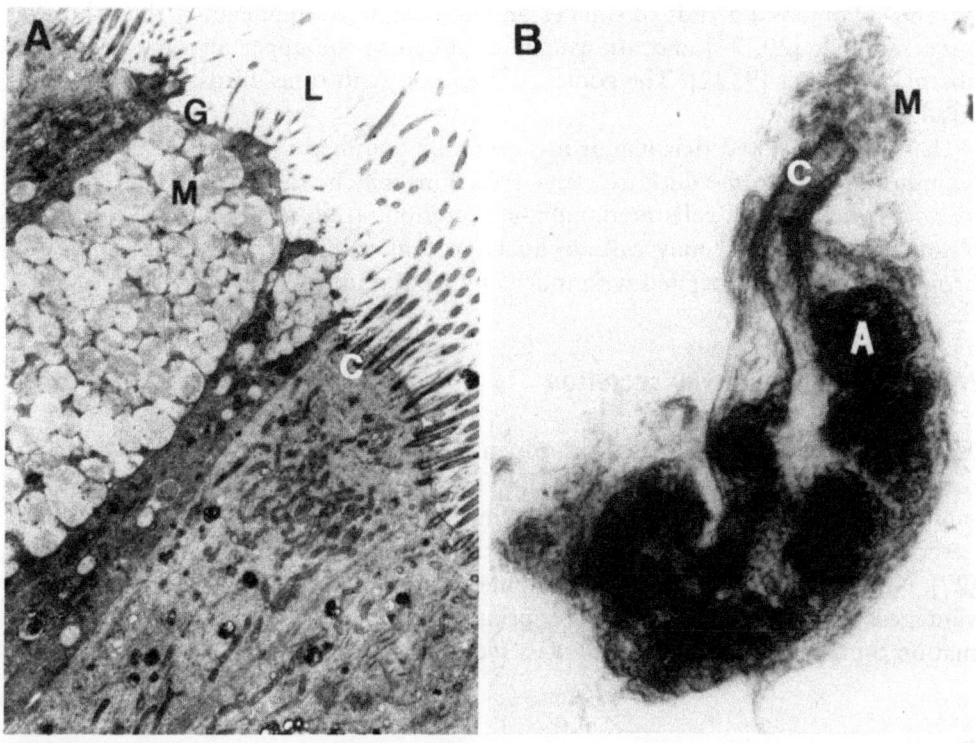

Figure 2
Airway secretory elements. (A) Goblet cell (G) in human bronchus (transmission electron micrograph). M, mucin granule (electron lucent); C, ciliated cell; L, lumen. (B) Isolated dog tracheal submucosal gland. A, acinus; C, ciliated duct; M, mucus. Image kindly supplied by Dr. Sanae Shimura.

Submucosal glands

The submucosal glands of larger mammals, including humans, are complex structures comprising secretory tubules, a collecting duct and a ciliated duct which opens onto the airway surface [16] (Fig. 2). The secretory tubules form acini which are lined with mucous and serous cells containing mucins staining acidic or neutral, respectively [17]. Pre-secreted mucus may be squeezed out of the tubules and ducts onto the airway surface by contraction of gland-associated myoepithelial cells. Isolated feline tracheal submucosal glands contract and expel mucus in response to cholinergic agonists [18] and electrical field stimulation [19], the latter predominantly *via* activation of cholinergic nerves.

The submucosal glands of smaller animals are more rudimentary than those of larger animals [20, 21] and are usually confined to the upper airways or may be absent altogether [9, 22]. The goose, in common with other birds, has no submucosal glands [23].

It should be noted that glands are generally confined to more proximal, cartilaginous airways from which excessive secretions may be cleared by cough. In small airways where goblet cells predominate, secretions from these cells are less easily cleared by cough and may assume an important role in the pathophysiology of bronchial diseases associated with mucus hypersecretion.

Measurement of airway secretion

Methods to measure lower airways' mucus secretion *in vitro* [24] and *in vivo* [25] have been reviewed recently and are only briefly summarised below. Similarly, two minimally invasive methods of sampling human nasal secretions have been described, namely the NIH nasal provocation model [26] and the nasal pool device [27]. No technique is perfect, and each method has inherent advantages and disadvantages. However, when used appropriately, these methods have provided information on the control of airway mucus secretion.

Direct measurement

Measurement of secretion from a single tracheo-bronchial submucosal gland can be made directly by: 1) recording the rate of filling of a micropipette inserted into the duct opening, or 2) the rate of appearance of fluid droplets (termed "hillocks") through tantalum dust spread on the airway surface. Both techniques can be used *in vivo* and *in vitro*, but are measures more of liquid secretion rather than being specific for mucus, because they do not account for the water content of the secretion.

Non-radiolabelling methods

Non-radiolabelling methods involve measurement of endogenous mucus markers, for example fucose [28, 29], or application of selective stains including Alcian blue and periodic acid-Schiff. Although these methods have value in giving an estimate of the magnitude of mucus secretion, they are not selective for the cellular source of secretion. It may be desirable to determine the cellular source of secretion to distinguish epithelial from glandular, or to separate serous from mucous acinar secretions. Lysozyme is localised in submucosal gland serous cells and is consequently a marker of serous secretion. In ferret trachea it can be detected by immunocytochemistry

[30] or bioassay [31]. However, other structures, for example the epithelium of human airways, also contain lysozyme, which limits its use as a marker of serous secretion. Monoclonal antibodies and plant lectins are useful, but continue to be hampered by specificity, particularly so for the lectins. Isolation of epithelial cells in culture [32], or individual submucosal glands [18], is useful in limiting the source of secretion but removes normal influences from surrounding structures. Morphometric measures, where loss of intracellular secretory material is used as an index of mucus output, have been used to determine secretion both from submucosal glands [33] and goblet cells [34]. Certain experimental preparations have a restricted source of mucus that enables the site of secretion to be defined. For example, the goose trachea has no submucosal glands, whereas the ferret trachea has extremely few goblet cells (see above); these preparations have been used to study, respectively, goblet cell and submucosal gland secretion.

Radiolabelling methods

Radiolabelled precursors of mucus are often used to quantify secretion *in vitro* and *in vivo*. Airway tissues are incubated with the radiolabel, which is incorporated into the mucus molecules and appears in the secreted product. The success of the technique depends upon the specificity of the radiolabel for mucus. Precursors may become incorporated in materials other than mucus, a problem most associated with carbohydrate labels (in particular glucose), and many precursors are metabolised. [^{35}S]-sulphate is not readily metabolised and is relatively selective for airway submucosal gland mucus in a number of animal species including cat, ferret and human [3, 35–37]. However, in bovine trachea, $^{35}SO_4$ labels a "novel" monomeric mucin rather than the characteristic large oligomeric mucins [38]. If this observation extends to mucins from other species, the use of $^{35}SO_4$ as a mucus marker may need re-evaluation. Tritiated labels are considered to be useful epithelial markers. For example, [^{3}H]-proline is a putative marker of epithelial secretion in cat trachea [35].

In the following sections, attention will be drawn to the technique used to quantify secretion so that an evaluation of the relevance of the measurement can be made.

Characteristics of basal and cholinergic-stimulated secretion

There may be no basal secretion in the airways under normal conditions. Dissection and subsequent protocols used to study basal secretion may themselves stimulate secretion. In the air-filled ferret *in vitro* whole trachea, a preparation containing abundant submucosal glands but virtually no goblet cells (see above), there is no detectable baseline secretion [39–41]. In the cat, a species with abundant glands (see above), the rate of filling of a micropipette with fluid from an individual gland is 1–9 nl/min [42, 43]. The process of introducing the pipette into the gland may have affected secretion. The rate was only slightly reduced by atropine, indicating that cholinergic influences, or tonic parasympathetic tone, contributed little to basal secretion, which is consistent with the lack of effect of tetrodotoxin on basal secretion in the cat trachea *in vivo* [45]. In contrast, in the goose trachea, a preparation lacking submucosal glands and where the principal source of mucus is the goblet cells, intra-tracheal atropine reduces basal secretion by 30% [23]. Thus, cholinergic activity may contribute to basal secretion in airways where the primary source of mucus is the epithelium.

In cat trachea *in vivo*, biophysical and biochemical differences between mucus secreted under "basal" conditions and mucus secreted in response to stimulation by the cholinomimetic pilocarpine, or to activation of cholinergic nerves, indicate different sources of mucus [45, 46]. Stimulated secretion (presumed to be predominantly from the submucosal glands) was consistent with a mucus glycoprotein, whereas basal secretion was mucin-like but not typical (denser and more resistant to reduction than typical mucus). The origin of this material is unknown, but may be the epithelium [47] or the glycocalyx. Thus, under resting conditions, normal airway surface fluid contains an atypical mucin which may be produced by the epithelium. In contrast, cholinergic stimulation recruits the high-capacity glandular system which produces a typical mucin [45–49] which, presumably, is better equipped to aid in the formation of a surface liquid appropriate to acute airway defence.

Cholinergic control of lower airways secretion

The cholinergic nervous system is the dominant neural pathway in the lower airways. In animal and human airways, cholinergic nerve fibres are closely associated with submucosal glands [50]. Autoradiographic mapping of muscarinic receptors [51–54] (Fig. 3) and localisation of muscarinic receptor mRNAs by *in situ* hybridisation [55] demonstrate dense labelling over the submucosal glands. Labelling of epithelium was less than the glands in ferret trachea and guinea pig lung and absent in human lung. In ferret trachea, muscarinic receptors are localised to both serous and mucous cells in the glands [56], which indicates that cholinergic stimulation induces secretion by both types of acini.

Mucus output

Muscarinic receptor agonists induce mucus secretion *in vitro* by airway tissue of a number of animal species including humans [3], ferret [39, 57], dog [58] and pig

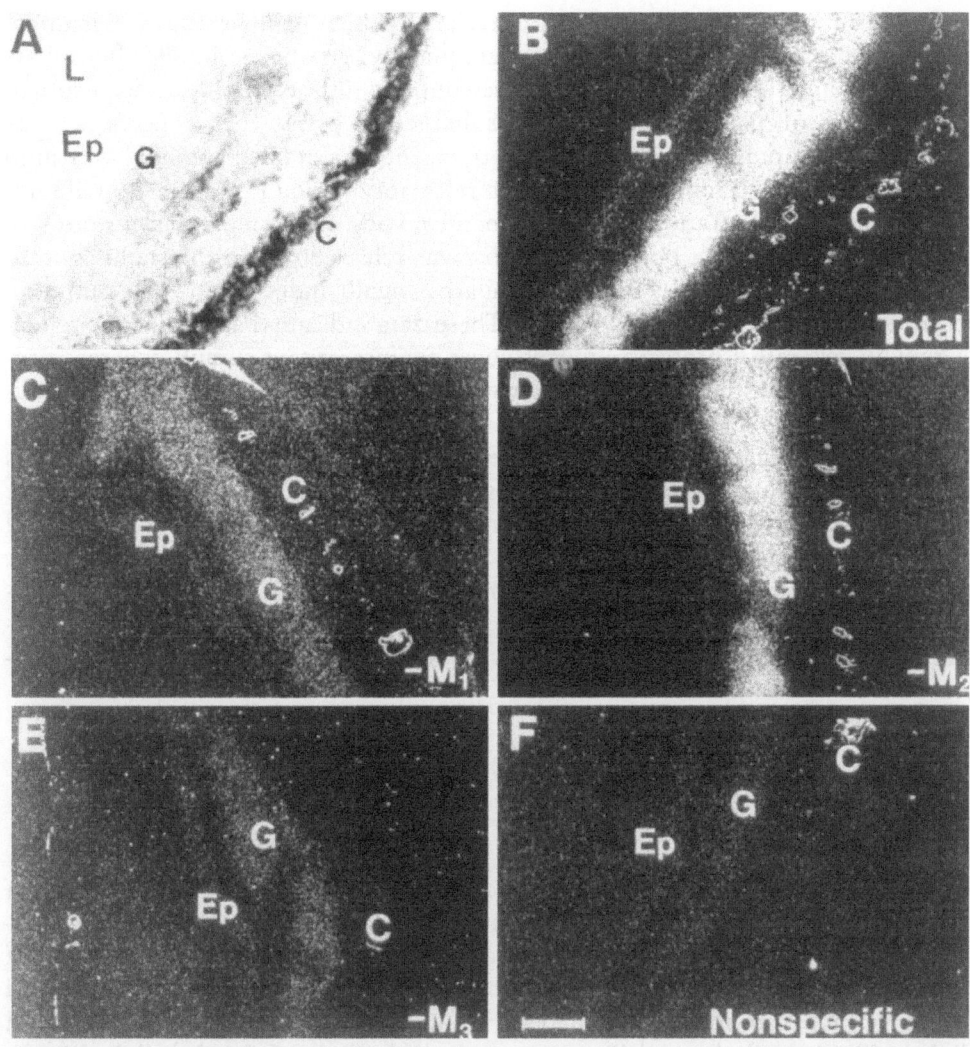

Figure 3
Localisation of muscarinic receptor types in ferret trachea. (A) Brightfield showing epitheli-um (Ep), submucosal gland (G) and cartilage (C). L, lumen. (B–F) Darkfield adjacent sections showing localisation of autoradiographic grains after incubation with the reversible and non-selective muscarinic antagonist [³H](–)QNB (1 nM), either alone (B) or in the presence of: (C) the M_1-selective antagonist telenzepine (100 nM) (i.e., minus M_1 to leave $M_2 + M_3$ receptors), (D) the M_2-selective antagonist methoctramine (50 nM) ($M_1 + M_3$ receptors), (E) the M_3-selective antagonist 4-DAMP (25 nM) ($M_1 + M_2$ receptors), or (F) atropine (1 μM). Antagonist displacement of [³H](–)QNB demonstrates a predominance of muscarinic M1 and M_3 receptors over the glands (compare C and E with B). Scale bar = 100 μm.

[59], and *in vivo* in cat [48, 49] and ferret [54], which demonstrates a functional role for the receptors localised autoradiographically (see above). Inositol phosphate generation underlies the stimulus response-coupling [60]. Brief (30 sec) exposure to acetylcholine of pig isolated tracheal submucosal gland cells triggers a large (10–25-fold), rapid (within 1 min) and transient (maintained for only 2–3 min) increase in mucus secretion followed by a refractory period [59]. Secretion did not increase in a concentration-dependent manner with increasing concentrations of acetylcholine. Instead, a bolus of secretion was released once a threshold concentration of acetylcholine was reached. Similarly, vagally induced lysozyme output in ferret trachea *in vivo* is short-lived [61]. These data indicate that cholinergic reflexes rapidly generate a protective mucus cover against irritant stimuli, whereas prolonged muscarinic stimulation limits airway flooding in response to continued irritation.

Electrical stimulation of cholinergic nerves in a variety of animal species induces tracheal mucus output, from unspecified secretory sources, either *in vitro* [57] or *in vivo* [42, 46, 54, 61]. Cholinergic nerve stimulation also induces airway goblet cell discharge (see below) and contributes to reflex mucus output (see below). Importantly, electrical field stimulation of human bronchi *in vitro* induces mucus secretion [3], a response which is blocked by atropine and by tetrodotoxin, demonstrating that cholinergic nerve activity will induce mucus secretion in human airways.

Electrolyte and water flow

Cholinoceptor agonists also induce a matched flow of Na^+ and Cl^- from submucosa to mucosa in ferret, cat and human airways [62, 63]. The secretion of NaCl is considered to be *via* the glands because (1) acetylcholine does not influence fluxes of these ions across rabbit trachea [64], a preparation lacking submucosal glands, and (2) Na^+ efflux from isolated submucosal glands of the cat has been demonstrated directly [8]. Apically located IP_3 receptors comprise part of the Ca^{2+}-linked intracellular signal transduction mechanism underlying acetylcholine-induced Cl^- secretion by acinar cells of cat and human airway submucosal glands [65]. IP_3/Ca^{2+}-activated Cl^- conductances are also involved in methacholine-induced serous cell secretion in cultured sheep submucosal gland cells [66]. The observations above are consistent with the concept that cholinergic stimulation induces an outpouring of a large volume of watery isosmotic "mucus" [67]. It should be noted, however, that the effect of cholinergic nerve stimulation on salt/water flow has not been studied. Unlike the response to exogenous administration of cholinomimetics, the final secretory response to parasympathetic nerve activity will be modified by VIP, VIP-related peptides and NO (see below).

Albumin flux

In contrast to its stimulatory action on mucus secretion and salt/water flow, cholinergic stimulation *in vivo* does not affect plasma exudation from the bronchial vasculature [68, 69]. However, methacholine markedly increases the transepithelial flow of albumin into the lumen of the ferret or rabbit whole trachea *in vitro* preparation [70, 71]. The latter observation indicates that although not involved in inducing plasma exudation, cholinergic stimulation concurrent with on-going exudation may facilitate plasma flow into the lumen. The net result of this is an increase in the macromolecular content of the airway surface liquid, with consequent changes in biochemical/biophysical characteristics. For example, plasma and mucins interact synergistically to produce a liquid of enhanced viscosity [72].

Muscarinic receptor subtypes and mucus output

Four muscarinic receptor subtypes are currently recognised pharmacologically (designated M_1–M_4) and five by genomic cloning (designated m1–m5) (see the chapter by Roffel et al. in this volume). Both M_1 and M_3 receptors have been localised by autoradiographic mapping to submucosal glands in human bronchi (in a ratio of approximately 1:2) [53] and ferret trachea [54] (Fig. 3). Functional studies in cat and pig trachea, using exogenous administration of cholinomimetics, demonstrate that the M_3 receptor antagonist 4-DAMP is the more potent of the receptor subtype antagonists currently available in inhibiting secretion induced by exogenous administration of cholinomimetics [73–75], which indicates that the muscarinic receptor mediating cholinergic mucus secretion is predominantly of the M_3 subtype. This suggestion was extended in ferret trachea *in vitro* and *in vivo* using electrical stimulation of cholinergic nerves: increased secretion was mediated *via* M_3 receptors on the secretory cells (Fig. 4). From these studies, the M_1 receptor, localised to the glands, does not appear to be involved with mucus secretion. One possibility is that it is involved in electrolyte and water secretion [73].

Cholinergic control of nasal secretion

Neural control of nasal secretion, including cholinergic control, has been reviewed in depth previously [26, 76, 77] and will only be discussed briefly herein. In contrast to the lower airways, sympathetic neural pathways contribute significantly to nasal secretion. Excitatory-NANC and i-NANC control is also demonstrable, even in humans [78]. However, as in the lower airways, parasympathetic pathways dominate. The nasal passages have a rich parasympathetic nervous supply although, intriguingly, it is easier to isolate the sympathetic than the parasympathetic or sen-

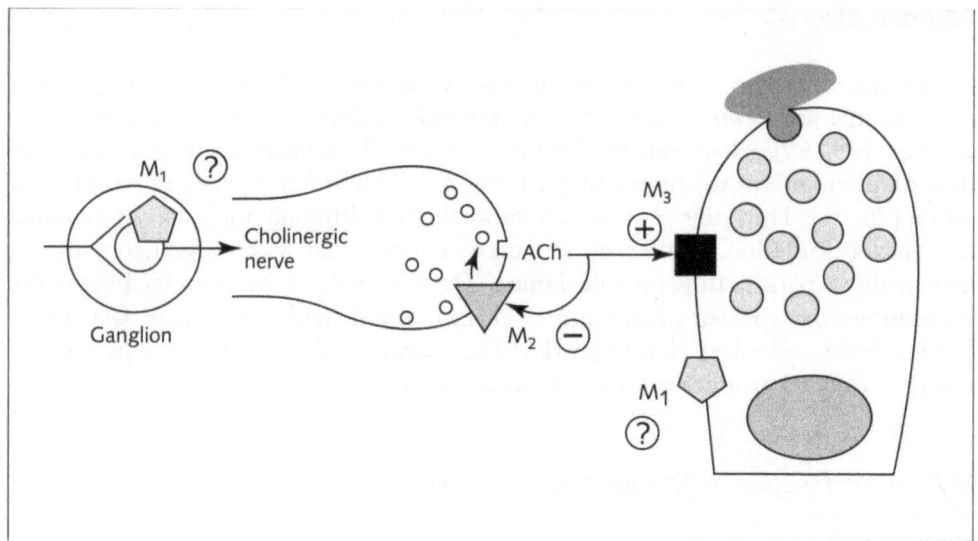

Figure 4

Cholinergic control of airway submucosal gland secretion. Traditionally, M_1 muscarinic (cholinergic) receptors are located on ganglia where they facilitate neurotransmission. They are also localised to airway submucosal glands. M_2 receptors are located on cholinergic nerve terminals and are autoinhibitory to acetylcholine (ACh) release. Muscarinic M_3 receptors are localised to the end-organ (in this case, the submucosal glands). Current information demonstrates that M_3 receptors mediate mucous secretion from the glands (+). The M_2 receptor is situated prejunctionally and regulates the magnitude of cholinergic mucous secretion (–). No evidence has been found for the ganglionic M_1 receptor in control of cholinergic gland secretion. The glandular-located M_1 receptor does not appear to be involved in control of mucous secretion. It may mediate ion and water secretion.

sory nerves to the nose, which is the exact opposite to the situation in the lower airways. Nasal challenge with cholinomimetics in animals or humans induces a marked increase in sampled liquid. Analysis of the liquid for mucus and plasma markers indicates that activation of muscarinic receptors causes secretion from serous and mucous cells in the submucosal glands, possibly also from epithelial cells, but has little or no effect on plasma exudation. In animals, electrical stimulation of parasympathetic nerves causes secretion which is largely blocked by atropine, which indicates predominant cholinergic control. In general, sympathetic nerve stimulation and adrenoceptor agonists cause a comparatively slow flow (< 100 mg/min) of protein-rich liquid, whereas parasympathetic nerve stimulation or cholinomimetics provoke a faster flow (150–500 mg/min) of liquid with a lower protein concentration.

Cholinergic control of goblet cell secretion

Electrical stimulation of the descending oesophageal nerves in the anaesthetised goose, a species lacking airway submucosal glands, or administration of acetylcholine, induces secretion of radiolabelled mucins from the tracheal goblet cells [23]. Three-quarters of the response is cholinergic. Similarly, electrical stimulation of the cervical vagus nerves in rats [79] or guinea pigs [34], both species with sparse glands, induces loss of intracellular mucus granules, considered to represent mucus secretion, from tracheal goblet cells. In the guinea pig, cholinergic pathways are the predominant neural control, although it is likely that there are facilitatory interactions with tachykininergic pathways. Cholinoceptor agonists also induce mucus secretion *in vitro* in guinea pig tracheal segments [80] and hamster tracheal epithelial cell cultures [32, 81], although in the latter studies high concentrations of cholinomimetics were used. In contrast to the stimulatory effect of cholinergic stimulation above, in species with abundant submucosal glands, for example the cat, neither nerve stimulation nor administration of cholinomimetic drugs induces tracheal goblet cell secretion [48, 49]. Thus, it seems that goblet cells in the upper airways are innervated in the absence of submucosal glands. Of interest would be whether, in species with an upper airways gland, goblet cells are innervated in the lower airways from which the gland is absent.

Direct administration of cigarette smoke into the cannulated trachea of guinea pigs induces airways goblet cell secretion *via* activation of nerves [82]. The "dose" of smoke delivered determines which type of nerve is activated: ten breaths of diluted smoke induces secretion *via* activation of parasympathetic ganglia by a component in the particulate phase of the smoke, presumably nicotine. In contrast, the vapour phase from 20 breaths of undiluted smoke activates tachykininergic nerves to induce secretion. Similarly, tracheal instillation of trimellitic anhydride (TMA), an industrial "curing" agent for epoxy resins which may cause occupational asthma, induces goblet cell degranulation *in vivo* in TMA-sensitised guinea pigs *via* activation of ganglia and capsaicin-sensitive nerves [83].

Neurally acting inflammatory mediators

Although many inflammatory mediators induce mucus secretion directly, a number of mediators affect secretion by influencing cholinergic-induced secretion. For example, serotonin has no direct effect on mucus secretion but potentiates neurogenic mucus secretion *in vivo* in dog trachea [84] and cat submucosal gland contraction *in vitro* [19, 92] *via* actions on cholinergic nerves. Similarly, endothelin-1 has no direct effect on secretion *in vitro* in ferret trachea but, in contrast to serotonin, inhibits secretion stimulated by methacholine [86]. Prostaglandins (PG) have varied actions on mucus volume output induced *in vitro* in the whole ferret trachea by methacholine: $PGF_{2\alpha}$ potentiated secretion, PGD_2 reduced secretion and PGE_1 had no effect on secretion [87].

Cholinergic reflex secretion

A number of autonomic reflexes induce airway mucus secretion, presumably as a protective response to airway irritation or in response to potentially injurious physiological responses elsewhere in the body, for example vomiting. In the cat, mechanical irritation of the larynx, nose and nasopharynx induces tracheal mucus secretion *via* a reflex involving vagal cholinergic (and probably sympathetic) motor pathways [88, 89]. Ammonia vapour stimulated a predominantly parasympathetic reflex *via* activation of cough receptors. Inhalation of inert dusts, unsurprisingly, induces tracheal secretion with part of the response due to a direct local action on the secretory cells and also *via* a vagal reflex [90]. Reflex gland secretion is linked to the cough reflex, presumably to produce a quantity of mucus sufficient to entrap the irritant and aid in its expectoration. Irritation of the stomach induces airway mucus secretion in the cat, also *via* a vagal reflex [91]. Increased activity in gastric sensory nerves may signal vomiting [92] and initiate cholinergic reflexes to protect the lung from aspirated fluid. The effects of drugs and foodstuffs on airway secretion may, in part, be the result of gastric reflexes. Hypoxaemia stimulates glandular secretion in dogs *via* a reflex linked to receptors in the carotid bodies [93]. If this reflex is also present in humans, it may exacerbate the symptoms of chronic bronchitis, asthma or CF by increasing the volume of luminal mucus.

Neuroregulation of cholinergic airway secretion

Previous sections of this chapter have demonstrated that cholinergic neural pathways are a potent drive to airway mucus secretion. It is not surprising, therefore, that regulatory mechanisms exist to limit cholinergic mucus output (Fig. 5).

M_2 receptors

First, there are the muscarinic M_2 receptors [94]. These are located on cholinergic nerves where they act as inhibitory autoreceptors which activate a negative feedback system whereby acetylcholine released from cholinergic nerves activates the receptor to inhibit further acetylcholine release. In ferret trachea *in vitro* and *in vivo*, M_2 receptors regulate the magnitude of cholinergic nerve-induced $^{35}SO_4$ output [54] (Figs. 4 and 5).

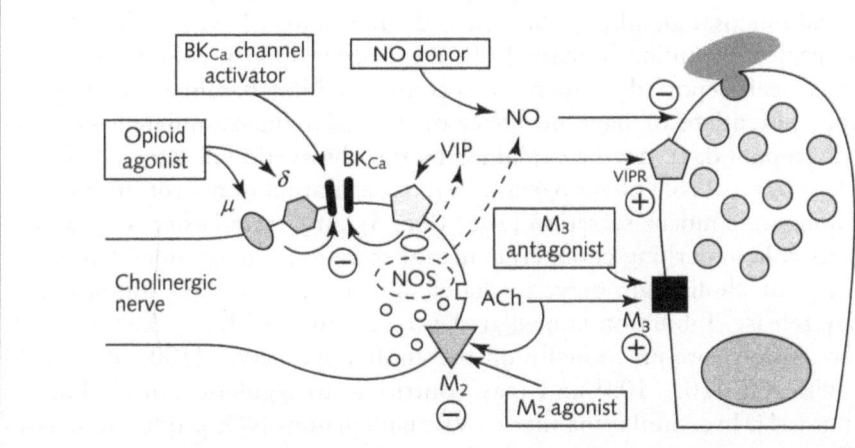

Figure 5

Neuroregulation of airways cholinergic mucus secretion and therapeutic targets. Acetylcholine (ACh) acts on postjunctional muscarinic M_3 receptors to stimulate secretion (+), and on prejunctional (autoinhibitory) M_2 receptors which regulate ACh release (–) (see Fig. 4). Cholinergic secretion should, therefore, be amenable to inhibition by M_3 receptor antagonists or M_2 agonists. Nitric oxide (NO), released by NO synthase (NOS) in association with cholinergic nerve activation, inhibits secretion, presumed to be via postjunctional functional antagonism (–). NO donors may, therefore, inhibit cholinergic secretion. Vasoactive intestinal peptide (VIP), co-released with ACh, induces secretion via postjunctional VIP receptors (VIPR). However, the small stimulatory action (+) is offset by its prejunctional inhibition of ACh release (–) and, hence, inhibition of cholinergic secretion. Agonists at μ or δ opioid receptors also regulate ACh release (–) and cholinergic secretion. The inhibitory action of both VIP and the opioid receptor agonists is mediated via opening of large conductance calcium-activated potassium channels (BK_{Ca}). Thus, BK_{Ca} channel activators should be effective inhibitors of airway cholinergic mucus secretion.

VIP and NO

VIP and NO are neurotransmitters of i-NANC bronchodilator neural pathways, although their relative involvement varies with species [95]. Their role in regulation of cholinergic mucus secretion has only recently been elucidated. VIP-containing nerves are found in intrinsic ganglia, smooth muscle, blood vessels and submucosal glands in the airways of a number of mammalian species, including humans [96, 97]. Depending upon species, VIP coexists with different neurotransmitters, including NO synthase and acetylcholine [97, 98]. By autoradiography, VIP receptors are localised to submucosal glands, epithelium and smooth muscle in the airways of a number of species, including humans [99]. Depending upon preparation, exogenous VIP has been reported to increase secretion, inhibit baseline secretion, or either inhibit, potentiate or have no effect on methacholine-stimulated secretion [2]. These discrepant data may be explained by the observation in ferret trachea *in vitro* that the principal role of endogenous VIP is regulation of neurotransmission, including cholinergic mucus secretion [100] (Fig. 5). Any stimulatory role for VIP is offset by its role in curbing cholinergic neural secretion. Endogenous VIP limits the magnitude of cholinergic nerve-mediated mucus secretion *via* inhibition of acetylcholine release. Inhibition is mediated *via* activation of BK_{Ca} channels, presumed to be situated pre-junctionally on the cholinergic nerves [100]. PHI/PHM co-exist(s) with VIP [102, 103] and may contribute to regulation of cholinergic mucus output [41]. In a similar manner to VIP, endogenous NO, produced in association with nerve activation, inhibits cholinergic $^{35}SO_4$ output in ferret trachea *in vitro* [103] (Fig. 5).

Opioids and K+ channels

In addition to the three endogenous regulatory mechanisms discussed above, exogenous administration of opioid agonists selective for μ or δ opioid receptors inhibits cholinergic nerve-induced $^{35}SO_4$ output in ferret trachea *in vitro* [104] and goblet cell discharge *in vivo* in guinea pig trachea [105] (Fig. 5). Interestingly, in a similar manner to VIP, opioid-inhibition is *via* activation of BK_{Ca} channels. Consistent with this is the observation that a putative BK_{Ca} channel activator inhibits cholinergic $^{35}SO_4$ output 104]. Thus, it is possible that activation of BK_{Ca} channels is a final common mechanism for a variety of neuroregulatory influences on cholinergic mucus secretion (Fig. 5). It should be noted, however, that although BK_{Ca} channels are the endogenous signalling mechanism for a number of cholinergic neuroregulators, activation of K_{ATP} channels will also inhibit cholinergic nerve-induced airway mucus secretion [104, 106].

Abnormal cholinergic control and airway mucus hypersecretion

Abnormal cholinergic control in respiratory disease and anticholinergic therapy are the subjects of the chapters by Chapman and Disse, respectively, in this book. However, for completeness in the present chapter, cholinergic abnormalities with specific reference to bronchial diseases associated with mucus hypersecretion (chronic bronchitis and asthma) are discussed below (Fig. 6).

Chronic bronchitis

Chronic bronchitis is characterised by cough and continued sputum production and is closely associated with chronic inhalation of pollutants, in particular cigarette smoke. The sputum production is associated with goblet cell hyperplasia and submucosal gland hypertrophy [107]. In experimental animals, repeated injection of cholinomimetics induces goblet cell hyperplasia [108, 109]. If true in humans, this observation indicates that chronic repeated stimulation of airway cholinergic nerves (for example by cigarette smoke – see below) could cause goblet cell hyperplasia, with associated mucus hypersecretion. *In vitro* in bronchi from patients with chronic bronchitis, basal mucus output is increased [110]. Not only is basal secretion increased, but the response to stimuli is exaggerated. Bronchi containing hypertrophied glands have an increased secretion in response to acetylcholine, and this is less effectively blocked by atropine compared with bronchi without gland hypertrophy. Similarly, VIP inhibits mucus secretion from normal human airways *in vitro* but has little inhibitory effect on secretion by airways of patients with chronic bronchitis [111]. Thus, the combination of increased responsiveness to acetylcholine and reduced inhibition by VIP could act in concert to induce exaggerated mucus production when cholinergic nerves are stimulated in patients with bronchial submucosal gland hypertrophy (Fig. 6).

Cigarette smoke

Cigarette smoke has a number of effects relevant to direct overproduction of mucus by cholinergic mechanisms (Fig. 6). Cigarette smoke provokes mucus secretion from submucosal glands [112] and goblet cells [82] *via* nicotine-stimulation of ganglia and activation of cholinergic nerves. Proteinases produced by inflammatory cells recruited into the airways by cigarette smoke [113, 114] will degrade VIP and so reduce its inhibitory effects on cholinergic mucus secretion.

Asthma

Asthmatics usually produce sputum during an asthmatic attack and this is associated with airways goblet cell hyperplasia and submucosal gland hypertrophy. The airways of patients dying in *status asthmaticus* are occluded by tenacious mucus plugs [115, 116], and partially-formed plugs are found in the airways of asthmatics who have died from causes other than asthma [117]. Cholinergic control of airway mucus secretion might arguably be augmented in asthma because a dysfunction in M_2 receptors (inhibitory autoreceptors; see above) in asthmatic airways [118, 119] would fail to regulate cholinergic nerve-induced mucus output [54] (Fig. 6). In addition, endogenous VIP-regulation may also be reduced. VIP immunoreactive nerves

191

are absent in the airways of asthmatics, whether mild or severe [120], although this has not been confirmed [121, 122]. VIP-like immunoreactivity is reduced in plasma of asthmatics compared with controls [123], and this correlated with reduced reversibility upon treatment. In contrast, VIP receptor number is unaltered in asthma [124]. If VIP is reduced in asthmatic airways, then endogenous VIP-regulation of cholinergic mucus secretion would be expected to be reduced, with consequent enhancement of secretion (Fig. 6). However, although inhibitory mechanisms may be dysfunctional in asthma, the refractoriness of acetylcholine-stimulated discharge of airway mucus [59] may limit continued and excessive secretion.

Conclusions

This chapter presents evidence that cholinergic nerve activity is a potent drive to alter the rate of mucus secretion which, together with activation of other processes including salt and water secretion, will affect the quantity, and possibly the composition, of the liquid lining of the airways. In the short term, increased mucus secretion is a protective mechanism developed by the airways to limit damage by inhaled airborne irritants. Chronic overproduction of mucus precipitates disease. Further characterisation of cholinergic pathways mediating mucus secretion, together with classification of receptor types involved and elucidation of inhibitory mechanisms, should identify critical basic mechanisms pertinent to airways hypersecretory disease. This information should, in turn, indicate therapeutic directions and target the development of drugs for appropriate inhibitory neural sites, for example BK_{Ca} channel activators (Fig. 5).

Acknowledgements
I thank the Cystic Fibrosis Research Trust, National Asthma Campaign, Clinical Research Committee of the Royal Brompton Hospitals and Pfizer Central Research (Sandwich, UK) for financial support, and all colleagues over the years who have contributed to the secretion studies.

Figure 6
Cholinergic mechanisms in the pathophysiology of airways mucus hypersecretion. Cholinergic nerve activity can be increased by the airway inflammation associated with asthma (initiated, for example, via allergenic activation of mast cells) and chronic bronchitis (initiated, for example, via inhalation of pollutants, most notably cigarette smoke). Neurally active inflammatory mediators, epithelial irritation and plasma nicotine trigger sensory nerves and initiate cholinergic reflexes with release of acetylcholine (ACh). Nicotine also triggers cholinergic nerves via ganglionic stimulation. Released ACh will induce mucus secretion. ACh may also induce secre-

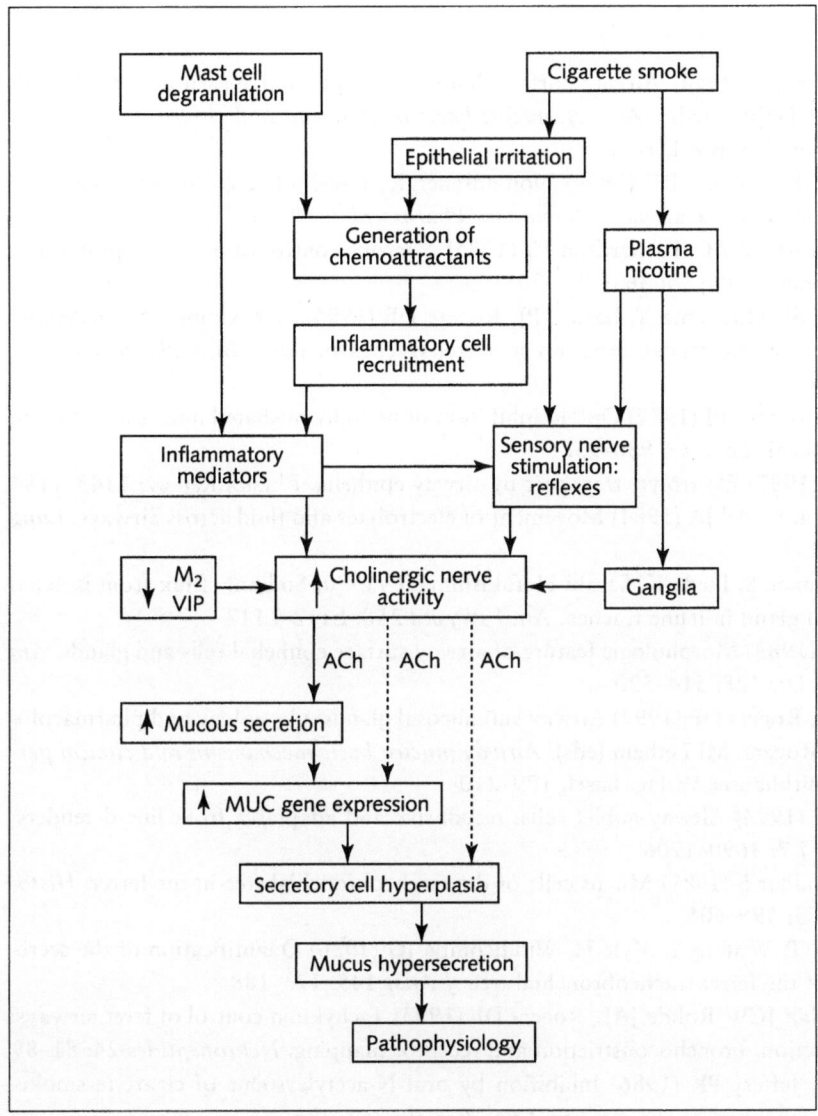

tory cell hyperplasia: via the increased mucus secretion, via induction of mucin (MUC) genes, via a direct effect on the secretory cells, or via a combination of these mechanisms. Impaired neuroregulation by dysfunctional muscarinic M_2 receptors and degradation of vasoactive intestinal peptide (VIP) (see text) will increase the post-junctional effects of ACh. Secretory cell hyperplasia is associated with mucus hypersecretion, which leads to the pathophysiological correlates of airflow limitation, mucus plugging, and infection. Solid arrows: experimental/clinical evidence available (see text); broken arrows: putative pathways awaiting demonstration.

References

1   Widdicombe JG (1998) Airway surface liquid: concepts and measurements. In: DF Rogers, MI Lethem (eds): *Airway mucus: basic mechanisms and clinical perspectives*. Birkhäuser Verlag, Basel, 1–17

2   Ramnarine SI, Rogers DF (1994) Non-adrenergic, non-cholinergic neural control of mucus secretion in the airways. *Pulmonary Pharmacol* 7: 19–33

3   Baker B, Peatfield AC, Richardson PS (1985) Nervous control of mucin secretion into human bronchi. *J Physiol* 365: 297–305

4   Ramnarine SI, Hirayama Y, Barnes PJ, Rogers DF (1994). Tachykinin NK_1 receptors mediate neurogenic mucus secretion in ferret trachea *in vitro*. *Br J Pharmacol* 113: 1183–1190

5   Rogers DF, Barnes PJ (1989) Opioid inhibition of neurally mediated mucus secretion in human bronchi. *Lancet* i: 930–932

6   Welsh MJ (1987) Electrolyte transport by airway epithelia. *Physiol Rev* 67: 1143–1184

7   Nathanson I, Nadel JA (1984) Movement of electrolytes and fluid across airways. *Lung* 162: 125–137

8   Sasaki T, Sanae S, Ikeda K, Sasaki H Takishima T (1990) Sodium efflux from isolated submucosal gland in feline trachea. *Am J Physiol* 258: L112–L117

9   Jeffery PK (1983) Morphologic feature of airway surface epithelial cells and glands. *Am Rev Respir Dis* 128: 514–520

10 Fung DCK, Rogers DF (1997) Airway submucosal glands: physiology and pharmacology. In: DF Rogers, MI Lethem (eds): *Airway mucus: basic mechanisms and clinical perspectives*. Birkhäuser Verlag, Basel, 179–210

11 Rogers DF (1994) Airway goblet cells: responsive and adaptable front line defenders. *Eur Respir J* 7: 1690–1706

12 Jacob S, Poddar S (1981) Mucus cells of the tracheobronchial tree in the ferret. *Histochemistry* 73: 599–605

13 Robinson NP, Venning L, Kyle H, Widdicombe JG (1986) Quantification of the secretory cells of the ferret tracheobronchial tree. *J Anat* 145: 173–188

14 Meini S, Mak JCW, Rohde JAL, Rogers DF (1993) Tachykinin control of feret airways: mucus secretion, bronchoconstriction and receptor mapping. *Neuropeptides* 24: 81–89

15 Rogers DF, Jeffery PK (1986) Inhibition by oral N-acetylcysteine of cigarette smoke-induced "bronchitis" in the rat. *Exp Lung Res* 10: 267–283

16 Meyrick B, Sturgess JM, Reid L (1969) A reconstruction of the human bronchial gland. *Thorax* 24: 729–736

17 Reid LM (1977) Secretory cells. *Fed Proc* 36: 2703–2707

18 Shimura A, Sasaki T, Sasaki H, Takishima T (1986) Contractility of isolated single submucosal gland from trachea. *J Appl Physiol* 60: 1237–1247

19 Shimura S, Sasaki T, Okayama H, Sasaki H, Takishima T (1987) Neural control of contraction in isolated submucosal gland from feline trachea. *J Appl Physiol* 62: 2404–2409

20   Goco RV, Kress MB, Branigan OC (1963) Comparison of mucus glands in the tracheal bronchial tree of man and animals. *Annals NY Acad Sci* 106: 555–571

21   Gatto LA, Aiello E (1981) Mucus-secreting glands and goblet cells in the trachea of the deer mouse Peromyscus leucopus. *Trans Am Micros Soc* 100: 355–365

22   Breeze RG, Wheeldon EB (1977) The cells of the pulmonary airways. *Am Rev Respir Dis* 116: 705–777

23   Phipps RJ, Richardson PS, Corfield A, Gallagher PK, Jeffery PK, Kent PW, Passatore M (1977) A physiological, biochemical and histological study of goose tracheal mucin and its secretion. *Philos Trans Royal Soc London (Biol)* 9: 513–543

24   Davis CW (1997) *In vitro* models for airways mucus secretion. *Pulmonary Pharmacol* 10: 145–155

25   Rogers DF (1997) *In vivo* preclinical test models for studying airway mucus secretion. *Pulmonary Pharmacol* 10: 121–128

26   Raphael GD, Baraniuk JN, Kaliner MA (1991) How and why the nose runs. *J Allergy Clin Immunol* 87: 457–467

27   Greiff L, Pipkorn U, Alkner U, Persson CGA (1990) The nasal pool device applies controlled concentrations of solutes on human nasal airway mucosa and samples its surface exudation/secretions. *Clin Exp Allergy* 20: 253–259

28   Boat TF, Cheng PW (1976) Mucous glycoproteins. In: JA Mangos, RC Talamo (eds): *Cystic fibrosis: projections into the future.* Stratton Intercontinental Medical Book Co, New York, 165–177

29   Rogers DF, Turner NC, Marriott C, Jeffery PK (1987) Cigarette smoke-induced "chronic bronchitis": a study *in situ* of laryngo-tracheal hypersecretion in the rat. *Clin Sci* 72: 629–637

30   Tom-Moy M, Basbaum CB, Nadel JA (1983) Localisation and release of lysozyme from ferret trachea: effects of adrenergic and cholinergic drugs. *Cell Tissue Res* 228: 549–562

31   Webber SE (1989) Receptors mediating the effects of substance P and neurokinin A on mucus secretion and smooth muscle tone of the ferret trachea: potentiation by an enkephalinase inhibitor. *Br J Pharmacol* 98: 1197–1206

32   Kim KC (1991) Biochemistry and pharmacology of mucin-like glycoproteins produced by cultured airway epithelial cells. *Exp Lung Res* 17: 533–545

33   Gashi AA, Borson DB, Finkbeiner WE, Nadel JA, Basbaum CB (1986) Neuropeptides degranulate serous cells of ferret tracheal glands. *Am J Physiol* 251: C223–C229

34   Tokuyama K, Kuo H-P, Rohde JAL, Barnes PJ, Rogers DF (1990) Neural control of goblet cell secretion in guinea pig airways. *Am J Physiol* 259: L108–L115

35   Davies JR, Corbishley CM, Richardson PS (1990) The uptake of radiolabelled precursors of mucus glycoproteins by secretory tissues in the feline trachea. *J Physiol* 420: 19–30

36   Gashi AA, Nadel JA, Basbaum CB (1987) Autoradiographic studies of the distribution of ^{35}sulfate label in ferret trachea: effects of stimulation. *Exp Lung Res* 12: 83–96

37   Pack RJ, Williams IP, Phipps RJ, Richardson PS (1984) A preparation for the study of secretory function of the human bronchus *in vitro*. *Eur J Respir Dis* 65: 239–250

38 Svitacheva N, Hovenberg HW, Davies JR (1998) Biosynthesis of mucins in bovine trachea: identification of the major radiolabelled species. *Biochem J* 333: 449–456

39 Webber SE, Widdicombe JG (1987) The actions of methacholine, phenylephrine, salbutamol and histamine on mucus secretion from the ferret *in vitro* trachea. Agents Actions 22: 82–85

40 Webber SE, Widdicombe JG (1987) The effect of vasoactive intestinal peptide on smooth muscle tone and mucus volume output from the ferret trachea. *Br J Pharmacol* 91: 139–148

41 Webber SE (1988) The effects of peptide histidine isoleucine and neuropeptide Y on mucus volume output from the ferret trachea. *Br J Pharmacol* 95: 49–54

42 Ueki I, German VF, Nadel JA (1980). Micropipette measurement of airway sub-mucosal gland secretion: autonomic effects. *Am Rev Respir Dis* 121: 351–357

43 Quinton PM (1979) Composition and control of secretions from tracheal bronchial submucosal glands. *Nature* 279: 551–552

44 Somerville M, Karlsson J-A, Richardson PS (1990) The effects of local anaesthetic agents upon mucus secretion in the feline trachea *in vivo*. *Pulmonary Pharmacol* 3: 93–101

45 Davies JR, Gallagher JT, Richardson PS, Sheehan JK, Carlstedt I (1991) Mucins in cat airway secretions. *Biochem J* 275: 663–669

46 Fung DCK, Beacock DJ, Richardson PS (1992) Vagal control of mucus glycoconjugate secretion into the feline trachea. *J Physiol* 453: 435–447

47 Varsano SC, Basbaum CB, Forsberg LS, Borson D, Caughey G, Nadel JA (1987) Dog tracheal epithelial cells in culture synthesise sufated macromolecular glyco-conjugates and release them from the cell surface upon exposure to extracellular proteinases. *Exp Lung Res* 13: 157–184

48 Florey H, Carlton H, Wells SA (1932) Mucus secretion in the trachea. *Br J Exp Pathol* 13: 269–284

49 Gallagher JT, Kent PW, Passatore M, Phipps RJ, Richardson PS (1975) The composition of tracheal mucus and the nervous control of its secretion in the cat. *Proc Royal Soc (Biology)* 192: 49–76

50 Laitinen A, Partanen M, Hervonen A, Laitinen LA (1985) Electron microscopic study on the innervation of the human lower respiratory tract: evidence of adrenergic nerves. *Eur J Respir Dis* 67: 209–215

51 Barnes PJ, Nadel, JA, Roberts JM, Basbaum CB (1983) Muscarinic receptors in lung and trachea: autoradiographic localization using [^{3}H]quinuclidinyl benzilate. *Eur J Pharmacol* 86: 103–106

52 Van Koppen CJ, Blankesteijn WM, Klaassen ABM, Rodrigues de Miranda JF, Beld AJ, Van Ginneken CAM (1988) Autoradiographic visualisation of muscarinic receptors in human bronchi. *J Pharmacol Exp Ther* 244: 760–764

53 Mak JCW, Barnes PJ (l990) Autoradiographic visualization of muscarinic receptor subtypes in human and guinea pig lung. *Am Rev Respir Dis* 141: 1559–1568

54   Ramnarine SI, Haddad E-B, Khawaja AM, Mak JCW, Rogers DF (1996) On muscarinic control of neurogenic mucus secretion in ferret trachea. *J Physiol* 494: 577–586

55   Mak JCW, Baraniuk JN, Barnes PJ (1992) Localization of muscarinic receptor subtype mRNAs in human lung. *Am J Respir Cell Mol Biol* 7: 344–348

56   Basbaum CB, Barnes PJ, Grillo MA, Widdicombe JH, Nadel JA (1983) Adrenergic and cholinergic receptors in submucosal glands of the ferret trachea: auto-radiographic localization. *Eur J Respir Dis* 64 (suppl. 128): 433–435

57   Borson DB, Charlin M, Gold BD, Nadel JA (1984) Neural regulation of $^{35}SO_4$-macromolecule secretion from tracheal glands of ferrets. *J Appl Physiol* 57: 457–466

58   Chakrin L, Baker AP, Christian P, Wardell J (1973) Effect of cholinergic stimulation on the release of macromolecules by canine trachea *in vitro*. *Am Rev Respir Dis* 108: 69–76

59   Dwyer TM, Szebeni A, Diveki K, Farley JM (1992) Transient cholinergic glycoconjugate secretion from swine tracheal submucosal gland cells. *Am J Physiol* 262: L418–L426

60   Hall IP (1992) Agonist-induced inositol phosphate responses in bovine airway submucosal glands. *Am J Physiol* 262: L257–L262

61   Wells UM, Robson A, Webber SE, Widdicombe JG (1992) An *in vivo* preparation for measurement of plasma protein and lysozyme output in the ferret tracheal lumen. *Pulmonary Pharmacology* 5, 183–189

62   Corrales RJ, Coleman DL, Jacoby DB, Leikauf GD, Hahn HL, Nadel JA, Widdicombe JH (1986) Ion transport across cat and ferret tracheal epithelia. *J Appl Physiol* 61: 1065–1070

63   Knowles M, Murray G, Shallal J, Askin F, Ranga V, Gatzy J, Boucher R (1984) Bioelectric properties and ion flow across excised human bronchi. *J Appl Physiol* 56: 868–877

64   Jarnigan F, Davis JD, Bromberg PA, Gatzy JT, Boucher RC (1983) Bioelectric properties and ion transport of excised rabbit trachea. *J Appl Physiol* 55: 1884–1192

65   Sasaki T, Shimura S, Wakui M, Ohkawara Y, Takishima T, Mikoshiba K (1994) Apically localized IP_3 receptors control chloride current in airway gland acinar cells. *Am J Physiol* 267: L152–L158

66   Griffin A, TM Newman, RH Scott (1996) Electrophysiological and ultrastructural events evoked by methacholine and intracellular photolysis of caged compounds in cultured ovine trachea submucosal gland cells. *Exp Physiol* 81: 27–43

67   De Sanctis GT, Rubin BK, Ramirez O, King M (1993) Ferret tracheal mucus rheology, clearability and volume following administration of substance P or methacholine. *Eur Respir J* 6: 76–82

68   Lundberg JM, Saria A (1982) Capsaicin-sensitive vagal neurons involved in control of vascular permeability in rat trachea. *Acta Physiol Scand* 115: 521–523

69   Belvisi MG, Barnes PJ, Rogers DF (1990) Neurogenic inflammation in the airways: characterisation of electrical parameters for vagus nerve stimulation in the guinea pig. *J Neurosci Methods* 32: 159–167

70   Webber SE, Widicombe JG (1989) The transport of albumin across the ferret *in vitro* whole trachea. *J Physiol* 408: 457–472

71 Price AM, Webber SE, Widdicombe JG (1990) Transport of albumin across the rabbit trachea *in vitro. J Appl Physiol* 68: 726–730

72 List SJ, Findlay BP, Forstner GG, Forstner JF (1978) Enhancement of the viscosity of mucin by serum albumin. *Biochem J* 175: 565–571

73 Yang CM, Farley JM, Dwyer TM (1988) Muscarinic stimulation of submucosal glands in swine trachea. *J Appl Physiol* 64: 200–209

74 Gater PR, Alabaster VA, Piper I (1989) A study of the muscarinic receptor subtype mediating mucus secretion in the cat trachea *in vitro. Pulmonary Pharmacol* 2: 87–92

75 Ishihara H, Shimura S, Satoh M, Masuda T, Nonaka H, Kase H, Sasaki T, Sasaki H, Takishima T, Tamura K (1992) Muscarinic receptor subtypes in feline tracheal submucosal gland secretion. *Am J Physiol* 262: L223–L228

76 Baraniuk JN (1991) Neural control of human nasal secretion. *Pulmonary Pharmacol* 4: 20–31

77 Widdicombe JG (1990) Nasal pathophysiology. *Respir Med* 84 (suppl A): 3–10

78 Baraniuk JN, Kaliner M (1991) Neuropeptides and nasal secretion. *Am J Physiol* 261: L223–L235

79 McDonald DM (1988) Neurogenic inflammation in the rat trachea. I. Changes in venules, leukocytes and epithelial cells. *J Neurocytol* 17: 583–603

80 Newman TTM, Robichaud A, Rogers DF (1996) Microanatomy of secretory granule release from guinea pig tracheal goblet cells. *Am J Respir Cell Mol Biol* 15: 529–539

81 Steel DM, Hanrahan JW (1997) Muscarinic-induced mucin secretion and intracellular signalling by hamster tracheal goblet cells. *Am J Physiol* 272: L230–L237

82 Kuo H-P, Rohde JAL, Barnes PJ, Rogers DF (1992) Cigarette smoke-induced airway goblet cell secretion: dose-dependent differential nerve activation. *Am J Physiol* 263: L161–L167

83 Hayes JP, Kuo H-P, Rohde JAL, Newman-Taylor AJ, Barnes PJ, Chung KF, Rogers DF (1995) Neurogenic goblet cell secretion and bronchoconstriction in guinea pigs sensitised to trimellitic anhydride. *Eur J Pharmacol (Env Toxicol Pharmacol Section)* 292: 127–134

84 Popovac, D., Chinn, R., Graf, P., Nadel, J. and Davis, B. (1979). Serotonin potentiates nervous stimulation of mucus gland secretion in canine trachea *in vivo. Physiologist* 22: 102 (abstract)

85 Shimura S, Sasaki T, Ishihara H, Sato M, Sasaki H, Takishima T (1992) Autonomic innervation to feline tracheal submucosal glands for mucus glycoprotein secretion. *Am J Physiol* 262: L15–L20

86 Yurdakos E, Webber SE (1991) Endothelin-1 inhibits prestimulated tracheal submucosal gland secretion and epithelial albumin transport. *Br J Pharmacol* 104: 1050–1056

87 Deffebach ME, Islami H, Price A, Webber SE, Widdicombe JG (1990) Prostaglandins alter methacholine-induced secretion in ferret *in vitro* trachea. *Am J Physiol* 258: L75–L80

88 Phipps RJ, Richardson PS (1976) The effects of irritation at various levels of the airway upon tracheal mucus secretion in the cat. *J Physiol* 261: 563–581

89 German VF, Ueki IF, Nadel JA (1980) Micropipette measurement of airway submucosal gland secretion: laryngeal reflex. *Am Rev Respir Dis* 122: 413–416

90 Peatfield AC, Richardson PS (1983) The action of dust in the airways on secretion into the trachea of the cat. *J Physiol* 342: 327–334

91 German VF, Corrales R, Ueki IF, Nadel JA (1982) Reflex stimulation of tracheal mucus gland secretion by gastric irritation in cats. *J Appl Physiol* 52: 1153–1155

92 Borison HL, Wang SC (1953) Physiology and pharmacology of vomiting. *Pharmacol Rev* 5: 193–230

93 Davis B, Chinn R, Gold J, Popovac D, Widdicombe JG, Nadel JA (1982) Hypoxaemia reflexly increases secretion from tracheal submucosal glands in dogs. *J Appl Physiol* 52: 1416–1419

94 Barnes PJ (1990) Muscarinic receptors in airways: recent developments. *J Appl Physiol* 68: 1777–1785

95 Lammers JW, Barnes PJ, Chung KF (1992) Nonadrenergic, noncholinergic airway inhibitory nerves. *Eur Respir J* 5: 239–246

96 Dey RD, Shannon Jr WA, Said SI (1981) Localization of VIP-immunoreactive nerves in airways and pulmonary vessels of dog, cat and human subjects. *Cell Tissue Res* 220: 231–238

97 Laitinen A, Partanen M, Hervonen A, Pelto-Huikko M, Laitinen LA (1985) VIP-like immunoreactive nerves in human respiratory tract: light and electronic microscopic study. *Histochemistry* 82: 313–319

98 Fischer AW, Hoffmann B (1996) Nitric oxide synthase in neurons and nerve fibers of lower airways and in vagal sensory ganglia of man. *Am J Respir Crit Care Med* 154: 209–216

99 Carstairs JR, Barnes PJ (1986) Visualisation of vasoactive intestinal peptide receptors in human and guinea pig lung. *J Pharmacol Exp Ther* 239: 249–255

100 Liu Y-C, Patel HJ, Khawaja AM, Belvisi MG, Rogers DF (1999). Neuroregulation by vasoactive intestinal peptide (VIP) of mucus secretion in ferret trachea: activation of BK_{Ca} channels and inhibition of neurotransmitter release. *Br J Pharmacol* 126: 147–158

101 Ghatei MA, Springall DR, Richards IM, Oostveen JA, Griffin RL, Cadieux A (1987) Regulatory peptides in respiratory tract of *Macaca fascicularis*. *Thorax* 42: 431–439

102 Lundberg JM, Fahrenkrug J, Hokfelt T, Martling CR, Larsson O, Tatemoto K (1984) Co-existence of peptide histidine isoleucine (PHI) and VIP in nerves regulating blood flow and bronchial smooth muscle tone in various mammals including man. *Peptides* 5: 593–606

103 Ramnarine SI, Khawaja AM, Barnes PJ, Rogers DF (1996) Nitric oxide inhibition of basal and neurogenic mucus secretion in ferret trachea *in vitro*. *Br J Pharmacol* 118: 998–1002

104 Ramnarine SI, Liu Y-C, Rogers DF (1998). Neuroregulation of mucus secretion by opioid receptors and K_{ATP} and BK_{Ca} channels in ferret trachea *in vitro*. *Br J Pharmacol* 123: 1631–1638

105 Kuo H-P, Rohde JAL, Barnes PJ, Rogers DF (1992) Differential inhibitory effects of opi-

oids on cigarette smoke, capsaicin and electrically-induced goblet cell secretion in guinea-pig trachea. *Br J Pharmacol* 105: 361–366

106 Kuo H-P, Rohde JAL, Barnes PJ, Rogers DF (1992) K$^+$ channel inhibition of neurogenic goblet cell secretion in guinea pig trachea. *Eur J Pharmacol* 215: 297–299

107 Reid L (1954) Pathology of chronic bronchitis. *Lancet* i: 275–278

108 Sturgess J, Reid L (1973) The effects of isoprenaline and pilocarpine on (a) bronchial mucus-secreting tissues and (b) pancreas, salivary glands, heart, thymus, liver and spleen. *Br J Exp Pathol* 54: 388–403

109 Kleinerman J, Sorensen J, Rynbrandt D (1976) Chronic bronchitis in the cat produced by chronic methacholine administration. *Am J Pathol* 82: A48

110 Sturgess J, Reid L (1972) An organ culture study of the effect of drugs on the secretory activity of the human bronchial submucosal gland. *Clin Sci* 43: 533–543

111 Coles SJ, Said SI, Reid LM (1981) Inhibition by vasoactive intestinal peptide of gluco-conjugate and lysozyme secretion by human airways *in vitro*. *Am Rev Respir Dis* 124: 531–536

112 Peatfield AC, Davies JR, Richardson PS (1986) The effect of tobacco smoke upon airway secretion in the cat. *Clin Sci* 71: 179–187

113 Koyama S, Rennard SI, Leikauf GD, Robbins RA (1991) Bronchial epithelial cells release monocyte chemotactic activity in response to smoke and endotoxin. *J Immunol* 147: 972–979

114 Robbins RA, Nelson KJ, Gossman GL, Koyama S, Rennard SI (1991) Complement activation by cigarette smoke. *Am J Physiol* 260: L254–L259

115 Houston JC, de Navasquez S, Trounce JR (1953) A clinical and pathological study of fatal cases of asthmaticus. *Thorax* 8: 207–213

116 Dunnill MS (1960) The pathology of asthma with special reference to changes in the bronchial mucosa. *J Clin Pathol* 13: 27–33

117 Dunnill MS (1975) The morphology of the airways in bronchial asthma. In: M Stein (ed): *New directions in asthma*. American College of Chest Physicians, Park Ridge, Illinois, 213–221

118 Ayala LE, Ahmed T (1989) Is there a loss of a protective muscarinic receptor mechanism in asthma? *Chest* 96: 1285–1291

119 Minette PAH, Lammers JWJ, Dixon CMS, McCusker MT, Barnes PJ (1989) A muscarinic agonist inhibits reflex bronchoconstriction in normal but not asthmatic subjects. *J Appl Physiol* 67: 2461–2465

120 Ollerenshaw S, Jarvis D, Woolcock A, Sullivan C, Scheibner T (1989) Absence of immunoreactive vasoactive intestinal polypeptide in tissue from the lungs of patients with asthma. *New Engl J Med* 320: 1244–1248

121 Howarth PH, Springall DR, Redington AE, Djukanovic R, Holgate ST, Polak JM (1995) Neuropeptide-containing nerves in endobronchial biopsies from asthmatic and nonasthmatic subjects. *Am J Respir Cell Mol Biol* 13: 288–296

122 Lilly CM, Bai TR, Shore SA, Hall AE, Drazen JM (1995) Neuropeptide content of lungs from asthmatic and nonasthmatic patients. *Am J Respir Crit Care Med* 151: 548–535

123 Cardell LO, Uddman R, Edvinsson L (1994) Low concentrations of VIP and elevated levels of other neuropeptides during exacerbations of asthma. *Eur Respir J* 7: 2169–2173

124 Sharma RK, Addis BJ, Jeffery PK (1995) The distribution and density of airway vasoactive intestinal polypeptide (VIP) binding sites in cystic fibrosis and asthma. *Pulmonary Pharmacol* 8: 91–96

The role of anticholinergics in asthma and COPD

Kenneth R. Chapman

University of Toronto and Asthma Centre, University of Health Network, Suite 4-011 ECW, 399 Bathurst Street, Toronto, Ontario, Canada M5T 2S8

Introduction

For centuries, physicians from various cultures and medical traditions have relied upon inhaled antimuscarinic agents to provide bronchodilator relief to their patients with obstructive respiratory diseases. No recorded history describes how the useful medicinal properties of alkaloid-containing plants were discovered, but careful records of antimuscarinic respiratory therapy are available from several cultures. In the ancient Ayurvedic literature of India and the pharmacopoeias of Europe and the United Kingdom in the 19th century, the preparation and administration of anticholinergic botanical derivatives are described. Many of the early observations of the clinical value of antimuscarinic bronchodilators have been validated in the modern era by pharmacological, physiological, clinical and biological studies. The pharmaceutical progeny of antimuscarinic botanicals are now the cornerstone of bronchodilator therapy in patients with chronic obstructive pulmonary disease and a useful supplemental bronchodilator for patients with asthma. In the following chapter, this rich medical history is reviewed briefly and the clinical role of modern pharmaceutical antimuscarinic bronchodilators is explored.

The history of anticholinergic bronchodilator therapy

The first use of inhaled antimuscarinic bronchodilator therapy is unrecorded and it seems most likely that the discovery was made in various locations and times. The most common botanical sources of anticholinergic therapy are certainly various species of the *Datura* genus, species that grow as uncultivated weeds worldwide. The leaves, stems and flowering tops of the plants are rich in atropine and related alkaloids. It is easy to imagine that the weeds' unusual properties were first observed when bundles of dried *Datura* weeds were used to kindle or supplement cooking fires. Inhalation of the smoke, particularly in a confined space, would deliver large

Muscarinic Receptors in Airways Diseases, edited by Johan Zaagsma, Herman Meurs and Ad F. Roffel
© 2001 Birkhäuser Verlag Basel/Switzerland

amounts of atropine to the lower airways with consequent systemic absorption. Just the right amount of alkaloid absorption might result in a pleasant euphoria. Somewhat more absorption might cause hallucinations — sometimes delightful, sometimes comical, but often terrifying. Even greater amounts of systemic absorption might result in death. Thus, men quickly learned to use *Datura* species for their hallucinogenic properties and sometimes to misuse them as poisons.

At least one historian has speculated that Homer's *Odyssey* offers the first recorded "history" of deliberate anticholinergic poisoning [1]. Circe, the nymph, is said to have fed Odysseus's crew a magical potion that turned them into swine. With the divine help of Hermes, Odysseus avoided this unhappy fate by ingesting the magical plant, moly, that protected him from Circe's potion and allowed him to rescue his crew. Does this myth recount in a somewhat fanciful fashion actual historical events? Could Circe have used *Datura* or related species to poison surreptitiously and thus to cause hallucinations? Could Odysseus have been protected from antimuscarinic poisoning by the ingestion of *Galanthus nivalus*, a natural source of the centrally acting anticholinesterase galanthine? Certainly the description of Odysseus's crew members trapped in animal form brings to mind later documented episodes of mass anticholinergic poisoning from *Datura* species. In 17th century New England, British troops set out to quell the Jamestown uprising. En route, the troops unknowingly added *Datura stramonium* plants to their cookpots to supplement their meager fare. Soon thereafter, a number of soldiers began to hallucinate:

> *"The Jamestown Weed (Datura stramonium) ... was gather'd very young for a boil'd salad, by some soldiers sent thither, to pacifie the troubles of Bacon; and some of them did eat plentifully of it, the effect of which was a very pleasant comedy; for they turned natural Fools upon it. One would blow up a feather in the air; and another stark naked was sitting up in a Corner like a monkey grinning and making Bows at them...A thousand such simple tricks they play'd, and ... returned to themselves ... not remembering anything that had passed..."[2]*

The hallucinogenic properties of *Datura* became well known and seem to have been exploited by almost all cultures. Many of the native peoples of the Americas, particularly in South America, inhaled the smoke of *Datura* species to achieve a state of euphoria during religious ceremonies [3]. On the Indian subcontinent, *Datura* species were well known to innocent pranksters and to criminals alike [4]. In renaissance Europe, history is replete with tales of political intrigue and scheming helped along by an occasional deadly nightshade poisoning. Even in the late twentieth century the medical literature abounds with case reports of thrill-seeking teenagers who smoked or ingested commercially produced stramonium-containing asthma cigarettes to obtain a "psychedelic" high.

It is plausible that at some time in this history of anticholinergic misuse, an asthma sufferer might have noted that inhalation of *Datura* smoke for whatever intended purpose had the pleasant side-benefit of relieving chest tightness. From such observations, respiratory therapy with *Datura* species found its way into various medical literatures. From the standpoint of Western medicine, the history of anticholinergic bronchodilator therapy began in Madras, India, at the turn of the 18th century [5]. There, the British surgeon William English decided to emulate the Ayurvedic medical practice of stramonium use and encouraged an army officer, General Gent, to use the therapy for his obstructive airway symptoms. The success of such treatments in India resulted in the use of the practice in the United Kingdom where the first reports of successful stramonium treatments were reported enthusiastically.

By the mid-eighteenth century, the place of stramonium was well established in Europe. In Henry Hyde Salter's great *Treatise on Asthma*, the preparation and use of stramonium were discussed in detail [6]. Even at that time, there were concerns about sub-optimal inhalation techniques and about the excipients sometimes added to patent preparations of the day. (The original Ayurvedic antimuscarinic remedies were prepared as dried pastes to which several non-stramonium ingredients had been added. The addition of crushed peppers is intriguing; the addition of arsenic is alarming). By the late eighteenth century, more than one astute clinician had suggested a particular benefit of antimuscarinic bronchodilator therapy for patients with tobacco-related emphysema, an observation now well validated by modern clinical and physiological research [7, 8].

The smoking of atropinic botanicals did not die with the last century nor did the practice disappear with the introduction of adrenergic agents such as ephedrine and epinephrine. Well into the late twentieth century it was possible to purchase without prescription a number of patent remedy "asthma cigarettes" that contained stramonium. As noted above, these were sometimes abused by teenagers in search of a cheap and legal hallucinogen [9–11]. Used for their legitimate purpose, the cigarettes were likely to be very effective. Ervenius and colleagues estimated that a typical asthma cigarette containing stramonium could liberate an homogenous aerosol of one micron droplets delivering up to half a milligram of atropine to the lungs [12]. In later studies, spirometry confirmed that useful bronchodilator effects were measurable and clinically significant [13].

Use by physicians of antimuscarinic bronchodilators decreased markedly in mid-century as adrenergic agents and theophylline became popular. Atropine inhalation was used occasionally for clinical and research purposes but atropine's erratic systemic absorption could lead to occasional problems of toxicity. The modern rediscovery of anticholinergic bronchodilator therapy resulted from the development of modern quaternary derivatives of atropine. There are several such derivatives of which two, ipratropium bromide and oxitropium bromide, are in common clinical use. The former is the best studied of this class of compounds and its pharmacodynamic and clinical properties can be considered representative.

Quaternary anticholinergic compounds

The quaternary anticholinergic compounds are poorly absorbed across biological membranes, so that when they are given by inhalation they produce the desired airway effects with a minimum of systemic toxicity. Ipratropium bromide is most commonly administered as a metered dose inhaler (MDI) or pressurized dry suspension aerosol. It is also widely available as a nebulizer solution. As an MDI, it is most commonly administered in a 20 μg per puff formulation, although higher concentrations are marketed in some countries. In the near future, the shift from chlorofluorocarbon propellants will result in the introduction of novel delivery formulations and devices.

Following inhalation, ipratropium produces bronchodilation more slowly than adrenergic bronchodilators [14]. Approximately 50% maximal effect is seen in 15 min, 80% in 30 min and peak effect is seen between 30 and 60 min following inhalation. The effect subsides gradually over the next 4 to 6 h, so that scheduled treatment is generally given four times daily. Only minute quantities of ipratropium are absorbed and can be measured in the circulation. The original pharmacokinetic studies quantified picogram amounts of circulating drug only after multiple puffs of radiolabelled drug were administered. This minimal systemic absorption accounts for a near absence of non-local side-effects. Ipratropium causes no tremor, a potentially useful property for a small number of patients who are intolerant of the skeletal muscle tremor triggered by β_2 agonist use. Similarly, ipratropium is essentially devoid of cardiovascular impact. One careful echocardiographic study showed that 160 μg of ipratropium (eight MDI puffs of 20 μg per puff) did not perturb systolic or diastolic blood pressure and left cardiac output unchanged [15]. By contrast, 400 μg of fenoterol triggered a 44% increase in cardiac output. In this study there was no evidence of tachycardia following anticholinergic inhalation. Indeed, there was a paradoxical and very small decrease in heart rate (3 beats per minute), an effect sometimes seen with sub-therapeutic doses of parenteral atropine and attributed to a small agonist effect at the muscarinic receptor. The absence of cardiovascular impact makes anticholinergic agents the logical bronchodilator alternative for patients intolerant of adrenergic agents for reasons of cardiovascular disease. There have been rare case reports of urinary retention in elderly men with benign prostatic hypertrophy.

The most commonly reported side-effects are local; drying of the mouth or a metallic taste in the mouth are reported occasionally. Another topical side-effect may be seen if ipratropium is inadvertently sprayed in the eye by an aberrant MDI actuation or, more commonly, by nebulization with an open facemask. Ipratropium introduced into the eye can cause mydriasis and, in predisposed individuals, can cause an abrupt increase in intraocular pressure [16]. This adverse effect is more likely if β_2 agonists are administered concurrently; these agents stimulate the production of aqueous humour. Serious adverse consequences appear to be rare, but the small risk of inadvertent ocular delivery of bronchodilator drugs is an additional argument for delivery of these agents by MDI plus spacer in preference to nebulizers.

Considerable research has addressed the effects of ipratropium and other quaternary anticholinergic compounds on mucociliary clearance and mucus secretion in the lower airways. In brief, ipratropium has negligible effects on ciliary activity, mucus secretion and mucociliary clearance rates. In patients with airways disease, long-term therapy is not associated with measurable changes in mucus composition and mucociliary clearance and there has been no change in the microscopic appearance of airway mucosa [17]. Most clinical studies describe no effect of ipratropium on cough or sputum production although one seven-week clinical study described a small but statistically significant reduction in sputum volume [18]. The near absence of effect on airway sputum is difficult to explain, given the airway drying effect of atropine and, for that matter, the nasal drying effect of ipratropium administered to the upper airway [19].

Uses in asthma – stable asthma

In clinical trials for product registration, ipratropium was most extensively given to patients with asthma and most typically to patients who would be considered to have undertreated and poorly controlled disease by today's more rigorous standards [20–22]. Most available trials have been acute bronchoprotection or acute bronchodilator trials. Amongst patients with relatively mild disease and relatively normal spirometry, ipratropium was found to have potentially useful bronchoprotective effects against a number of specific and non-specific stimuli. This includes blunting of the bronchoconstriction triggered by the inhalation of cold air, methacholine and antigen [23]. Exercise-induced bronchoconstriction is also blunted [24]. However, as compared to β_2 selective agonists, ipratropium and related compounds are less effective. In the setting of exercise, ipratropium's bronchoprotective effect is additive to that of salbutamol.

In asthma patients with persistent disease and airflow obstruction at rest, single-dose comparative studies show that ipratropium improves spirometry but does so more slowly and less completely than β_2 selective agents [25]. However, the combination of the two agents, given together from a single MDI canister or as separate sequential inhalations from two MDIs, can produce greater bronchodilator effect than either monotherapy. The effect is thought to be additive rather than synergistic. When inhaled bronchodilators are given chronically, there is evidence of mild tolerance to the bronchodilator (and bronchoprotective) effects of β_2 agonists but not to the effects of anticholinergics. (This difference is thought to be the consequence of the adrenergic drugs' agonist mechanism of action *versus* the anticholinergic antagonist mechanism of action). As a consequence, the bronchodilator superiority of adrenergic drugs is less marked in chronic administration studies than in short-term or single-administration studies. Rebuck and colleagues, for example, showed that inhaled fenoterol was superior to inhaled ipratropium when given acutely to 22 stable theophylline-treated patients [26]. The combination of these two agents was better than either agent alone. However, after one month taking each of the regimens four times per day, the bronchodilator advantage of the adrenergic agent over the anticholinergic agent was less easily discerned. Throughout the study, a combination of adrenergic and anticholinergic bronchodilators was superior to either monotherapy.

Although administering the combination of an anticholinergic bronchodilator with an adrenergic bronchodilator can produce greater airway opening than administering either agent singly, such a combination strategy is of lesser importance now that asthma management strategies focus on preventive, anti-inflammatory therapy. For most patients with asthma, small amounts of short-acting β_2 agonist are sufficient for as-needed relief of symptoms. For patients who have frequent need for short-acting β_2 agonists, the daily use of a moderate dose of inhaled corticosteroids can produce disease control and a minimum of bronchodilator need. Relatively few patients with moderate to severe asthma require combination drug strategies and for these patients the combination is more likely to be a long-acting β_2 agonist added to an inhaled corticosteroid [27, 28]. Anticholinergic agents are less and less used in the maintenance therapy of stable asthma but are now more commonly used for the treatment of acute asthma.

Anticholinergic bronchodilators have been said to have useful properties under certain clinical circumstances or for certain types of patients. For example, their benefit is said to be greater in older patients, perhaps explaining their preferential use in the treatment of chronic obstructive pulmonary disease [29]. This benefit is attributed to the decreased sensitivity of the adrenergic nervous system that occurs with normal aging. Older patients are said to be less responsive to adrenergic bronchodilators so that, in a relative sense, the importance of antimuscarinic bronchodilators is greater. In a more extreme instance of unresponsive adrenergic receptors, patients with asthma who are inadvertently given a beta-blocker will be refractory to adrenergic bronchodilators. Anticholinergic agents are the bronchodilator treatment of choice in this setting. The anticholinergic bronchodilators have been reported to be of benefit for patients with asthma who suffer nocturnal awakenings. Ipratropium given before bedtime can reduce the frequency of such awakenings, presumably by blunting the increase in vagal tone said to occur during sleep. However, long-acting β_2 agonists, inhaled corticosteroids or both are markedly effective in reducing the frequency of nocturnal awakenings, making this property of antimuscarinic agents redundant. Finally, patients with so-called psychogenic asthma are said to be particularly responsive to ipratropium [30, 31]. It has long been known that bronchoconstriction can be triggered by emotional stimuli in susceptible asthmatic patients. The inhalation of a bronchoconstrictor produces less airway narrowing if patients are informed that they are receiving a bronchodilator by inhalation. Similarly, the converse is true when the bronchodilators are administered. These cortical influences on airway caliber are mediated largely, if not entirely, by the cholinergic vagus pathway. Whether anticholinergic bronchodilators are more effective in this setting than other bronchodilators or treatment strategies is unknown.

Acute asthma

Although most patients with stable asthma are treated so as to minimize their bronchodilator need, acutely ill patients with asthma require aggressive bronchodilator therapy. Thus, the role of anticholinergic bronchodilators remains important in this setting. Early investigations with nebulized ipratropium bromide compared the anticholinergic agent to inhaled salbutamol. Both agents provided bronchodilator benefit and in the small-scale studies the benefit appeared to be similar [32]. However, this misperception of equality between agents was the consequence of Type II error. The benefits of anticholinergic bronchodilator acute asthma appear to replicate the benefits described in the foregoing section on stable asthma. That is, antimuscarinic agents as monotherapy provide less bronchodilator benefit than adrenergic agents. However, when anticholinergic agents and adrenergic agents are combined, the effect of the two drugs is greater than either agent alone. Again, the effect appears to be additive rather than synergistic. One study typical of this benefit would be that reported by Rebuck and colleagues [33]. They described 149 patients with acute asthma treated with one of three nebulizer regimens in randomized double-blind fashion. Patients received either ipratropium bromide 500 µg, fenoterol 1.25 mg or the two drugs in combination. Although significant improvements in FEV_1 were seen with all three regimens, the effect was greatest for the combination, less for the adrenergic agent alone and least for the anticholinergic agent alone. In a finding echoed by subsequent studies, the benefit of combination bronchodilator therapy with inhaled anticholinergic and adrenergic agents appeared to be greatest amongst patients with most severe airflow obstruction at presentation. The effect does not appear to be the consequence of under-dosing with either single agent therapy. Bryant reported that when patients with acute asthma received initial therapy with fenoterol, they benefited more from subsequent anticholinergic bronchodilator than from additional doses of fenoterol [34]. Similar findings have been reported for children suffering acute asthma. Beck and colleagues, for example, reported that children receiving maximal doses of salbutamol, but continuing to suffer from severe bronchoconstriction, benefited from the inhalation of ipratropium [35].

Not all studies have been able to confirm bronchodilator benefit of adding inhaled ipratropium or another anticholinergic agent to inhaled adrenergic bronchodilators in the setting of acute asthma. Fitzgerald and colleagues reported that the combination of ipratropium and salbutamol could not be shown to be clinically superior to salbutamol alone in the treatment of acutely ill adults with asthma [36]. These investigators did note small trends towards greater bronchodilation and fewer hospitalizations among patients randomized to combination bronchodilator therapy rather than salbutamol alone. However, these trends were not statistically significant. Perhaps the definitive answer to the question of whether combination anticholinergic plus adrenergic bronchodilator therapy is superior to adrenergic therapy alone was provided by a pooled analysis of studies by Lanes and colleagues [37]. These investigators combined the results of three randomized double-blind clinical trials undertaken in the United States, Canada and New Zealand to assess emergency room nebulizer therapy of acute asthma. All studies compared nebulized salbutamol in a dosage of 2.5 mg to the combination of 0.5 mg of ipratropium bromide plus 2.5 mg of salbutamol. Bronchodilator outcomes were measured at 45 and 90 minutes following nebulization and patients were then followed for up to 48 hours after discharge to learn the outcome of their asthma exacerbation and its acute treatment. A total of 1,064 patients were randomized and 90 percent of patients were available for outcome measurement at 90 minutes. There was a significantly better FEV_1 response amongst patients receiving combination therapy, although this benefit was small and was estimated at 43 ml. Despite this small bronchodilator benefit, patients receiving combination therapy had a lower risk for three adverse clinical outcomes: the need for additional treatment, the risk of exacerbation and a risk of hospitalization. Thus, combination bronchodilator therapy appears to offer additional clinical benefit beyond simply improved spirometric outcomes. Similar results have been reported in a paediatric population. Schuh and colleagues undertook a randomized, double-blind and placebo-controlled trial in one hundred twenty children aged 5 to 17 years who were brought to the emergency room for treatment of acute asthma [38]. All patients were treated with nebulized salbutamol at initial presentation. One group of children received three administrations of nebulized ipratropium bromide (250 μg per dose); another group received one administration of ipratropium bromide (250 μg); a third group received no ipratropium. Spirometry improved most in the group of children receiving three doses of ipratropium, slightly less amongst those receiving one dose of ipratropium and least among those receiving no ipratropium. The spirometric benefits of added antimuscarinic bronchodilator therapy were most evident amongst patients whose baseline FEV_1 was less than or equal to 30 percent of the predicted value. Similar to the adults described by Lanes and colleagues, the children receiving ipratropium had a lesser risk of hospitalization than children who did not receive ipratropium. As a consequence of such findings, consensus statements on the treatment of acute asthma have recommended the addition of ipratropium to adrenergic bronchodilator amongst patients with more severe episodes of disease [39].

Chronic obstructive pulmonary disease

International guidelines for the treatment of chronic obstructive pulmonary disease increasingly recognize the advantages of anticholinergic bronchodilator therapy for these diseases [40, 41]. The treatment of tobacco-related chronic obstructive pulmonary disease is often frustrating to clinicians. Patients demonstrate little acute bronchodilator responsiveness and typically demonstrate minimal corticosteroid responsiveness [42]. The bronchodilator superiority of anticholinergic agents over adrenergic agents in older patients with COPD is modest and some investigators have suggested that aggressive dosing with inhaled adrenergic agents can produce the same spirometric benefit as can be achieved with anticholinergic agents [43]. Typical acute bronchodilator comparisons report greater spirometric response to two MDI puffs of ipratropium than 2 two MDI puffs of salbutamol [44]. Similarly, ipratropium produces greater bronchodilator benefit than oral theophylline [45]. However, acute bronchodilator comparisons are not a realistic model for actual clinical use. COPD patients typically use their inhaled bronchodilators on a regular basis over long periods of time. When clinical trials study the effects of repeated dosing, the bronchodilator superiority of anticholinergic agents is demonstrated more clearly as tolerance develops to the administration of adrenergic agents.

The regular and frequent use of β_2 agonist bronchodilators in asthma has alarmed a number of observers who have noted the association between the frequency of β_2 agonist administrations and the risk of asthma death [46]. Most now regard this association as merely that and consider the frequency of β_2 agonist use to be simply a useful marker of asthma control. Nonetheless, there is some laboratory evidence that the regular use of short-acting β_2 agonists in mild asthma can cause increases in specific and non-specific bronchial hyperreactivity [47, 48]. Finally, there have been reports that the long term use of bronchodilators, either adrenergic or anticholinergic, can lead to an accelerated decline in lung function amongst patients with asthma or COPD [49]. Several of these reports arose from repeated analyses of a single data set with results that were inconsistent from analysis to analysis [50–52]. Several completed studies (as described below) now reassure that long-term regular ipratropium administration does not accelerate lung function decline in COPD.

An intriguing retrospective analysis of clinical trials was reported by Rennard and colleagues [53]. These investigators described changes in baseline or pre-bronchodilator spirometry amongst patients enrolled in seven different but comparable clinical trials comparing ipratropium to β_2 agonist over a three-month treatment period. Long-term therapy with ipratropium improved pre-bronchodilator endpoints modestly. On average, baseline FEV_1 increased 28 ml while baseline vital capacity increased 131 ml. In contrast, three months of β_2 agonist therapy did not alter baseline spirometry. These improvements in baseline spirometry, coupled with a lack of tolerance to the bronchodilator effects of the antimuscarinic agent, meant that chronic bronchodilator benefit was clearly superior with the anticholinergic agent as compared to the adrenergic agent. A longer period of ipratropium treatment was tracked in the Lung Health Study. The Lung Health Study sought to determine whether early intervention with an effective bronchodilator would slow the rate of decline in lung function in early COPD. Cigarette smokers with mild spirometric abnormalities of obstruction were recruited to a five-year trial in which they would be randomized to one of three groups: a smoking cessation intervention plus three times per day ipratropium by MDI; a smoking cessation intervention with placebo inhaler; or no intervention. Ten North American centres studied 5,887 men and women aged 35 to 60 years. Smoking cessation was encouraged by behavior modification and nicotine replacement therapy. FEV_1 decline was significantly less in the two intervention groups, a difference attributable to smoking cessation for those cigarette smokers who were able to abstain. Ipratropium produced an acute bronchodilator benefit but there was no subsequent change in the rate of lung function decline with continued ipratropium use. These findings appear to be in contrast to the three-month improvements in baseline spirometry reported by Rennard and colleagues in a more severely obstructed patient population. However, the Lung Health Study does reassure that there is no acceleration in lung function decline with regular anticholinergic bronchodilator administration.

Trials of therapy in the obstructive lung disease typically use spirometry as the primary endpoint, particularly for pivotal product registration studies. However, in the setting of COPD, acute spirometric changes are minimal and there is no evidence that chronic bronchodilator administration alters long-term spirometric progression of the disease. Of greater consequence to patients are changes in symptoms, quality of life and exercise tolerance. In general, ipratropium bromide therapy or oxitropium therapy appears to lead to improved quality of life and decreased dyspnea, although changes in exercise tolerance are more difficult to demonstrate [54–56].

Although anticholinergic monotherapy may be superior to adrenergic monotherapy for patients with COPD, the marked dyspnea and disability of many COPD sufferers must prompt the obvious consideration that combination bronchodilator therapy may be best. Combination bronchodilator therapy is increasingly practical with the growing availability of combination MDIs and pre-mixed, combination nebulized solutions. A variety of studies have shown that the combination of ipratropium plus salbutamol or fenoterol produces greater acute bronchodilator benefit than either ipratropium or the adrenergic agent alone [30, 57]. Of particular interest is the chronic administration study of combination inhaled therapy reported by Petty. In a three-month trial comparing ipratropium monotherapy, salbutamol monotherapy and the combination inhaled regularly in MDI format by patients with COPD, the spirometric benefits were greatest amongst patients using the combination therapy. Moreover, the patients inhaling this combination therapy were least likely to suffer an exacerbation requiring a burst of systemic corticosteroids. Somewhat more corticosteroid therapy was given to patients in the anticholinergic monotherapy group and the greatest amount of corticosteroid was given to patients in the adrenergic monotherapy group. Thus, spirometric changes appear to be paralleled by differences in clinical outcomes. Indeed, Friedman and colleagues have analyzed the cost-effectiveness of combination ipratropium and salbutamol *versus* either agent alone as the inhaled bronchodilator therapy of COPD [58]. A total of 1,067 patients were randomized to receive ipratropium alone (two puffs of 20 µg/ puff) four times daily, salbutamol alone (two puffs of 100 µg/puff) four times daily or the combination (two puffs of 20 µg ipratropium and 100 µg salbutamol per puff) four times daily. Spirometric outcomes were best in the combination therapy group. More important, compared with salbutamol, patients receiving ipratropium and ipratropium plus salbutamol experienced significantly fewer COPD exacerbations and patient-days of exacerbation. The increased frequency of exacerbations observed in the salbutamol group was associated with a significant increase in the number of patient-hospital-days and antibiotic and corticosteroid use. As a result, the total cost of treatment over the study period was significantly less for ipratropium ($ 156 per patient) and ipratropium plus salbutamol ($ 197 per patient) than for salbutamol ($ 269 per patient),

Anticholinergic agents in exacerbations of COPD

One might expect that the response to bronchodilator therapy in exacerbations of COPD would parallel responses in stable COPD or in exacerbations of asthma. That is, one would expect better responses to a combination of anticholinergic and adrenergic bronchodilators than to either agent used alone. However, no study of acute COPD exacerbation has demonstrated such a benefit. Typical of the few available studies would be that by Rebuck and colleagues [33]. The investigators studied the emergency room bronchodilator treatment of acute exacerbations of obstructive lung disease. In 51 patients with exacerbations of COPD, nebulized ipratropium (500 µg) was an effective bronchodilator but was no more effective than fenoterol (1.25 mg). The combination of both agents was not better than either single-agent treatment. Backman and Hellstrom studied 40 patients hospitalized with an exacerbation of COPD and randomized to fenoterol (0.5 mg three times daily) or ipratropium (0.2 mg three times daily) [59]. Although they could demonstrate small spirometric improvements over one week, they could not show superiority of one regimen over another. The only study to suggest a benefit of adding anticholinergic therapy to adrenergic therapy in COPD exacerbation was the report by Shrestha and colleagues [60]. They administered isoetharine by nebulizer hourly (5.0 mg per administration) to patients with COPD as initial and continuing treatment. In addition, they assigned the patients at random to receive ipratropium or placebo MDI puffs (60 µg in the first hour, 40 µg in the second hour and 40 µg in the fourth hour).

Although there were no spirometric differences between groups, the group receiving combination therapy had a shorter stay in the emergency department.

The difficulty in determining the optimal bronchodilator regimen for the treatment of COPD exacerbations is the nature of the exacerbations themselves. It is difficult to define an exacerbation and in practical terms it means a patient-perceived worsening of dyspnea or cough. Although such events are commonly treated as respiratory tract infections accompanied by worsening airflow obstruction, there are many potential non-respiratory explanations for any COPD patient's symptomatic deterioration. Even for exacerbations that are primarily respiratory in nature, the deterioration in lung function may be complicated by respiratory muscle fatigue, metabolic disturbances, acidosis and hypoxemia. In short, it may be unrealistic to hope that any one bronchodilator regimen will show measurable superiority in a setting where spirometric responsiveness is poor. The lack of perceptible benefit of combination therapy for exacerbations has led to differing recommendations. Some have argued for the use of anticholinergic therapy alone, given that it is least likely to cause side-effects and will be less expensive than combination regimens. Others have argued that combination therapy is most likely to be optimal in the stable state and should be used during exacerbations until the stable state is achieved. Given the diagnostic uncertainty of the acute setting and the demonstrable superiority of combination regimens in acute asthma, it may be safest to offer the combination to all acutely ill patients with obstructive processes.

The future of anticholinergic bronchodilator therapy

Two major developments are likely to alter our use of antimuscarinic bronchodilator therapy in the future. First, newer anticholinergic agents may be selective for muscarinic receptor subtypes as described elsewhere in this monograph. Second, the introduction of longer-acting agents could increase our use of these agents in both COPD and asthma.

One agent, tiotropium, seems to offer both receptor subtype selectivity and a long duration of bronchodilator and bronchoprotective action. It has undergone extensive clinical study to assess its potential benefit in the treatment of COPD. Tiotropium bromide is a quaternary ammonium compound similar in structure to ipratropium bromide. Like ipratropium, tiotropium binds non-selectively to M_1, M_2 and M_3 receptors. Tiotropium is five times more potent than either ipratropium or atropine in inhibiting nerve-induced contraction of guinea pig trachea and dissociates much more slowly from the receptors to which it is bound. It may, however, dissociate more rapidly from M_2 receptors than from either M_1 or M_3 receptors [61]. Thus, it may in this fashion have a degree of functional receptor selectivity. In clinical trials of patients with COPD, the inhalation of 10 to 80 µg of tiotropium produces rapid bronchodilation that is maximal between 1 in 4 h after inhalation [62]. After a single dose, significant bronchodilator benefit extends to 32 h. Such a long duration of action could allow once-daily dosing. Tiotropium could also be used in patients with asthma. In a study of 12 patients with asthma, O'Connor compared bronchoprotective effects of three doses of tiotropium (10, 40 and 80 μg) to placebo [63]. Each dose of tiotropium produced mild bronchodilation such that FEV_1 increased between 6 and 11% from baseline. These bronchodilator benefits were sustained for 24 h. There was significant and dose-dependent protection against methacholine challenge appearing as early as two hours following tiotropium administration.

Conclusion

For centuries, antimuscarinic bronchodilators have been inhaled for the treatment of obstructive lung disease. In the modern era, crude botanical preparations have given way to safe and effective quaternary ammonium compounds that produce bronchodilator benefit without risk of significant systemic side effects. The best studied of the anticholinergic bronchodilators, ipratropium bromide, is the single-agent bronchodilator of choice in the treatment of COPD. In acute administration it produces the greatest bronchodilator effect of any single agent. In chronic administration studies, ipratropium's spirometric benefits are unblunted by regular administration. These spirometric benefits are paralleled by improvements in quality of life and symptom scores. When ipratropium is combined with adrenergic agents in the chronic therapy of COPD, patients have enjoyed further spirometric benefit beyond single-agent therapy. Such spirometric benefits seem to be paralleled by a decrease in exacerbation rate and thus in the personal and societal costs of COPD. In the treatment of asthma, anticholinergic agents are used infrequently in the stable ambulatory patient. However, in the setting of acute asthma, anticholinergic bronchodilators can be added to adrenergic bronchodilators to produce more rapid improvements in flow rate. Moreover, such spirometric benefits are accompanied by decreased hospitalization rates, a finding seen in both adult and paediatric studies.

References

1   Plaitakis A, Duvoisin RC (1983) Homer's moly identified as *Galanthus nivalis* L.: physiologic antidote to stramonium poisoning. *Clin Neuropharmacol* 6: 1–5
2   Johnson CE (1967) Mystical force of the nightshade. *Int J Neuropsychiatry* 3: 268–275
3   Schultes RE (1983) Peruvian and Chilean psychoactive plants mentioned in Ruiz's Relacion (1777–1788). *J Psychoactive Drugs* 15: 303–312
4   Guerra F (1974) Sex and drugs in the 16th century. *Br J Addict* 69: 269–289

5   Gandevia B (1975) Historical review of the use of parasympatholytic agents in the treatment of respiratory disorders. *Postgrad Med J* 51: 13–20

6   Salter HH (1864) Treatment of the asthmatic paroxysm (continued) – treatment by stimulants. In: *On asthma: its pathology and treatment*. Blanchard and Lea, Philadelphia, 134–144

7   Fothergill JM (1882) *Chronic bronchitis: its form and treatment*. Bailleire Tindall & Co., London

8   Waters ATM (1862) *Researches on the nature, pathology and treatment of emphysema of the lungs*. Churchill, London

9   Ballantyne A, Lippiet P, Park J (1976) Herbal cigarettes for kicks. *BMJ* 2: 1539–1540

10   Goldsmith SR, Frank I, Ungerleider JT (1968) Poisoning from the ingestion of a stramonium-belladonna mixture: flower power gone sour. *JAMA* 204: 169–170

11   Gowdy JM (1972) Stramonium intoxication: review of symptomatology in 212 cases. *JAMA* 221: 585–587

12   Ervenius O, Holmstedt B, Wallen O (1958) Atropin- und Stramoniumzigaretten. *Naunyn-Schmiedebergs Archiv Pharmakol Exper Pathol* 234: 343

13   Trechsel K, Bachofen H, Scherrer M (1973) Die bronchodilatorische Wirkung der Asthmazigarette. *Schweiz Med Wochenschr* 103: 415–418

14   Pakes GE, Brogden RN, Heel RC, Speight TM, Avery GS (1980) Ipratropium bromide: a review of its pharmacological properties and therapeutic efficacy in asthma and chronic bronchitis. *Drugs* 20: 237–266

15   Chapman KR, Smith DL, Rebuck AS, Leenen FH (1985) Hemodynamic effects of inhaled ipratropium bromide, alone and combined with an inhaled beta 2-agonist. *Am Rev Respir Dis* 132: 845–847

16   Reuser T, Flanagan DW, Borland C, Bannerjee DK (1992) Acute angle closure glaucoma occurring after nebulized bronchodilator treatment with ipratropium bromide and salbutamol. *J R Soc Med* 85: 499–500

17   Gross NJ (1988) Ipratropium bromide. *N Engl J Med* 319: 486–494

18   Ghafouri MA, Patil KD, Kass I (1984) Sputum changes associated with the use of ipratropium bromide. *Chest* 86: 387–393

19   Bronsky EA, Druce H, Findlay SR, Hampel FC, Kaiser H, Ratner P, Valentine MD, Wood CC (1995) A clinical trial of ipratropium bromide nasal spray in patients with perennial nonallergic rhinitis. *J Allergy Clin Immunol* 95 (Suppl): 1117–1122

20   British Thoracic Society (1997) The British guidelines on asthma management 1995 review and position statement. *Thorax* 52: S1–S21

21   Ernst P, Fitzgerald JM, Spier S (1996) Canadian asthma consensus conference: Summary of recommendations. *Can Respir J* 3: 89–100

22   Sheffer AL, Bousquet J, Busse WW, Clark TJH, Dahl R, Evans D, Fabbri LM, Hargreave FE, Holgate ST, Magnussen H et al (1992) *International consensus report on diagnosis and management of asthma*. Bethesda: U.S. Department of Health and Human Services, Publication #92-3091

23   Gross NJ, Skorodin MS (1984) Anticholinergic, antimuscarinic bronchodilators. *Am Rev Respir Dis* 129: 856–870

24   Magnussen H, Nowak D, Wiebicke W (1992) Effect of inhaled ipratropium bromide on the airway response to methacholine, histamine and exercise in patients with mild bronchial asthma. *Respiration* 59: 42–47

25   Ruffin RE, Fitzgerald JD, Rebuck AS (1977) A comparison of the bronchodilator activity of Sch 1000 and salbutamol. *J Allergy Clin Immunol* 59: 136–141

26   Rebuck AS, Gent M, Chapman KR (1983) Anticholinergic and sympathomimetic combination therapy of asthma. *J Allergy Clin Immunol* 71: 317–323

27   Greening AP, Ind PW, Northfield M, Shaw G (1994) Added salmeterol versus higher-dose corticosteroid in asthma patients with symptoms on existing inhaled corticosteroid. *Lancet* 344: 219–224

28   Pauwels RA, Löfdahl CG, Postma DS, Tattersfield AE, O'Byrne P, Barnes PJ, Ullman A (1997) Effect of inhaled formoterol and budesonide on exacerbations of asthma. *N Engl J Med* 337: 1405–1411

29   Ullah MI, Newman GB, Saunders KB (1981) Influence of age on response to ipratropium and salbutamol in asthma. *Thorax* 36: 523–529

30   Ingram RH, Jr. (1996) Bronchodilating effects of combined therapy with clinical dosages of ipratropium bromide and salbutamol for stable COPD. *Chest* 109: 293–293

31   Rebuck AS, Marcus HI (1979) SCH 1000 in psychogenic asthma. *Scand J Resp Dis* (Suppl) 103: 186–191

32   Leahy BC, Gomm SA, Allen SC (1983) Comparison of nebulized salbutamol with nebulized ipratropium bromide in acute asthma. *Br J Dis Chest* 77: 159–163

33   Rebuck AS, Chapman KR, Abboud R, Pare PD, Kreisman H, Wolkove N, Vickerson F (1987) Nebulized anticholinergic and sympathomimetic treatment of asthma and chronic obstructive airways disease in the emergency room. *Am J Med* 82: 59–64

34   Bryant DH (1985) Nebulized ipratropium bromide in the treatment of acute asthma. *Chest* 88: 24–29

35   Beck R, Robertson C, Galdes-Sebaldt M, Levison H (1985) Combined salbutamol and ipratropium bromide by inhalation in the treatment of severe acute asthma. *J Pediatr* 107: 605–608

36   Fitzgerald JM, Grunfeld A, Pare PD, Levy RD, Newhouse MT, Hodder R, Chapman KR (1997) The clinical efficacy of combination nebulized anticholinergic and adrenergic bronchodilators vs nebulized adrenergic bronchodilator alone in acute asthma. Canadian Combivent Study Group. *Chest* 111: 311–315

37   Lanes SF, Garret JE, Wentworth CE III, Fitzgerald JM, Karpel JP (1998) The effect of adding ipratropium bromide to salbutamol in the treatment of acute asthma – A pooled analysis of three trials. *Chest* 114: 365–372

38   Schuh S, Johnson DW, Callahan S, Canny G, Levison H (1995) Efficacy of frequent nebulized ipratropium bromide added to frequent high-dose albuterol therapy in severe childhood asthma. *J Pediatr* 126: 639–645

39   Beveridge RC, Grunfeld AF, Hodder RV, Verbeek PR (1996) Guidelines for the emergency management of asthma in adults. CAEP/CTS Asthma Advisory Committee. Canadian Association of Emergency Physicians and the Canadian Thoracic Society. *Can Med Assoc J* 155: 25–37

40   Chapman KR, Bowie DM, Goldstein RS, Hodder RV, Julien M, Kesten S, Newhouse MT, Pare PD (1992) Guidelines for the assessment and management of chronic obstructive pulmonary disease. Canadian Thoracic Society Workshop Group. *Can Med Assoc J* 147: 420–428

41   Celli BR, Snider GL, Heffner J, Tiep B, Ziment I, Make B, Braman D, Olsen G, Phillips Y, Stoller J et al (1995) Standards for the diagnosis and care of patients with chronic obstructive pulmonary disease. *Am J Respir Crit Care Med* 152: S77–S121

42   Callaghan CM, Dittus RS, Katz BP (1991) Oral corticosteroid therapy for patients with stable chronic obstructive pulmonary disease. *Ann Intern Med* 114: 216–223.

43   Easton PA, Jadue C, Dhingra S, Anthonisen NR (1986) A comparison of the bronchodilating effects of a beta-2 adrenergic agent (albuterol) and an anticholinergic agent (ipratropium bromide), given by aerosol alone or in sequence. *N Engl J Med* 315: 735–739

44   Braun SR, McKenzie WN, Copeland C, Knight L, Ellersieck M (1989) A comparison of the effect of ipratropium and albuterol in the treatment of chronic obstructive airway disease. *Arch Intern Med* 149: 544–547

45   Bleecker ER, Johns M, Britt EJ (1988) Greater bronchodilator effects of ipratropium compared to theophylline in chronic airflow obstruction. *Chest* 94 (Suppl 1): 3S

46   Spitzer W, Suissa S, Ernst P, Horwitz RI, Habbick B, Cockcroft D, Boivin J-F, McNutt M, Buist AS, Rebuck AS (1993) The use of beta-agonists and the risk of death and near death from asthma. *N Engl J Med* 326: 501–506

47   Vathenen AS, Knox AJ, Higgins BG, Britton JR, Tattersfield AE (1988) Rebound increase in bronchial responsiveness after treatment with inhaled terbutaline. *Lancet* 1: 554–558

48   Bhagat R, Swystun VA, Cockcroft DW (1996) Salbutamol-induced increased airway responsiveness to allergen and reduced protection versus methacholine: Dose response. *J Allergy Clin Immunol* 97: 47–52

49   Van Schayck CP, Dompeling E, Van Herwaarden CL, Folgering H, Verbeek AL, van der Hoogen HJ, Van Weel C (1991) Bronchodilator treatment in moderate asthma or chronic. *BMJ* 303: 1426–1431

50   Van Schayck CP, Dompeling E, Van Weel C, Akkermans RP (1992) Bronchodilator treatment in asthma: Continuous or on demand? Reply. *BMJ* 304: 504–504

51   Van Schayck CP, Rutten-van Molken MP, van Doorslaer EK, Folgering H, Van Weel C (1992) Two-year bronchodilator treatment in patients with mild. *Chest* 102: 1384–1391

52   Van Schayck CP, Graafsma SJ, Visch MB, Dompeling E, Van Weel C, Van Herwaarden CL (1990) Increased bronchial hyperresponsiveness after inhaling salbutamol during 1 year is not caused by subsensitization to salbutamol. *J Allergy Clin Immunol* 86: 793–800

53 Rennard SI, Serby CW, Ghafouri M, Johnson PA, Friedman M (1996) Extended therapy with ipratropium is associated with improved lung function in patients with COPD – A retrospective analysis of data from seven clinical trials. *Chest* 110: 62–70
54 Colice GL (1996) Nebulized bronchodilators for outpatient management of stable chronic obstructive pulmonary disease. *Am J Med* 100: 11S–18S
55 Ikeda A, Nishimura K, Koyama H, Tsukino M, Mishima M, Izumi T (1996) Dose response study of ipratropium bromide aerosol on maximum exercise performance in stable patients with chronic obstructive pulmonary disease. *Thorax* 51: 48–53
56 Tsukino M, Nishimura K, Ikeda A, Hajiro T, Koyama H, Izumi T (1998) Effects of theophylline and ipratropium bromide on exercise performance in patients with stable chronic obstructive pulmonary disease. *Thorax* 53: 269–273
57 Ikeda A, Nishimura K, Izumi T (1996) Bronchodilating effects of combined therapy with clinical dosages of ipratropium bromide and salbutamol for stable COPD. *Chest* 109: 294–294
58 Friedman M, Serby CW, Menjoge SS, Wilson JD, Hilleman DE, Witek TJ, Jr. (1999) Pharmacoeconomic evaluation of a combination of ipratropium plus albuterol compared with ipratropium alone and albuterol alone in COPD. *Chest* 115: 635–641
59 Backman R, Hellstrom PE (1985) Fenoterol and ipratropium bromide for treatment of asthma and chronic bronchitis. *Curr Ther Res* 38: 135–140
60 Shrestha M, O'Brien T, Haddox R, Gourlay HS, Reed G (1991) Decreased duration of emergency department treatment of chronic obstructive pulmonary disease exacerbations with the addition of ipratropium bromide to beta-agonist therapy. *Ann Emerg Med* 20: 1206–1209
61 Takahashi T, Belvisi MG, Patel H, Ward JK, Tadjkarimi S, Yacoub MH, Barnes PJ (1994) Effect of Ba 679 BR, a novel long-acting anticholinergic agent, on cholinergic neurotransmission in guinea pig and human airways. *Am J Respir Crit Care Med* 150: 1640–1645
62 Maesen FP, Smeets JJ, Sledsens TJ, Wald FD, Cornelissen PJ (1995) Tiotropium bromide, a new long-acting antimuscarinic bronchodilator: a pharmacodynamic study in patients with chronic obstructive pulmonary disease (COPD). Dutch Study Group. *Eur Respir J* 8: 1506–1513
63 O'Connor BJ, Towse LJ, Barnes PJ (1996) Prolonged effect of tiotropium bromide on methacholine-induced bronchoconstriction in asthma. *Am J Respir Crit Care Med* 154: 876–880

Novel perspectives in anticholinergic therapy

Bernd Disse

Boehringer Ingelheim Clinical Research Institute, Respiratory Medicine, Boehringer Ingelheim GmbH, D-55216 Ingelheim/Rhein, Germany

Introduction

Muscarinic receptors are widely distributed throughout the body and, when activated, serve a variety of important regulatory functions [1, 2]. Considering their physiological importance, comparatively few muscarinic agonists or antagonists constitute part of established pharmacotherapy.

Muscarinic antagonists serve in bronchodilation in obstructive lung disease, premedication in anaesthesia, induction of mydriasis or cycloplegia in ophthalmology, treatment of exocrine gland hypersecretion in, e.g., allergic and non-allergic rhinitis, relaxation of gastro-intestinal spasms, treatment of urinary urge incontinence, duodenal ulcer and bradycardia. Stimulatory activity of the central nervous system (CNS) by direct or indirect muscarinic agonism is a therapeutic opportunity in dementia. The limited range of indications so far may be explained by the difficulty of selectively influencing an organ or body function without inducing the variety of muscarinic or anti-muscarinic effects possible.

The following effects of muscarinic stimulation can be seen after systemic administration of choline esters or the alkaloid pilocarpine:

- vasodilation, with injured endothelium vasoconstriction,
- depression of sino-atrial and atrio-ventricular rate of conduction and of atrial and ventricular force of contraction,
- release of catecholamines from the adrenal medulla and activation of sympathetic ganglia,
- enhanced secretory activity in the gastrointestinal system, including gallbladder and biliary ducts, and increases in tone and peristaltic activity,
- increase in ureteral peristalsis in the urinary tract, contraction of the detrusor muscle of the urinary bladder and relaxation of the sphincters,
- sweating and salivation,
- miosis and spasm of accommodation,
- submucosal gland secretion and bronchospasm in the lungs,
- a cortical arousal reaction in the CNS.

Muscarinic Receptors in Airways Diseases, edited by Johan Zaagsma, Herman Meurs and Ad F. Roffel
© 2001 Birkhäuser Verlag Basel/Switzerland

Antimuscarinic effects are classically described following moderate systemic doses of atropine, as this alkaloid is a very potent and specific antimuscarinic agent with no selectivity for the subtypes of muscarinic receptors and, in addition, high oral absorption and penetration into the CNS. The spectrum of pharmacological activities results from inhibition of the above-mentioned muscarinic effects, the severity and sequence depending on physiological cholinergic tone, muscarinic receptor density, transduction efficacy and junctional auto-receptors inhibiting acetylcholine release.

Following systemic atropine administration:

– bronchodilation is a sensitive early event,
– followed by dry mouth and inhibition of gastric and duodenal secretion,
– bradycardia may occur at low threshold doses as well as inhibition of sweating,
– tachycardia/palpitations and more pronounced dryness of mucus membranes at higher doses, mydriasis and accommodation disturbances,
– inhibition of intestinal motility, biliary and urinary retention,
– at very high doses dry hot skin, restlessness, hallucinations and delirium.

Atropine and the related alkaloid scopolamine have been used in carefully titrated doses for many of the indications mentioned above. The high incidence of unwanted side-effects triggered the quest for more selective antimuscarinics.

Inhaled ipratropium bromide, the standard for comparison in anticholinergic therapy of obstructive airways disease

Oral, nasal and lung absorption of quaternary anticholinergics

A breakthrough for anticholinergic bronchodilators was the synthesis of congeners of the naturally occurring alkaloids atropine and scopolamine, in which the tertiary nitrogen of the tropine or scopine moiety is quaternized to an ammonium salt. The charged molecules remain specific antimuscarinics. They retain anticholinergic potency or even show an increase, favourably combined with low oral absorption and no penetration of the blood-brain barrier as reviewed previously [3–5].

Table 1 demonstrates a similar sequence of sensitivity for antimuscarinic effects in the dog following *systemic* administration of ipratropium or atropine: bronchoprotective effects at the lowest dose, then inhibition of salivation; at higher doses, increase in heart rate. Following *oral* administration, these effects need 60-fold higher doses of ipratropium compared to only 8-fold higher for atropine. Following *inhalation*, the ratio of inhibition of salivation to bronchoprotective effect increases to 300 for ipratropium and 90 for atropine.

Table 1 - From atropine to ipratropium bromide: pharmacological studies in dogs

		Ipratropium		Atropine	
		ED_{50}[1]	(f)[2]	ED_{50}[1]	(f)[2]
i.v.[1]	*Inhibition of bronchospasm (ACh)*	*0.15*		*0.22*	
i.v.[1]	Inhibition of salivation (pilocarpine stimulated)	1.5	(10)	2.5	(11)
s.c.[1]	Heart rate increase	11	(73)	16	(73)
p.o.[1]	Heart rate increase	640		135	
	{ratio dose p.o/dose s.c.}	{58}		{8.4}	
		EC_{50}[3]	(f)[2]	EC_{50}[3]	(f)[2]
Inhal.[3]	*Inhibition of bronchospasm (ACh)*	*0.14*		*0.32*	
Inhal.[3]	Inhibition of salivation (pilocarpine stimulated)	40	(290)	30	(94)
Inhal.[3]	Heart rate increase (EC 9–40%)	30	(210)	10	(31)

[1]ED_{50} *(µg/kg) i.v., s.c., p.o.;* [2]*ratio of dose of extrapulmonary effect/dose of bronchoprotective effect;* [3]EC_{50} *(mg/ml) of nebulized inhaled solution; ACh, acetylcholine; data from [6, 7]*

Human pharmacokinetics explain these results by showing an oral bioavailability of only 2% for ipratropium [8]. The apparent systemic availability of ipratropium from the nasal cavity with intranasal administration or following inhalation is higher: 7.8% of an intranasal dose and 6.9% of an inhaled dose, as determined by comparison of unchanged drug urinary excretion following systemic and nasal or lung administration, respectively [8, 9]. This systemically available fraction in both cases results from the fact that only a fraction of the nominal dose of about 5 to 30% can be topically deposited in the nasal cavity or in the lungs (in this case the inhalable fraction or fine particle fraction of the aerosol dose), which then shows appreciable absorption from the nasal cavity or the peripheral lung, respectively. The swallowed major part of the nominal dose contributes little because of the low oral absorption.

Efficacy of ipratropium as a bronchodilator

In COPD, inhaled ipratropium (Atrovent®) showed superior or comparable efficacy as a bronchodilator compared to β_2-agonists [10–12]. Ipratropium bronchodilates or protects from cholinergic stimuli for up to 6 hours [12, 13]. Ipratropium has an excellent safety record from controlled clinical trials. The lung health study documents up to 5 years of ipratropium treatment of 1961 of the 5887 participants [14]. In these mild "patients" or symptomatic smokers, ipratropium showed small but sustained bronchodilation, reversing on withdrawal of the drug. No safety limitations were identified over this 5-year period of controlled observation. In a meta-analysis of 1445 patients with COPD, Rennard and colleagues [15] described a potential advantage of ipratropium over the β_2-agonist: with ipratropium there was an increase of 28 ml in FEV_1 or 131 ml in FVC from the pre-study baseline to the three-month pre-bronchodilator value (trough level, 12 h after last administration), whereas a small loss of lung volumes was seen with β_2-agonist treatment. The effect of ipratropium on the baseline must be explained by mechanisms other than by acute bronchodilation.

Is there an influence of quaternary anticholinergics on mucus secretion and mucociliary clearance of the lungs?

Mucus secretion from airway submucosal glands is at least partly regulated by M_3 and M_1-receptor activation [16–19] (see also the chapter by Rogers in this volume). However, results of studies on a potential influence of antimuscarinics on the bronchial secretions of patients with chronic bronchitis are ambiguous: inhaled (1200–2400 µg BID) as well as intramuscular (600–1200 µg) atropine had no effect on sputum volume and sputum viscosity in patients with COPD in a study by Lopez-Vidriero and colleagues [20]. However, in the same study oral treatment with 600 µg QID for 5 weeks reduced the sputum volume. Taylor and colleagues [21] did not see an influence on sputum volume after ipratropium 200 µg TID for 4 weeks, nor did Pavia and colleagues [22] using oxitropium at 200 µg TID. Ghafouri and colleagues [23] reported a reduction of sputum volume after 7 weeks of inhaled ipratropium at 40 µg QID and Tamaoki and colleagues [24] reported a sputum volume reduction by 1/3 after 7 weeks of an 8-week trial with oxitropium 200 µg TID. Interestingly, in this study, there was a progressive decline of sputum volume starting with week 3 up to week 7 of treatment with oxitropium (Fig. 1). The slow decline with time may indicate that acute blockade of cholinergic-induced secretion is not likely to explain this effect.

Muscarinic agonists like acetylcholine stimulate the beat frequency of airway ciliated cells above their baseline frequency; *in vitro* this effect can be blocked by antimuscarinics [25, 26]. *In vivo* in humans lung mucociliary clearance is preferably measured as the overall effect of ciliary activity and secretion. The contributions of the individual components cannot be easily assessed. By inhalation as well as following oral, i.m. or i.v. administration (400–600 µg), atropine and its active enantiomer hyoscine have been shown to acutely reduce the mucociliary clearance in

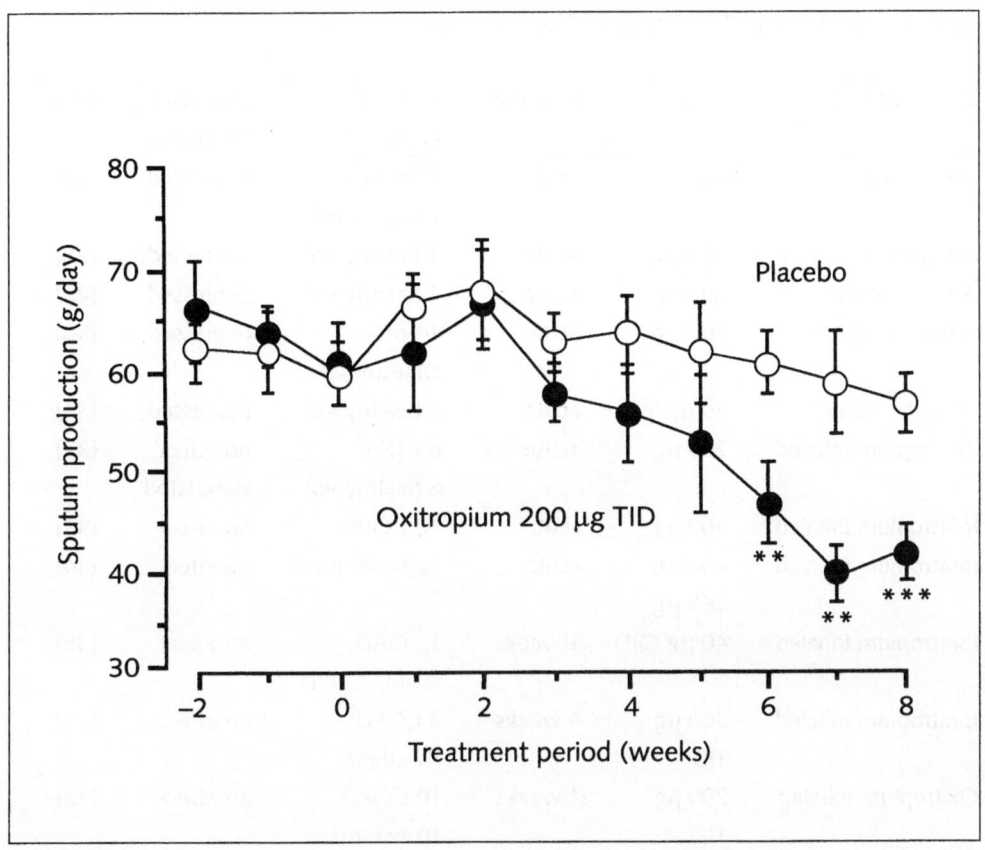

Figure 1
Time course of sputum production (wet weight) following inhalation of 200 µg oxitropium
TID (●) or placebo (O) in patients with chronic bronchitis or diffuse panbronchiolitis. Drug
treatment started at week zero and lasted throughout week eight. Shown are the means ±
SE of 17 (●) and 16 (O) patients. **p<0.01, ***p < 0.001 indicate significant differences from
placebo. Adapted from Tamaoki and colleagues [24].

healthy volunteers, measured as the increase in the particle retention time of a radiolabelled inhaled tracer (Tab. 2).

The quaternary anticholinergics differed from atropine in that they were demonstrated not to reduce the mucociliary clearance in several acute and chronic studies with normal volunteers and patients with asthma and COPD (Tab. 2). The highest anticholinergic doses used in studies of this type was oxitropium MDI 200 µg TID for 4 weeks [22].

Table 2 - Effect of anticholinergics on lung mucociliary clearance

Drug and route	Dose	Duration	Probands, Patients[1]	Mucociliary Clearance	Refs.
Hyoscine p.os.	8 µg/kg	acute	8 healthy vol. 2 obstructed	depressed	[27]
Atropine i.v.	600 µg	acute	4 healthy vol.	depressed	[28]
Atropine p.os.	not given	acute	4 healthy vol.	depressed	[29]
Atropine i.m.	400 µg	acute	14 pre- anaesthesia	depressed	[30]
Atropine inhaled	25 µg/kg	acute	8 healthy vol.	depressed	[31]
Ipratropium inhaled	200 µg	acute	6 COPD 6 healthy vol	no effect stimulated	[32]
Ipratropium inhaled	200 µg	acute	10 COPD	no effect	[33]
Ipratropium inhaled	40, 80, 160 µg	acute	12 healthy vol.	no effect	[34]
Ipratropium inhaled	40 µg QID	1 week	12 OAD, mainly COPD	no effect	[35]
Ipratropium inhaled	200 µg TID	4 weeks	8 COPD 7 Asthma	no effect	[21]
Oxitropium inhaled	200 µg TID	4 weeks	10 COPD 10 Asthma	no effect	[22]

[1]number of probands or patients on active drug

Systemic antimuscarinic effects of quaternary anticholinergics

Salivary secretion is not influenced by conventional doses of ipratropium, as shown in a study by Thomas and colleagues [36]: ipratropium in higher doses of 120 or 240 µg was directly sprayed into the oral cavity of healthy volunteers. If there were any effect at all, then it would be a tendency to increase secretion which may have been observed. Nevertheless, dryness of the mouth and bitter taste are reported side-effects of ipratropium which occur with low incidence, but some regularity in clinical trials [10, 37]. Effects on lung function and pulse rate were assessed in a cumulative dosing study by Ikeda and colleagues [38] in patients with COPD. Six puffs of oxitropium MDI (600 µg) or 30 puffs of ipratropium MDI (600 µg, highest dose tested) were without effect on the pulse rate. Fourteen puffs of oxitropium (1400 µg) significantly increased the pulse rate by 10 beats/minute. Using an estimate of the equipotent dose to 600 µg oxitropium, i.e. 1200 µg ipratropium [38, 39], a therapeutic window of f > 30 for the conventional dose of 40 µg ipratropium can be calculated.

Because of the favourable qualities outlined above, ipratropium became available world-wide in several formulations for inhalation (metered-dose inhaler (MDI), inhalation solution, powder inhalation system) and gained wide acceptance in the treatment of obstructive airways disease. The compound can be regarded as first-line therapy in COPD [40], which is also reflected in its prominent role in treatment guidelines [41, 42].

In summary, ipratropium bromide, the standard to be surpassed, is a specific antagonist at muscarinic acetylcholine-receptors, bronchodilates or protects against cholinergic bronchospasms in COPD, but is less broadly used in asthma. Administered by MDI or powder inhalation at a dose between 40–200 µg (500 µg by nebulizer) there is selectivity to the airways because of the topical administration, offering a huge therapeutic window. Furthermore, selectivity is also due to its characteristic of not passing the blood-brain barrier. The compound is not subtype-selective; its pharmacokinetic and pharmacodynamic half-life allows for administration 3–4 times daily, which is in agreement with a duration of action of 6–8 h. Besides bronchodilation there is only weak evidence that the compound may reduce bronchial secretions after weeks of therapy. The mucociliary clearance was not reduced in several acute and chronic studies.

Other structurally and conceptually related quaternary anticholinergics, e.g., atropine methonitrate, oxitropium or flutropium, have been studied and are introduced and available in several countries. They mainly differ in potency and slightly in their duration of action.

Concepts for improving anticholinergic therapy in airways diseases

The following concepts have been advanced to improve anticholinergic/antimuscarinic therapy for obstructive airways disease over the standard ipratropium bromide.

(1) More efficacious antimuscarinic agents through (a) higher potency/receptor affinity or improved lung bioavailability or (b) M_3-receptor subtype selectivity
(2) Antimuscarinic agents with an improved therapeutic window through M_3-receptor subtype selectivity
(3) Antimuscarinic agents for oral use through (a) special patterns of subtype selectivity or (b) lung/airway affinity and selectivity
(4) Long-acting antimuscarinic agents.

More efficacious drug through higher potency/receptor affinity

Can the efficacy be improved (in other words, can a higher level of bronchodilation or a higher level of inhibition of other muscarinic effects, e.g., secretion, be achieved) by increased *in vivo* potency of new compounds? This may be brought about by higher affinity to muscarinic receptors. In this context, it has to be considered that atropine and congeners are already highly potent compounds, with affinity constants or apparent K_D-values in the sub-nanomolar range (Tab. 3). Ipratropium showed dose-dependent bronchodilation in patients with COPD in a range from 5 to about 240 µg (dependent on lung delivery of the device used). At doses above 40 µg (delivered by metered aerosol) the dose-response curve becomes flat; Higgins and colleagues [11] showed that in patients with COPD, cumulative doses of ipratropium bromide of 5, 105, 855 and 1855 µg induced dose-dependent increases in FEV_1, plateauing beyond the 105 µg dose. Newnham and colleagues [48] showed an additional but non-significant increase in FEV_1, if 40 µg of ipratropium were followed by 200 µg (mean ΔFEV_1 from baseline in l, n = 27, (95% confidence intervals): 40 µg: 0.13 (0.10–0.16); 240 µg: 0.17 (0.14–0.20)).

Oxitropium (Oxivent®) is about twice as potent as ipratropium [38, 39] and Figure 2 shows that a dose of 100–200 µg is plateauing. In addition the diagram shows that increasing the dose from 20 µg (equivalent to 40 µg of ipratropium) also increases the duration of action. Takishima and colleagues [49] reported statistically significant improvement of lung function and COPD symptoms by treatment with oxitropium 200 µg TID compared to ipratropium 40 µg QID in a four-week trial in patients with COPD. So, oxitropium may be perceived as the more potent or high-dose form of ipratropium with a (at the doses used) slightly longer duration of action. A further improvement along these lines seems unlikely as oxitropium is already dosed into the plateau of its pharmacodynamic activity. A compound with similar efficacy at a lower dose (lower than 200 µg) is unlikely to offer a substantial clinical benefit.

Better lung bioavailability would be achievable by a more efficient device for inhalation than the conventional MDI. This would result in similar efficacy at a lower nominal or delivered dose, but the same lung (deposited) dose. Such highly efficient devices as described by Pavia and Moonen will become available in the future [50].

More efficacious drug through M_3-receptor subtype selectivity

Five human muscarinic receptor genes have so far been identified [51] and 4 subtypes (M_1–M_4) have been recognised pharmacologically and by using specific cDNA probes (see also the chapter by Roffel et al. in this volume). Only M_1-, M_2- and M_3-

Table 3 - Affinity-constants (K_i or K_D) (nM) of antimuscarinics in membrane preparations from insect cells or hamster ovary cells expressing human M_1 to M_5-receptors or to rat organ M_1 to M_3-receptors

	Atropine (1)	Pirenzepine (1)	Darifenacin (1)	Zamifenacin (1)	Tiquizium (1)	Ipratropium (2)	AQ-RA 721 (2)	DAC 5889 (2)
M_1	0.28	7.1	13	55	4.1	0.50	3.2	5.0
M_2	0.76	303	77	183	4.0	0.50	79	389
M_3	0.19	75	2.0	10	2.8	0.63	6.3	39.8
M_4	0.13	17	22	68	3.6			
M_5	0.24	66	5.4	34	8.2			

	YM-46303 (3)	Atropine (4)	Glycopyr-rolate(4)	Ipratropium (5)	Tiotropium (5)
M_1	0.67			0.18	0.04
M_2	5.9	4.3	1.9	0.20	0.02
M_3	0.39	4.9	1.7	0.20	0.01

(1) Inhibition constants K_i (nM) = $IC_{50}/1 + (L)/K_D$, ligand for displacement (L) was ^3H-NMS [43] at membrane preparations from recombinant insect cells expressing the human muscarinic receptors M_1–M_5 [43]

(2) Inhibition constants K_i (nM) at homogenates from rat cortex, heart or salivary gland [44]

(3) Inhibition constants K_i (nM) at homogenates from rat cortex, heart or salivary gland [45]

(4) Inhibition constants K_i (nM) at homogenates from rat ventricle or submandibular gland [46]

(5) Dissociation constants K_D (nM) = k_{-1}/k_1, kinetically determined, using radiolabelled compounds and membrane preparations from chinese hamster ovary cells expressing the human muscarinic receptors M_1–M_3 [47]

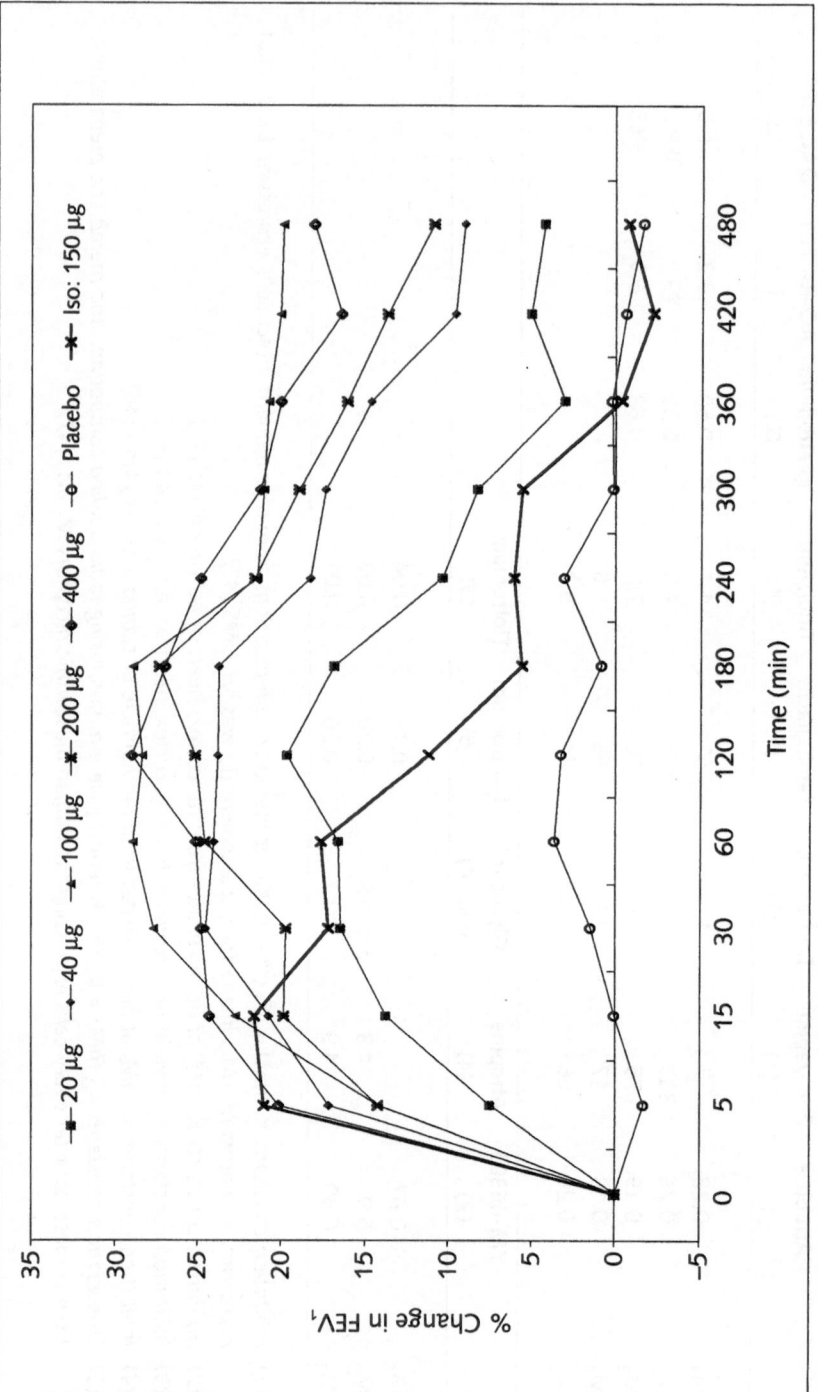

Figure 2

Time course and dose-dependency of bronchodilation (expressed as % change in FEV₁ from study day baseline) following inhalation of single doses of oxitropium (indicated in the figure), placebo or 150 μg isoproterenol in a crossover study in fourteen patients with moderate to severe COPD. Data from an unpublished study performed for Boehringer Ingelheim by Skorodin and Gross.

receptors have importance in the human lung (reviewed by Barnes [52]). Their precise function and role in airway physiology, disease and therapy is not fully established; however, the following have been defined:

M_1-receptors are localised to parasympathetic ganglia, submucosal glands and the alveolar wall (probably to type II pneumocytes); upon stimulation they facilitate nicotinic neurotransmission and by this means may enhance cholinergic reflexes. They induce mucus secretion and, potentially, alveolar surfactant secretion.

M_3-receptors are located in airway smooth muscle, at submucosal glands, airway epithelium and vascular endothelium; activation induces smooth muscle contraction, increased mucus secretion and vasodilation [52, 53].

M_2-receptors are localised to postganglionic nerves and serve as prejunctional inhibitory autoreceptors. A dysfunctional state in asthma or post-viral infection and blockade by inflammatory mediators or drugs may lead to increased junctional acetylcholine release, causing, or contributing to, bronchial hyperresponsiveness [54–58] (see also the chapter by Adamko et al. in this volume). It has further been hypothesized that the efficacy of non-subtype-selective antagonists like ipratropium in blocking post-junctional M_3-receptors may be weakened by the concomitant M_2-receptor blockade and increased junctional acetylcholine release [59]. Takahashi and colleagues [60] determined the ACh release from guinea pig and human airways evoked by electrical field stimulation; on incubation with non-subtype-selective anticholinergics they found about 40% enhancement. However, this moderate increase in junctional ACh is unlikely to limit the efficacy (= maximum effectiveness) of a potent anticholinergic agent like an atropine congener. Therefore, the maximum effectiveness at high doses of ipratropium or oxitropium should still indicate the maximum achievable.

An argument against the use of M_3-selective antimuscarinics may be derived from the presence of M_2-receptors in bronchial smooth muscle. These receptors have been described as inhibiting both the β-agonist-induced cAMP accumulation, probably *via* G_i-proteins, and the activation of large conductance calcium-dependent potassium channels (K_{Ca}, maxi-K), which are also activated by β-agonists *via* G_s-proteins (reviewed by Roffel and colleagues [61]; see also the chapter by Eglen and Watson in this volume). The functional importance of this counteraction of β-agonist activity by cholinergic activation has been demonstrated occasionally *in vitro* with guinea pig and rabbit tracheal preparations and in Basenji-greyhound dogs *in vivo*, but clear evidence of the role of smooth muscle M_2-receptors and their importance is missing in man [62].

Therefore, it is unclear whether an M_3-subtype-selective compound would have a major advantage. Although it is unlikely that this type of compound may provide a higher efficacy than high doses of ipratropium or oxitropium at acute dosing, advantages for chronic use, e.g., of a lower relative dose, cannot be excluded.

Improved bronchoselectivity/fewer side-effects through M_3-subtype selectivity

It has been proposed that non-subtype-selective drugs like ipratropium may induce paradoxical bronchoconstriction [52], which is occasionally reported after inhalation of ipratropium or oxitropium. For this to occur, e.g., ipratropium in lower doses would have to block M_2-neuronal autoreceptors before effective blockade of the postjunctional M_3-receptor at the smooth muscle is achieved with higher doses (or after more time has elapsed). The evidence comes from study results with guinea pig and dog airways *in vitro* and *in vivo* and human airways *in vitro* [54, 59, 60, 63].

Paradoxical bronchoconstriction is clinically defined as a fall in airflow (fall in $FEV_1 \geq 15\%$, other definitions are also used) in the first 30 min after inhalation of a drug with or without symptoms. Cases are reported with almost any drug available for inhalation. In two prospective studies such events (fall in $FEV_1 \geq 10\%$) occurred in 1.6% of 1450 salbutamol-treated patients and (fall in $FEV_1 \geq 15$ %) in 2.8% of 900 orciprenaline-treated patients, but in 5.7% of the orciprenaline-placebo group [64, 65]. The higher incidence with placebo is not surprising, as a putative constrictive event is not immediately counteracted by the β-agonist. Cocchetto and colleagues [66] conclude in a review that the active agent should not be regarded as the stimulus, but that properties of the carrier and/or the device are more likely to induce the adverse effects (preservatives, pH, osmolarity, excipients, cooling).

The incidence of such paradoxical reactions is unlikely to be higher with ipratropium. Spontaneous event reporting to Boehringer Ingelheim has recorded 114 cases allocated to the preferred term codes "bronchospasm aggravated" or "bronchospasm" in association with the use of ipratropium over a period of 10 years from 1985. Even when one takes into account the underreporting of spontaneous suspected adverse reactions, this figure does not at all indicate an appreciable incidence with ipratropium, considering the millions of patients treated. In published studies the following is documented: Beasley and colleagues [67] studied 22 patients with asthma, who inhaled preserved ipratropium nebulizer solution. Six developed paradoxical bronchoconstriction. These six sensitive patients subsequently on separate occasions inhaled isotonic saline or non-preserved ipratropium solution (Fig. 3) and in two additional periods increasing concentrations of benzalkonium chloride or edetic acid. None developed bronchospasm with saline, those on ipratropium bronchodilated, but all reacted to the preservatives. Bryant and Rogers [68] exposed 25 patients with stable asthma and 25 patients with acute asthma in four periods to nebulizer inhalation treatment: (1) preserved ipratropium, (2) acidic ipratropium, (3) neutralised ipratropium, and (4) saline. Only one patient with acute asthma (none of the stable asthmatics) reacted with a fall of 12% of predicted FEV_1 on exposure to preserved solution. This patient did not react to non-preserved solution.

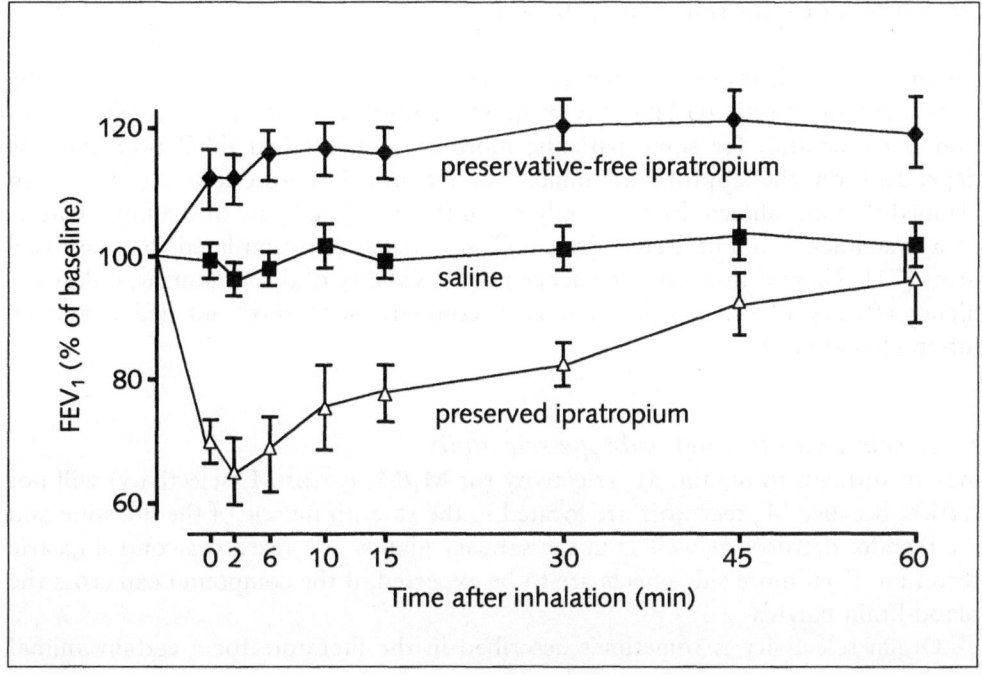

Figure 3
Change in FEV$_1$ (means ± SE, expressed as % of baseline) following inhalation of 4 ml of nebulised preserved ipratropium bromide solution (△) in six out of 22 asthmatic patients, who had shown a bronchoconstrictive response of > 20%. The same six patients after inhalation of preservative-free ipratropium (●), or saline (■) in this non-randomised single-blinded study. Adapted from Beasley and colleagues [67].

O'Callaghan and colleagues [69] report on 8 infants (average age 9 months) hospitalised for an acute episode of wheeze. The infants showed a fall in specific conductance of 20% one minute after inhalation of non-preserved nebulized ipratropium; 20 min after inhalation they showed improvement of 33% above baseline. This fall was not seen in a group of wheezy infants following inhalation of ipratropium bromide from a MDI/spacer [70].

In conclusion, most cases of paradoxical bronchoconstriction can presumably be explained by a reaction to the formulation rather than to the active ingredient. Only for infants with, probably, a history of bronchiolitis it cannot be excluded that they showed a pharmacological paradoxical bronchoconstriction. This means that, as far as paradoxical bronchoconstriction is concerned, room for improvement through subtype selectivity of an inhaled drug is limited.

Antimuscarinics for oral administration

An antimuscarinic for oral (systemic) use has to avoid antimuscarinic side-effects by sufficient lung selectivity. This may be quite an advantage, because oral administration is easier and, for some patients, more convenient than inhalation and less dependent on the appropriate inhalation technique. Furthermore, it has been claimed that an inhaled drug can only reach the ventilated part of the lung, whereas a systemically administered drug will also reach non-ventilated, but perfused areas [71]. There is no clinical evidence for the validity of this argument, unless the higher efficacy of oral corticosteroids in comparison to those inhaled would be interpreted as such.

Lung selectivity through subtype selectivity

may be difficult to obtain. M_3-selectivity (or M_1/M_3 *versus* M_2-selectivity) will not suffice, because M_3-receptors are located in the smooth muscle of the intestine and the bladder detrusor as well as in the salivary glands. M_1-receptors control gastric secretion. Even more side-effects are to be expected, if the compound can cross the blood-brain barrier.

Organ selectivity is sometimes described in the literature for a certain animal species or even in man without real explanation, unless receptor subtype- or isoenzyme-selectivity matches the organ. Other explanations, like a peculiar pattern of pharmacokinetics, tissue distribution or different glycosylation-patterns of the receptor protein in different organs have been advanced, but rarely with convincing evidence provided. Special profiles of subtype-selectivity have been proposed, e.g., in the case of antimuscarinics, Doods [44] hypothesized that a selectivity profile of $M_1 > M_3 >> M_2$ may be appropriate for an oral anticholinergic bronchodilator. This was based on animal experimental evidence, showing that M_1-antagonists strongly block reflex bronchospasms (by blocking facilitating ganglionic M_1-receptors); by this mechanism M_1-receptor antagonism should contribute to the overall protective efficacy of M_3-blockade. Concerning the most prominent anticholinergic side-effect, dry mouth, the following consideration applies: Wellstein and Pitschner [72] explained one possible mechanism for the well-known cholinomimetic activity of atropine at low doses on salivary flow (stimulation) as a blockade of inhibitory prejunctional M_1-receptors and increased ACh release. Their explanation is based on the dose range, being in agreement with M_1-receptor occupancy and the finding that the effect was prevented by pirenzepine. In consequence, an $M_1 > M_3 >> M_2$-antagonist should have secretion-stimulating M_1-effects, counteracting inhibition through M_3-receptor antagonism and perhaps turning out neutral on salivary secretion.

The problem with these kinds of concepts is the possibility of significant species differences in pharmacokinetics, receptor protein glycosylation pattern and distribution of subtypes in tissues. Therefore, a clinical development carries a high risk.

M_1-selective antagonists could theoretically be useful as oral antimuscarinic bronchodilators. For example, pirenzepine was profiled for oral treatment of peptic ulcers with minimal cardiovascular, ocular and CNS side-effects and was shown to reduce airway resistance in 6 of 8 patients with obstructive airways disease by Sertl and colleagues [73]. Lammers and colleagues [74] found inhaled doses of pirenzepine and ipratropium that were equipotent for inhibition of SO_2-induced reflex bronchospasm to be quite different in antagonising methacholine-induced bronchospasms (direct M_3-receptor effect); 7 µg ipratropium were quite effective, whereas 70 µg of pirenzepine did not raise the methacholine PC_{40}. So the importance of this concept depends on whether blockade of synaptic facilitation of vagal transmission *via* M_1-receptors in the airways is sufficiently important in patients.

By airway or lung tropism of a compound

The mucolytic/surfactant stimulator drug ambroxol has been described as reaching higher concentrations in lung tissue than in whole blood following systemic administration [75]. Patients eligible for thoracic surgery were given an intravenous infusion of 1 g of ambroxol in 3 h. Tissue biopsies from healthy lung lobar parenchyma obtained 2 to 12 h after the start of drug infusion showed 17- to 25-fold higher levels than the blood samples drawn at corresponding time points. These results suggest that ambroxol has a pulmonary tropism which would be a useful quality, especially for an oral antimuscarinic bronchodilator. However, no such example has been described in other pharmaceutical classes of compounds, including the bronchodilators.

Long duration of action through long plasma, tissue or receptor half-life

A long duration of action is of high value as it will contribute to patient convenience and compliance. Ipratropium or oxitropium have to be inhaled 3–4 times a day, in severe disease even more frequently, and depending on cultural background this may be problematic, as it may force patients to take medication in public. Furthermore, a long-acting compound should preferably have a low variability in its effectiveness from peak to trough (value before next administration), thus avoiding more frequent administration than necessary and prescribed.

For the β_2-agonist bronchodilators, a long duration of action through long plasma half-life has been achieved by oral administration of a highly β_2-selective compound (clenbuterol), or a slow release pro-drug concept (bambuterol). The inhaled β_2-agonist bronchodilators salmeterol and formoterol achieve their long duration of action by long target-tissue binding (salmeterol, binding to exoreceptor) or by local mechanisms not completely understood in the case of formoterol [112]. For antimuscarinics, long-lasting systemic exposure is unlikely to be an option. Therefore, the only real option is long-lasting topical effects in the target tissue or at the receptor level, which are likely to be achieved only by inhalation.

Novel compounds

Many attempts have been undertaken to develop new antimuscarinic bronchodilators according to the principles described above. The available information, especially from clinical studies, is limited and partly stems from published abstracts. Ongoing clinical development is only reported for tiotropium.

AQ-RA 721 and DAC 5889

AQ-RA 721 (Fig. 4) has been described as an $M_1 > M_3 >> M_2$-selective compound and DAC 5889 has the "oral use" pattern of $M_1 >> M_3 >> M_2$-selectivity (Tab. 3) [44]. In preclinical studies with rabbits, both compounds showed bronchoprotective effects with methacholine as the agonist at a 1.5 log units lower dose than for antibradycardic effects. They did not potentiate vagally mediated bronchoconstriction in guinea pigs at low doses (≤ 1 µg/kg i.v.), which is seen following ipratropium under these conditions. In these experiments bronchoprotective doses were reported at about 10–100 µg/kg i.v. In anaesthetised guinea pigs, DAC 5889 showed a "therapeutic window" of about $f = 30$ for bronchoprotective effects *versus* antibradycardic and inhibition of salivation effects. A clinical development of the compounds has not been reported.

Zamifenacin

Zamifenacin has been described as an M_3-selective compound ($M_3 >> M_1 = M_2$) with a so far unexplained selectivity for smooth muscle *versus* salivary gland in guinea pigs (Tabs. 3 and 4). Furthermore, unexplained selectivity for intestinal *versus* tracheal smooth muscle led to a clinical development in irritable bowl syndrome at a dose of about 40 mg [79].

Revatropate

Revatropate (U.K.-112,166; Fig. 4) has been described as an $M_1 \geq M_3 >> M_2$-selective compound and also showed *in vivo* (in animals) an 80-fold selectivity of bronchoprotection over cardiac effects (Tab. 4) [78]. In guinea pig and dog experiments, paradoxical bronchospasms, which have been recorded at low threshold doses with

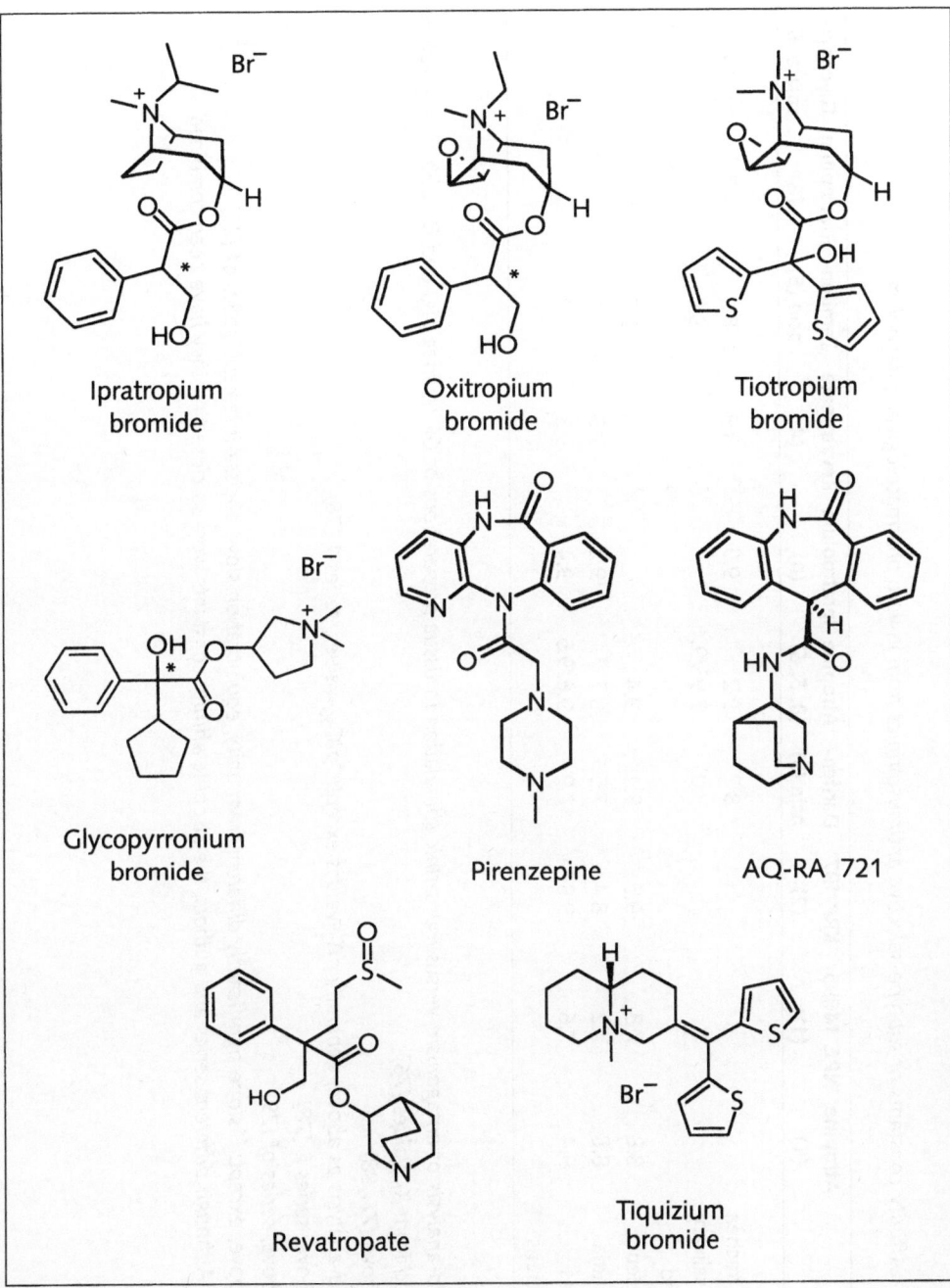

Ipratropium bromide

Oxitropium bromide

Tiotropium bromide

Glycopyrronium bromide

Pirenzepine

AQ-RA 721

Revatropate

Tiquizium bromide

Figure 4
Chemical structures of some antimuscarinics.

Table 4 - Affinity constants of subtype-selective antimuscarinics from in vitro pharmacological antagonism

	Atropine (1)	NPC 14695 (1)	KRP-197 (2)	Darifen-acin (3)	Atropine (3, 5, 6)	Ipratropium (4)	Revatropate (4)	Zamifen-acin (5)	Tiotropium (6)	Glycopyr-rolate (6)
M$_3$ GP trachea	8.8			8.7	9.2	9.5	8.9	8.1	9.5**	9.4
GP salivary gland				7.0*	7.9*/9.1			8.0		
GP ileum	8.8	7.8	9.5	9.4	9.4			9.3		
M$_2$ GP atria	8.8	7.2	8.4	7.5	8.7	9.2	7.3	7.1		
M$_1$ R vas deferens	9.4	7.6	9.3	7.9	9.6/9.5	9.4	9.3	7.4		

(1) Schild-analysis of antagonism versus muscarinic stimulation in tissue preparations of GP (guinea pig) and R (rabbit), pA$_2$-values calculated from K_b-values [76]

(2) As above [77, 88]

(3) Schild-analysis as above, generally pA$_2$-values except: *pIC$_{50}$-values, series 1 [78]

(4) As above, series 2 [78]

(5) As above, series of [79]

(6) As above, except: **slope significantly different from unity, equilibration slow, abcissa-intercept, series of [26]

(3, 5, 6) Although different series and authors, the atropine affinity constants were so close that they have been combined ipratropium in these models, were not observed. The compound was promoted to clinical development. In a Phase II study including 42 patients with COPD, 320 μg inhaled revatropate was compared to 80 μg ipratropium and placebo in a 3-period crossover-study. All compounds were administered from an MDI. The efficacy and safety profiles of the two compounds were comparable, including peak effect and duration of action [80]. In this acute setting, advantages of the M_3-selectivity were not evident and ongoing clinical development is not reported, so the properties during chronic use have not been revealed.

Darifenacin

Darifenacin (UK-88,525) was described as an $M_3 >> M_1 \geq M_2$-selective compound (Tabs. 3 and 4) [78]. Because of higher activity in the urinary bladder and ileum than in the trachea, the compound has been promoted to clinical development in urinary incontinence and irritable bowel syndrome. Beaumont et al. [81] reported a rather short half-life of less than 2 h, a saturation of plasma clearance and a complicated pattern of metabolism, properties which do not favour clinical development for COPD.

Pirenzepine and telenzepine

Pirenzepine and telenzepine (Fig. 4) possess M_1-selectivity (Tab. 3) [111, 43] and have been used for oral treatment of peptic ulcers with minimal systemic antimuscarinic side effects. Cazzola and colleagues [82, 83] showed that in healthy volunteers a higher intravenous dose of pirenzepine (Gastrozepine®, 10 mg) increased FEF_{50}, but not FEV_1 (Tab. 5). FEF_{50} predominantly reflects small airways' patency. Atropine (1 mg) increased the FEV_1. In patients with COPD, both drugs increased FEV_1 and FEF_{50}, but atropine was more effective (Tab. 5). In two crossover studies with 18 patients with COPD and 18 with asthma, Ceyhan and colleagues [84] compared single doses of, in each case, 100 μg inhaled pirenzepine administered by nebulizer with 125 or 250 μg of ipratropium. Concerning FEV_1 as endpoint pirenzepine was about 50% less effective than ipratropium in COPD and was not effective in asthma. Cazzola and colleagues [85] also studied the effect of 5 mg oral telenzepine *versus* placebo and found 14% increase in FEV_1 and 26% of FEF_{50} *versus* baseline (time average over 6 h) in patients with COPD. The median of peak increase in FEV_1 following telenzepine was higher (33%) than the median of the effect 15 min after 200 μg salbutamol (20%). Ukena and colleagues [86] studied a 5-day OD treatment regimen of 3 mg oral telenzepine *versus* placebo in a crossover study including 21 patients with COPD shown to be responsive to bronchodilators. Disappointingly, no significant bronchodilation was observed for FEV1. The only drug-

Table 5 - Bronchodilation (% change) by pirenzepine or atropine

	Healthy volunteers		COPD	
	FEV_1	FEF_{50}	FEV_1	FEF_{50}
Pirenzepine 10 mg i.v.	−4%	+28%	+21%	+10%
Atropine 1 mg i.v.	+14%	+3%	+21%	+88%

Data from [83], % calculated. Peak change in interval 5 to 120 min selected.

related side-effect was dry mouth in 9 patients during the telenzepine period and 2 on placebo (of 21). No further clinical studies or development have been reported for M_1-selective antagonists. The lower efficacy compared to ipratropium probably makes the class unattractive, although the selective effect on small airways has not been fully evaluated. Potential usefulness in this respect is not established, but cannot be rejected either, based on present data.

Other subtype-selective antimuscarinics

NPC 14695 (3-(4-benzyl-piperazinyl)-1-cyclobutyl-1-hydroxy-1-phenyl-2-propane) was reported to be in preclinical development for urinary incontinence and potentially for obstructive airways disease [76, 87]. The compound has a weak selectivity of $M_3 \geq M_1 > M_2$ (Tab. 4). In guinea pigs the compound was active as bronchodilator in a dose range of 40–1200 µg/kg i.v. or s.c. and was also orally absorbed. The compound is claimed to distinguish smooth muscle M_3 from glandular M_3-receptors. A clinical development for airways disease has not been reported. Miyachi and colleagues [77, 88] describe a series of imidazol derivatives. The most prominent compound of the series, KRP-197 (4-(2-methylimidazol-1-yl)-2,2-diphenylbutyramide), was characterised by a binding affinity to M_3-receptors of 0.32 nM and a M_2/M_3-selectivity ratio of 13 (Tab. 4). Naito and colleagues published data on a series of biphenyl carbamates. Their most prominent compound YM-46303 (quinuclidin-4-yl biphenyl-2-ylcarbamate) had a binding affinity of 0.39 nM to M_3-receptors and a selectivity ratio of 15 (Tab. 3). The latter two compounds are foreseen for altered smooth muscle tone and contractility-related disorders, especially urinary urge incontinence. YM-46303 inhibited bladder contraction of rats at a dose of 100 µg/kg i.d. A clinical development is not yet reported.

J-104129 (K_i 4.2 nM, 120-fold selectivity M_3 over M_2) and J-106366 are reported as M_3-selective compounds in preclinical development for smooth muscle contraction-related disorders [114]. The active dose range is 80 µg/kg i.v. or 300 µg/kg orally (Pharma Projects 1988, Merck &Co). For comparison, ipratropium is effective at 1.5 µg/kg i.v. or inhaled.

Glycopyrrolate

Glycopyrrolate (Robinul®, Fig. 4) is a synthetic quaternary ammonium compound. It has primarily been used during anaesthesia to reduce cholinergic reflex events like salivation, gastric acid secretion or bradycardia and to prolong the effectiveness of anticholinesterases [89]. Glycopyrrolate is not subtype-selective and its potency *in vitro* is similar to ipratropium (Tabs. 3 and 4). The compound has occasionally been studied by inhalation from a nebuliser in asthma [90, 113] and COPD [91, 92] in doses from 80 to 2000 µg and combined with orciprenaline or salbutamol. In asthma, a plateau of bronchodilatory effect seems to be reached with 240–480 µg [113]. The duration of action with the higher doses is claimed to exceed 8 h, depending on how the duration of action is defined (see below, Tab. 6). In COPD a dose of 1000 µg may provide bronchodilation for 8 h [92].

Tiquizium

Tiquizium (Fig. 4) is also a quaternary ammonium compound used as an antispasmodic agent for gastrointestinal, biliary and urinary system disorders [93]. Tiquizium is not subtype-selective and is described as having about one order of magnitude lower affinity than atropine for cloned human muscarinic receptors (Tab. 3), but a similar potency on animal tissues *in vitro* [94]. In a pilot study of 7 patients with COPD, 1000 µg tiquizium inhaled by nebulizer showed plateauing bronchodilation. At a dose of 800 µg the duration of action was 6–8 h (Tab. 6). Further development is not reported.

Tiotropium

Preclinical results/in vitro
The most promising of the new developments is tiotropium bromide (Spiriva™, Fig. 4). This compound is composed of a quaternized scopine moiety connected by an ester bond to di-thienylglycolic acid. The first part resembles oxitropium, the latter part has a unique structure. Apparent dissociation constants ranging from 0.1 to 0.3 nM for the subtypes of muscarinic receptors, combined with low affinity for other neurotransmitter receptors (H_1: 81 nM, unlikely to be of biological relevance), identify the compound as a high-affinity specific antimuscarinic (Tabs. 3, 4 and 7)

Table 6 - Trough to peak (responses)-ratio of bronchodilators (T/P%)

Agent, dose, device, duration and ref.	Baseline FEV$_1$ (l)	6 h T/P%	8 h T/P%	12 h T/P%	24 h T/P%
Ipratropium, 36 µg, MDI [13]	1.02	47			
Ipratropium, 36 µg, MDI [12]	1.11	47			
Oxitropium 200 µg, MDI [95]	1.10	46	30		
Oxitropium 200 µg, MDI [S]	1.12	59	40		
Glycopyrrolate 1000 µg, nebulizer [92]	1.27	63	50		
Tiquizium 800 µg, nebulizer [94]	1.06	40	27		
Salmeterol 50 µg, MDI + spacer [96] vs.	0.83			41	
Ipratropium 36 µg, MDI + spacer [96]	0.81	26			
Salmeterol 50 µg, MDI, BID [97] 1st/84th day	1.28			60/59	
Ipratropium 36 µg, MDI, QID [97] 1st/84th day	1.16	47/41			
Tiotropium 18 µg, DPI, OD, after 1 week:	1.16				59
[98] 2 weeks:					62
4 weeks:					87
Tiotropium 18 µg, DPI, OD, at day 8–92 [99]	1.01				57–55

SD, single dosing study; MDI, metered dose inhaler;DPI, dry powder inhaler
[13]: 25 patients with COPD, SD , FEV$_1$ from Fig. 1, T/P % (group peak response) are
from group means.
[12]: 72 patients with COPD, SD, FEV$_1$ from Fig. 1, T/P % (group peak response) are from
group means.
[95]: 24 patients with stable chronic obstruction (mainly COPD), SD,% change in FEV$_1$
from Fig. 1, T/P % (group peak response) are from group means.
[S] Skorodin M, Gross N, unpublished study for Boehringer Ingelheim: 14 patients with
COPD, SD,% change in FEV$_1$, T/P% (group peak response) are from group means.
[92]: 11 patients with COPD, SD, FEV$_1$ from Fig. 1, T/P % (group peak response) are from
group means.
[94]: 7 patients with COPD, SD, FEV$_1$ from Table 3, T/P % (group peak response) are from
group means
[96]: 16 patients with COPD, SD crossover, FEV$_1$ from Fig. 1, T/P % (group peak response)
are from group means.
[97]: 135 COPD patients salmeterol; 133 ipratropium; FEV$_1$-changes from Fig. 1, 1st day
(1st dose) and 84th day T/P % (group peak response) are from group means; 84th day data
presumably represent changes from the study baseline.
[98]: 33 COPD patients in dose group 18 µg; peak: mean 6 h average FEV$_1$-response from
Fig. 3, mean trough FEV$_1$ over study baseline from Fig. 2.
[99]: 279 COPD patients; peak: mean 3 h average FEV$_1$-response and trough over study
baseline from the table.

Table 7 - Binding affinity of antimuscarinics and dissociation half-lives of the receptor-drug complexes

| | K_D (nM) | | $t_{1/2}$ (h) | |
	Tiotropium JD	Ipratropium JD	Tiotropium JD	Ipratropium JD
Hm1	0.041	0.183	14.6	0.11
Hm2	0.021	0.195	3.6	0.035
Hm3	0.014	0.204	34.7	0.26

Membrane preparations from chinese hamster ovary –K1 cells expressing human muscarinic receptors. Binding affinity was kinetically determined at 23°C. Dissociation half-lives from 3–5 triplicate experiments [47].

[26]. Reversible antagonism of methacholine-induced contraction of isolated tracheal rings from guinea pigs yielded in the Schild-analysis an abcissa-intercept of 9.5. The slope, significantly higher than unity (2.53), was explained by non-equilibrium conditions, although the incubation lasted one hour. More reliable K_D-values were obtained from kinetic measurements as k_{off}/k_{on}, which describe tiotropium more adequately as a muscarinic ligand in the 10^{-11} M range (Tab. 7) [47]. The slow equilibration is explained by very slow dissociation of tiotropium from human muscarinic receptors, with a half-life of 34.7 h for the M_3-receptor compared to 16 min for ipratropium.

Antimuscarinics have no influence on the baseline beat frequency of airway ciliated cells *in vitro*. However, muscarinic agonists stimulate the frequency, and this can be inhibited by antimuscarinics. In line with the slow *in vitro* dissociation, much longer lasting inhibition of muscarinic effects at guinea pig ciliated cells [26] and at guinea pig as well as human tracheal smooth muscle [60] was reported following incubation and washout of tiotropium than with ipratropium or atropine.

Another important quality of the molecule emerges when comparing the dissociation half-lives of tiotropium from the muscarinic receptor subtypes (Tab. 7). For both tiotropium and ipratropium, the complex with M_3- or M_1-receptors dissociates slower than with M_2-receptors (about 8 or 3.5 times, respectively), but only for tiotropium is this difference likely to matter: With dissociation half-lives in the range of hours, this pattern may result in kinetic control of a subtype-selective blockade, $M_3 \geq M_1 > M_2$, which is unlikely for ipratropium with its half-life of minutes. Functional evidence of subtype-selective blockade was provided by *in vitro* studies of acetylcholine release from guinea pig airways after electrical field stimulation [60]. Ipratropium, atropine and tiotropium increased ACh-release to a similar extent (30–40%), which reflects blockade of the inhibitory prejunctional M_2-autoreceptor. This response was lost after 2 h of washout of all 3 drugs. In contrast, persistent inhibition of the postjunctional cholinergic twitch response was seen with a washout half-life of 9 h for tiotropium and only 81 min for ipratropium, which reflects inhibition of the M_3-mediated smooth muscle contraction.

Preclinical results/in vivo

As predicted from *in vitro* experiments, inhaled tiotropium provided dose-dependent protection against methacholine-induced bronchospasms in anaesthetized dogs [26]. Inhalation of the aerosol generated from a 1 mg/ml aqueous solution was 100% protective with nearly no decline of efficacy in the 6-h observation period. The fully protective dose of ipratropium generated from a 10 mg/ml-solution declined to half the maximal response within 2 h. In unrestrained conscious guinea pigs, inhaled tiotropium was compared to ipratropium (0.1 mM, 3 min inhalation). The ratio of the bronchoconstricting dose of methacholine after tiotropium *versus* a baseline methacholine dose reached 8 as peak value. A factor of 2 was still present 96 h after inhalation of the drug. The same nominal dose of ipratropium reached a factor of 2 as peak value, lasting for 12 h [61].

The preclinical safety of this antimuscarinic agent of a new generation was demonstrated in toxicological studies by the oral, intravenous and inhalation route for up to 2 years.

Clinical results/Phase I

Tiotropium was developed pharmaceutically for inhalation, first administered in man from a piezoelectric ultrasonic atomizer (single breath) [100], and later as a lactose-blended dry powder from a single capsule inhaler (HandiHaler®). In Phase I trials with healthy volunteers, tiotropium behaved as a typical quaternary anticholinergic agent with low oral absorption of the drug (< 2%) and following inhalation no relevant drug-related effects were observed on vital signs, blood pressure, heart rate, respiratory rate, ECG, pupillometry and routine laboratory test. Dry mouth, the most sensitive anticholinergic side-effect, was reported after higher doses. In patients with COPD, the estimated overall systemic availability was 7% of the inhaled nominal dose [101]. These values correspond well to ipratropium with an apparent systemic availability of 2% after oral administration and 6.9% following inhalation [8]. 20 µg of inhaled tiotropium bromide, similar to many inhaled drugs, has an almost i.v.-like absorption profile with peak plasma concentrations 5 min post-dosing of about 18 pg/ml and a subsequent rapid decline, in less than 1 h, to very low levels in the 4 pg/ml range [26, 101]. At these low levels, below relevant systemic muscarinic receptor occupancy, the drug is eliminated with a terminal plasma half-life of 5–6 days. Peak as well as trough (i.e., –2 to 0 h before dosing) plasma levels did not show clinically relevant accumulation with OD dosing up to three months in patients with COPD [101].

Clinical results/single dose studies

In 12 patients with mild asthma and documented hyperresponsiveness to inhaled methacholine (demographics: mean FEV_1 = 91.6% of predicted, range 80–121, mean PC_{20} methacholine = 1.7 mg/ml, range 0.04–4.9), inhaled tiotropium bromide in doses of 10, 40 and 80 µg produced significant bronchodilation, measured by an increase in FEV_1 between 5.5 and 11% from baseline, that was sustained for 24 h [102]. The response in terms of clinically significant protection from methacholine reached impressive levels at the dose of 40 µg of over 7 doubling concentrations and lasted for more than 48 h (Fig. 5). This may be compared to a protective effect of 200 µg of oxitropium bromide of up to 6 h [103].

The bronchodilator response and safety of tiotropium in patients with COPD were studied in a pilot dose-response study by Maesen and colleagues [100]. The results were similar to a subsequent larger double-blind placebo-controlled crossover study by the same group [104]. In these and following studies COPD patients were included if they

- conformed to the ATS definition of COPD [105],
- had an age > 40 years and ≥ 10 pack-year smoking history, FEV_1 < 65% of predicted,
- blood eosinophils < 600/mm³, and no history of asthma, atopy or allergic rhinitis.

In the study by Maesen and colleagues, patients with a mean baseline FEV_1 = 1.34 l (44% of predicted) inhaled single doses of 10, 20, 40 and 80 µg tiotropium bromide from a dry powder inhaler. Significant bronchodilation, determined by FEV_1-response, was observed at all doses. The effect was dose-dependent, 10 µg producing half the improvement of the 80 µg dose (Fig. 6). Peak improvement in FEV_1 ranged 20–25% of test-day baseline compared to 11% for placebo. During early morning all group-values dropped and the difference between drug and placebo became smaller, but widened again more than 24 h after administration. The duration of the effect exceeded the 32-h observation period. There was no evidence of systemic anticholinergic side-effects.

Clinical results/ multiple and chronic dosing studies

The appropriate dose for such a long-acting drug can only be defined in steady state. 169 patients with COPD (mean baseline FEV_1 = 1.08 l, 42% of predicted value, mean age 66 years, 57% male) were randomized to receive placebo, 4.5, 9, 18 or 36 µg tiotropium by dry powder inhaler OD for 28 days in a double-blind, parallel group comparison (tiotropium declaration as cation with 18 µg equivalent to 20 µg of the bromide) [98]. The FEV_1-response as primary efficacy variable was measured three times in the 4 h prior to drug intake at 12.00 noon (trough) and after inhalation hourly for 6 h (peak in the interval and 6-h average) during the weekly clinic visits; 162 patients completed all visits. All doses of tiotropium produced significantly greater responses than placebo for peak, average and trough FEV_1, FVC and serial PEF. The gross effects of the highest dose on FEV_1 were most pronounced, but there was no statistical significance between active doses (Fig. 7). The time-effect profile of tiotropium showed a flat maximum within the first 6 h. Average 6-h values after 1 week of treatment were generally higher than after the first dose. The difference between peak or average value in the 6 h after administration and trough became smaller with increasing duration of treatment, indicating a stabilisation of lung function with chronic treatment (Tab. 6). The response gradually returned to the pre-treatment baseline over the 3-week period following discontinuation of tiotropium. The overall safety profile for the tiotropium doses was similar to placebo. Dry mouth was the most frequently reported adverse event related to tiotropium treatment, 2 of 34 (4.5 µg group), 2 of 33 (18 µg group) and 3 of 34 (36 µg group), none with placebo or 9 µg, always described as mild, except for one patient as moderate in the highest dose group.

In a subgroup of a long-term phase III trial, 17 patients underwent frequent pulmonary function testing to characterise the onset of the pharmacodynamic steady state [106]. The major proportion of spirometric improvement at peak was achieved within 24 h, whereas the trough value showed additional improvement on the second day and further slight increases for up to one week.

Preliminary data after three months of tiotropium therapy are published as abstracts from two randomised double-blind, parallel group, controlled multi-center trials comparing 18 µg of tiotropium OD with placebo (Casaburi et al., 1999 [99]: 279 patients tiotropium, 191 patients placebo) or with 40 µg ipratropium-MDI QID (Van Noord et al., 2000 [107]: 191 patients tiotropium, 97 patients ipratropium). Also in these studies, with broad patient exposure for prolonged periods of time, tiotropium provided a significant increase in FEV_1 and FVC within 30 min after the first dose. Steady state at trough was reached after one week. Spirometric improvement (FEV_1) at peak (3 h average) ranged 20% and trough 11% above the study baseline throughout the 13-week treatment period. The comparison to ipratropium showed superiority of tiotropium already 3 h after administration, more pronounced at 6 hours and especially at the 24-h trough value in the morning of a clinical visit. After 13 weeks of therapy, ipratropium had shown an increase of baseline FEV_1 24 h after the last administration of 30 ml, whereas tiotropium was 160 ml higher.

Evaluation/duration of action

Different methods have been used to determine the duration of action of a bronchodilator. It has mostly been investigated whether clinically significant bronchodilation is still present at the end of the administration interval. Clinically significant bronchodilation has been defined as 15 or 12% of baseline FEV_1 [108, 109]. This

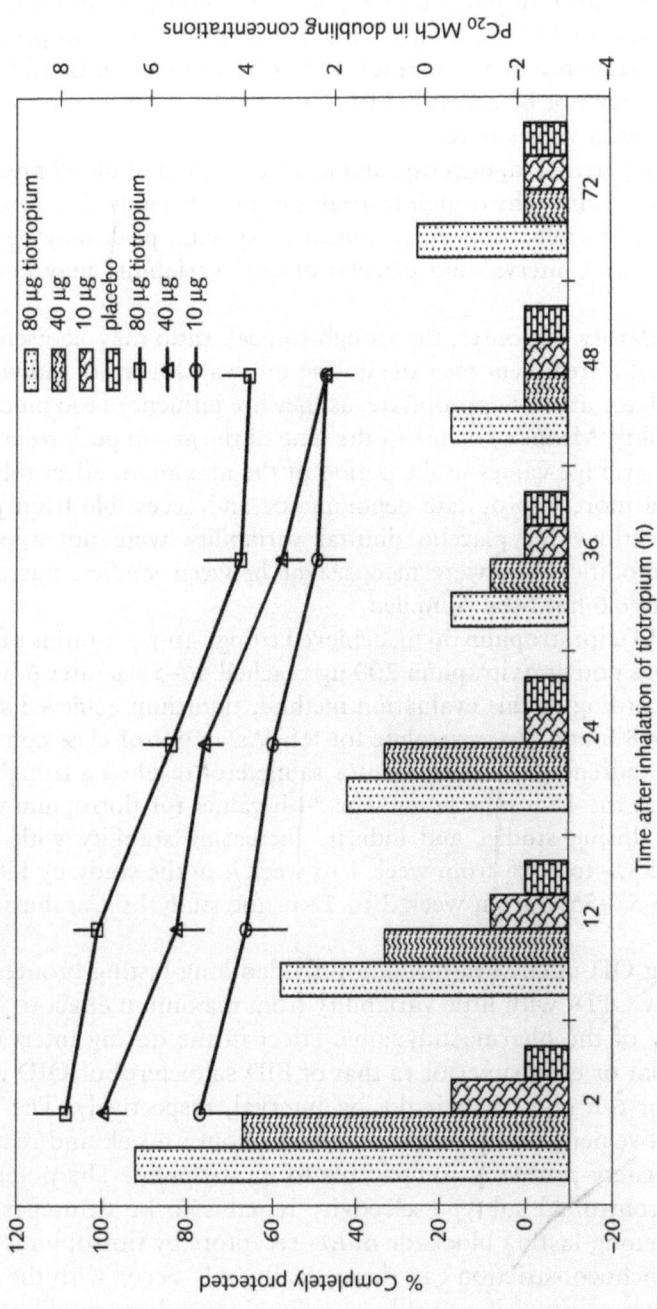

Figure 5
Dose-dependency and duration of tiotropium-mediated protection against methacholine challenge in 12 mild asthmatic patients with airway hyperreactivity. Bars represent the percentage of patients completely protected (defined as < 20% fall of FEV_1 on inhalation of the highest challenge dose of 128 mg/ml methacholine, left ordinate), whereas lines represent the increase in methacholine PC_{20} expressed as doubling concentrations compared to each patient's placebo value (right ordinate). Doubling concentrations are shown as means ± SE. Note that only four patients participated at 72 h. Adapted from O'Connor and colleagues [102].

definition has been used and may be appropriate to define the duration of effect in asthma, considering that a peak response of, e.g., 30% of 3 liter (450 ml) can then be monitored until decline to 15% (225 ml). This definition is less appropriate in COPD, where the peak response may just reach 15% of 1 l (150 ml). In addition, these individual values may not be accessible from the published literature, which mostly provides group mean values only.

In the treatment of hypertension, lowering and low fluctuation of blood pressure in the dosing interval is the aim. The trough-to-peak ratio has been used to describe a drug profile. Methods of calculation differ considerably, e.g., peak may be true peak or the maximum in an interval and placebo diurnal variability may be corrected for [110].

To assess bronchodilatory responses, the trough-to-peak ratio may be useful to indicate the variability of a treatment over the dosing interval. Means of individual peak values as denominator are not appropriate, as they are influenced too much by individual value variability. Means of values at the time of the group peak response, or means of individual average values in the period of the maximum effect following administration, is a more appropriate denominator and accessible from published literature. Corrections for placebo diurnal variability were not applied, because the resulting modifications were inconsistent between studies. Based on these considerations Table 6 has been compiled.

In patients with COPD, ipratropium 36 µg achieved trough-to-peak ratios of 26–47% after 6 h. The more potent oxitropium 200 µg reached 46–59% after 6 h and 30–40% after 8 h. According to this evaluation method, tiquizium achieved sufficient stabilisation for 6–8 h and glycopyrrolate for 8 h. As an out-of-class comparison, the long-acting β_2-adrenergic bronchodilator salmeterol reached a trough-to-peak ratio of 41–60% at the 12 h time point. The 24-h values for tiotropium were derived from multiple dosing studies and indicate increasing stability with prolonged exposure from 59% to 87% from week 1 to week 4 in the study by Littner and colleagues [98], or 57–55% from week 2 to 13 in the study by Casaburi and colleagues [99].

Summarizing, 18 µg OD inhaled tiotropium provided long-lasting bronchodilation in patients with COPD, with little variability from maximum effect to 24-h trough. The variability of the pharmacodynamic effect in the dosing interval of OD tiotropium is similar or even superior to that of BID salmeterol or QID ipratropium at the 12-h or 6-h end of their dosing interval, respectively. The pulmonary function improvement reaches steady state in about a week and reaches a stable plateau. The safety profile is comparable to ipratropium. The potential benefit of kinetically controlled subtype-selectivity remains to be demonstrated clinically, but given the long-lasting blockade of M_3-receptors by tiotropium, M_2-based paradoxical bronchoconstriction can theoretically only occur with the first dose. The compound has a promising profile as maintainance bronchodilator in COPD.

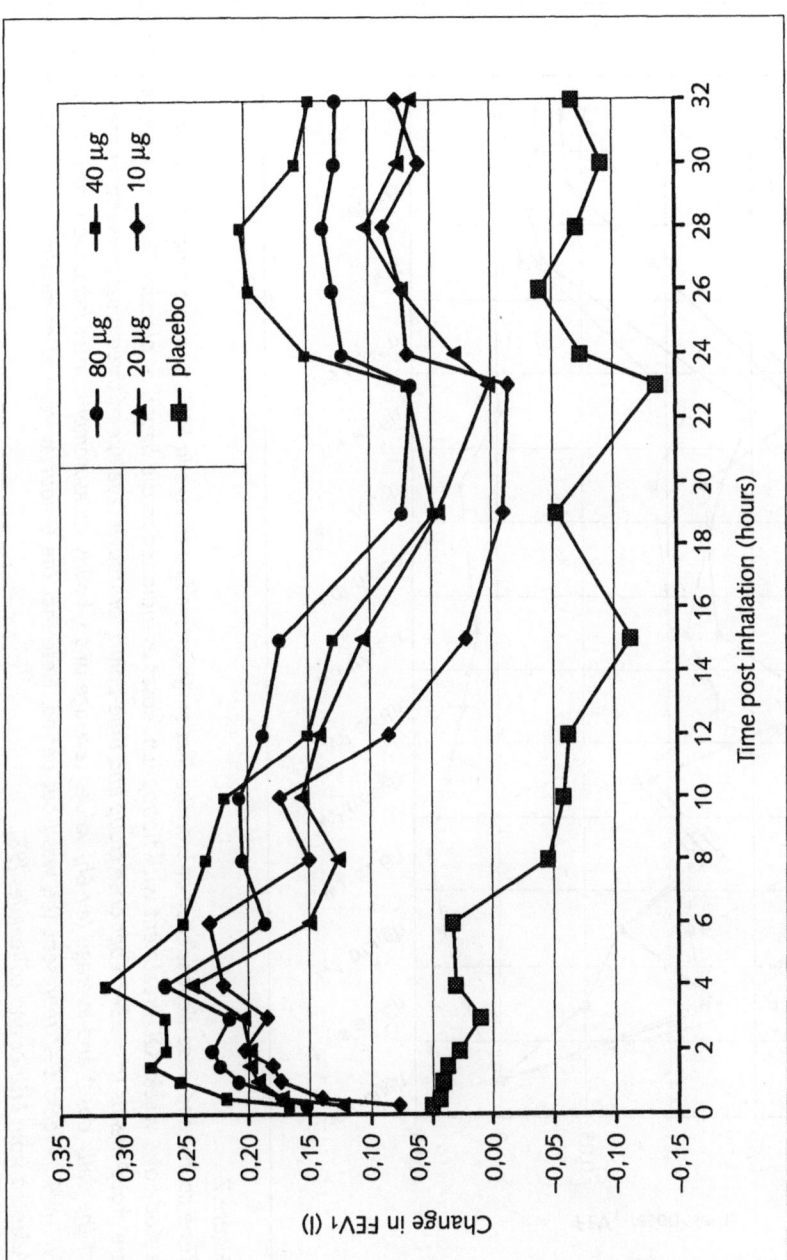

Figure 6

Time course and dose-dependency of bronchodilation (expressed as mean change in FEV$_1$ from test day baseline) following inhalation of single doses of tiotropium or placebo (indicated in the figure) in patients with COPD. Note that for this analysis 19 patients were evaluable after exclusion of test days which followed tiotropium 20, 40 or 80 µg, because of carry-over effects. Adapted from Maesen and colleagues [104].

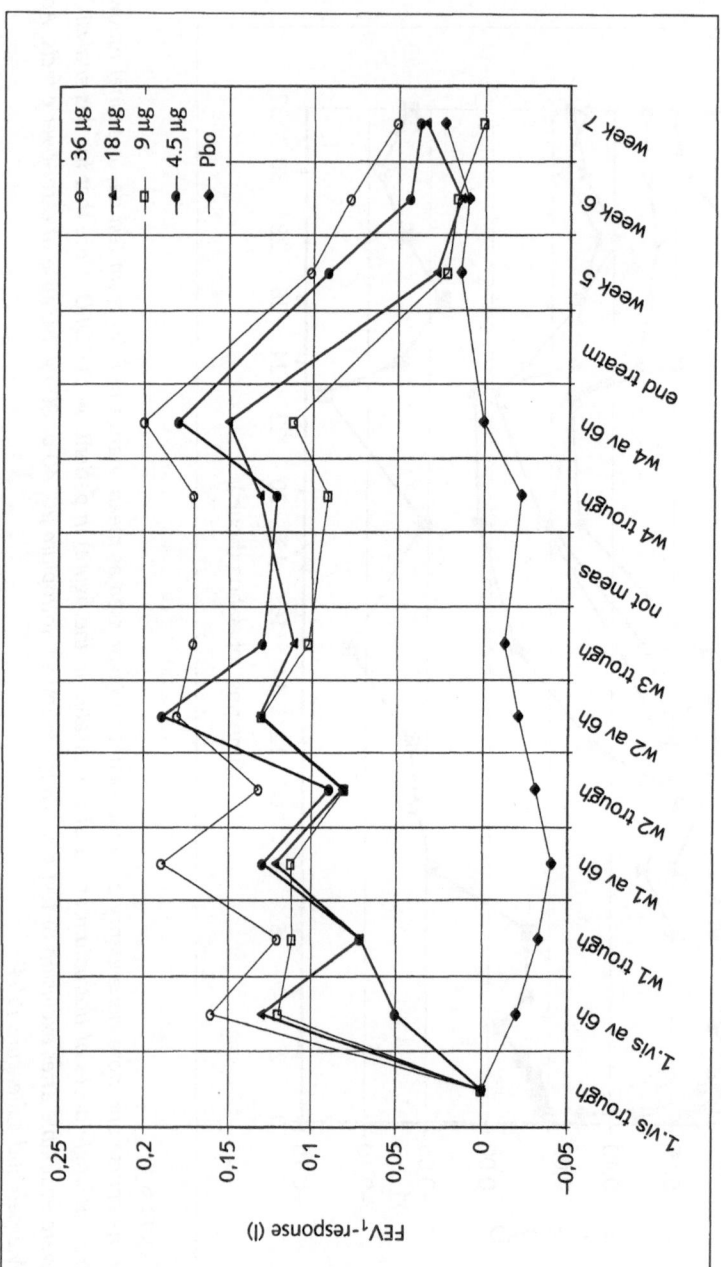

Figure 7
Time course and dose-dependency of bronchodilation (expressed as change in FEV₁ from pre-study (1.vis) baseline trough values) in a study of 4 weeks OD treatment with tiotropium (doses as indicated in the figure) or placebo (Pbo) in patients with COPD. Shown are group means (n = 33–35 per dose group and time-point) for trough (average of three measurements in the 4 h prior to inhalation of the daily dose) and average (av 6h) values (average of six hourly measurements after inhalation) as determined at weekly visits (w1–w4). Weeks 5–7 represent the wash-out period. Note that the 6-hour average after inhalation was not measured in week 3. Adapted from Littner and colleagues [98].

Conclusion

The standard anticholinergic bronchodilator recommended in treatment guidelines for COPD, inhaled ipratropium bromide, as MDI conventionally dosed 36 µg QID or higher, up to now has not been surpassed by novel approaches. Several compounds designed to have special profiles of muscarinic receptor subtype-selectivity or airway smooth muscle selectivity so far did not show improved efficacy or broadened therapeutic ratio in the clinic. Compounds appropriate for oral use did not emerge from these concepts either.

Oxitropium and glycopyrrolate, which are dosed higher than ipratropium, allow for TID administration. Advanced clinical development in COPD is reported for tiotropium, a specific quaternary antimuscarinic agent, topically selective when inhaled, for which slow dissociation from the M_3-receptor is likely to be responsible for the long duration of action. With once-daily inhalation, tiotropium achieved efficient bronchodilation with long duration and comparatively low variability between peak effect and trough at the end of the 24-h dosing interval. This profile was sustained with chronic therapy. Results beyond spirometry, e.g., influence on exercise performance, undisturbed sleep, quality of life and other outcomes are eagerly awaited.

References

1   Goyal RK (1989) Muscarinic receptor subtypes – physiology and clinical implications. *N Engl J Med* 321: 1022–1029
2   Eglen RM, Sharath SH, Watson N (1996) Muscarinic receptor subtypes and smooth muscle function. *Pharmacol Rev* 48: 531–565
3   Gross NJ, Skorodin MS (1984) State of the art, anticholinergic, antimuscarinic bronchodilators. *Am Rev Respir Dis* 129: 856–870
4   Gross NJ (1988) Ipratropium bromide. *N Engl J Med* 319: 486–494
5   Bauer R, Banholzer R (1993) Pharmacology of quaternary anticholinergic drugs. In: Gross NJ (ed): *Anticholinergic therapy in obstructive airways disease.* Franklin Scientific Publications, London, 105–115
6   Bauer R, Kuhn F-J, Stockhaus K, Wick H (1976) Allgemeine Pharmakologie und sekretionshemmende Wirkung von (8r)-3α-Hydroxy-8-isopropyl-1αH,5αH-tropaniumbromid(±)-tropat (Ipratropiumbromid). *Arzneim-Forsch (Drug Res)* 26: 974–980
7   Bauer R, Püschmann S, Wick H (1976) Wirkung von (8r)-3α-Hydroxy-8-isopropyl-1αH,5αH-tropaniumbromid-(±)-tropat (Ipratropiumbromid) auf Spasmen des Tracheobronchialbaumes und die Bronchialsekretion, die Speichelsekretion, EKG und Herzfrequenz. *Arzneim-Forsch (Drug Res)* 26: 981–985
8   Ensing K, de Zeeuw RA, Nossent GD, Koeter GH, Cornelissen PJG (1989) Pharmaco-

kinetics of ipratropium bromide after single dose inhalation and oral and intravenous administration. *Eur J Pharmacol* 36: 189–194

9   Tennant W, MacGreggor T, Adelglass J, Dockhorn R, Korpalski D, Wecker M (1997) Pharmakokinetic assessment following administration of Atrovent® nasal spray 0.06% (ANS) in a pediatric population with naturally aquired common colds. *Am J Respir Crit Care Med* 155 (part 2): A713

10   Tashkin DP, Ashutosh K, Bleeker ER, Britt EJ, Cugell DW, Cummiskey JM, DeLorenzo L, Gilman MJ, Gross GN, Gross NJ et al. (1986) Comparison of the anticholinergic bronchodilator ipratropium bromide with metaproterenol in chronic obstructive pulmonary disease. *Am J Med* 81 (Suppl 5A): 81–90

11   Higgins BG, Powell RM, Cooper S, Tattersfield AE (1991) Effect of salbutamol and ipratropium bromide on airway calibre and bronchial reactivity in asthma and chronic bronchitis. Eur Respir J 4: 415–420

12   Braun RS and Levy SF (1991) Comparison of ipratropium bromide and albuterol in chronic obstructive pulmonary disease: a three-center study. *Am J Med* 91 (Suppl 4A): 28S–32S

13   Braun SR, McKenzie WN, Copeland C, Knight L, Ellersieck M (1989) A comparison of the effect of ipratropium and albuterol in the treatment of chronic obstructive airway disease. *Arch Int Med* 149: 544–547

14   Anthonisen NR, Connett JE, Kiley JP, Altose MD, Bailey WC, Buist AS, Conway WA, Enright PL, Kanner RE, O'Hara P et al (1994) Effects of smoking intervention and the use of an inhaled anticholinergic bronchodilator on the rate of decline of FEV_1. The lung health study. *JAMA* 272: 1497–1504

15   Rennard SI, Serby CW, Ghafouri Mo, Johnson PA, Friedman M (1996) Extended therapy with ipratropium is associated with improved lung function in patients with COPD. *Chest* 110: 62–70

16   Yang CM, Farley JM, Dwyer TM (1988) Muscarinic stimulation of submucosal glands in swine trachea. *J Appl Physiol* 64: 200–209

17   Mak JCW, Barnes PJ (1990) Autoradiographic visualization of muscarinic receptor subtypes in human and guinea pig lung. *Am Rev Resp Dis* 141: 1559–1568

18   Mak JCW, Baraniuk JN, Barnes PJ (1992) Localisation of muscarinic receptor subtype mRNAs in human lung. *Am J Respir Cell Mol Biol* 7: 344–348

19   Ishihara H, Shimura S, Satoh M, Masuda T, Nonaka H, Kase H, Sasaki T, Sasaki H, Takishima T, Tamura K (1992) Muscarinic receptor subtypes in feline tracheal submucosal gland secretion. *Am J Physiol* 262: L223–L228

20   Lopez-Vidriero MT, Costello J, Clark TJH, Das I, Keal EE, Reid L (1975) Effect of atropine on sputum production. *Thorax* 30: 543–547

21   Taylor RG, Pavia D, Agnew JE, Lopez-Vidriero MT, Newman Sp, Lennard-Jones T, Clarke SW (1986) Effect of four weeks' high dose ipratropium bromide treatment on lung mucociliary clearance. *Thorax* 41: 295–300

22   Pavia D, Lopez-Vidriero MT, Agnew JE, Taylor RG, Eyre-Brook A, Lawton WA, Pellow PGD, Clarke SW (1989) Effect of four-week treatment with oxitropium bromide on

lung mucociliary clearance in patients with chronic bronchitis or asthma. *Respiration* 55: 33–43

23 Ghafouri MA, Patil KD, Kass I (1984) Sputum changes associated with the use of ipratropium bromide. *Chest* 86: 387–393

24 Tamaoki J, Chiyotani A, Tagaya B, Sakai N, Konno K (1994) Effect of long term treatment with oxitropium bromide on airway secretion in chronic bronchitis and panbronchiolitis. *Thorax* 49: 545–548

25 Corssen G, Allen CR (1959) Acetylcholine: its significance in controlling ciliary activity of human respiratory epithelium *in vitro*. *J Appl Physiol* 14: 901–904

26 Disse B, Reichl R, Speck G, Traunecker W, Rominger KL, Hammer R (1993) BA 679 BR, a novel long-acting anticholinergic bronchodilator. *Life Sci* 52: 537–544

27 Pavia D, Thomson ML (1971) Inhibition of mucociliary clearance from human lung by hyoscine. *Lancet* 1: 449–450

28 Yeates DB, Aspin N, Levison H, Jones MT, Bryan AC (1975) Mucociliary tracheal transport rates in man. *J Appl Physiol* 39: 487–495

29 Foster WM, Bergofsky EH, Bohning DE, Lippman M Albert RE (1976) Effect of adrenergic agents and their mode of action on mucociliary clearance in man. *J App Physiol* 41: 146–152

30 Annis P, Landa J, Lichtiger M (1976) Effects of atropine on velocity of tracheal mucus in anesthetized patients. *Anesthesiology* 44: 74–77

31 Groth ML, Langenback EG, Foster WM (1991) Influence of inhaled atropine on lung mucociliary function in humans. *Am Rev Respir Dis* 144: 1042–1047

32 Ruffin RE, Wolff RK, Dolovich MB, Rossman CM, Fitzgerald JD, Newhouse MT (1978) Aerosol therapy with Sch 1000. Short-term mucociliary clearance in normal and bronchitic subjects and toxicology in normal subjects. *Chest* 73: 501–506

33 Nakhosteen JA, Wichtmann G, Petro W, Konietzko N (1980) Beeinflußt Ipratropiumbromid die mukoziliare Klärfunktion? *Prax Pneumol* 34: 570–574

34 Francis RA, Thomson ML, Pavia D, Douglas RB (1977) Ipratropium bromide: Mucociliary clearance rate and airway resistance in normal subjects. *Br J Dis Chest* 71: 173–178

35 Pavia D, Bateman JRM, Sheahan NF, Clarke SW (1979) Effect of ipratropium bromide on mucociliary clearance and pulmonary function in reversible airways obstruction. *Thorax* 34: 501–507

36 Thomas VE, O'Connel F, Harrison AJ, Fuller RW (1992) Ipratropium bromide delivered orally by metered dose inhaler does not decrease salivary flow in normal subjects. *Br J Clin Pharmac* 34: 266–268

37 Cugell DW (1986) Clinical pharmacology and toxicology of ipratropium bromide. *Am J Med* 81 (Suppl 5A): 18–22

38 Ikeda A, Nishimura K, Koyama H, Izumi, T (1995) Comparative dose-response study of three antcholinergic agents and fenoterol using a metered dose inhaler in patients with chronic obstructive pulmonary disease. *Thorax* 50: 62–66

39 Bauer R (1985) Zur Pharmakologie des Bronchospasmolytikums Oxitropiumbromid. *Arzneim-Forsch (Drug Res)* 35: 435–440

40 Ferguson GT, Cherniack RM (1993) Management of chronic obstructive pulmonary disease. *N Engl J Med* 328: 1017–1022

41 American Thoracic Society (1995) Standards for the diagnosis and care of patients with chronic obstructive pulmonary disease. *Am J Respir Crit Care Med* 152: S77–S120

42 Siafakas NM, Vermeire P, Pride NB, Paoletti P, Gibson J, Howard P, Yernault JC, Decramer M, Higenbottam, T, Postma DS, Rees, J (1995) Optimal assessment and management of chronic obstructive pulmonary disease (COPD). *Eur Respir J* 8: 1398–1420

43 Moriya H, Takagi Y, Nakanishi T, Hayashi M, Tani T, Hirotsu I (1999) Affinity profiles of various muscarinic antagonists for cloned human muscarinic acetylcholine receptor (MACHR) subtypes and MACHRS in rat heart and submandibular gland. *Life Sci* 64: 2351–2358

44 Doods HN (1992) Selective muscarinic antagonists as bronchodilators. *Drug News Perspect* 5: 345–352

45 Naito R, Takeuchi M, Morihira K, Hayakawa M, Ikeda K, Shibanuma T, Isomura Y (1998) Selective muscarinic antagonists. II. Synthesis and antimuscarinic properties of biphenylylcarbamate derivatives. *Chem Pharm Bull* 46: 1286–1294

46 Gomez A, Bellido I, Sanchez de la Cuesta F (1995) Atropine and glycopyrronium show similar binding patterns to M2 (cardiac) and M3 (submandibular gland) muscarinic receptor subtypes in the rat. *Br J Anaesth* 74:549–552

47 Disse B, Speck GA, Rominger KL, Witek ThJ, Hammer R (1999) Tiotropium (SpirivaTM): Mechanistical considerations and clinical profile in obstructive lung disease. *Life Sci* 64: 457–464

48 Newnham DM, Dhillon DP, Winter JH, Jackson CM, Clark RA, Lipworth BJ (1993) Bronchodilator reversibility to low and high doses of terbutaline and ipratropium bromide in patients with chronic obstructive pulmonary disease. *Thorax* 48: 1151–1155

49 Takishima T, Sekizawa K, Tamura G, Inoue H (1991) Anticholinergics in treatment of COPD – site of bronchodilatation. In: Saunders KB (ed): *Pathway to successful management.* Wells Medical, Kent, Research and Clinical Forums 13, 2 Part 2: 49–59

50 Pavia D, Moonen D (1999) Preliminary data from phase II studies with Respimat, a propellant-free soft mist inhaler. *J Aerosol Med* 12 (Suppl 1): S33–S39

51 Hulme EC, Birdsall NJM, Buckley NJ (1990) Muscarinic receptor subtypes. *Ann Rev Pharmacol* 30: 633–673

52 Barnes PJ (1993) Muscarinic receptor subtypes in airways. *Life Sci* 52: 521–527

53 Roffel AF, Elzinga CRS, Zaagsma J (1990) Muscarinic M_3-receptors mediate contraction of human central and peripheral airway smooth muscle. *Pulmonary Pharmacol* 3: 347–351

54 Fryer AD, MacLagan J (1987) Ipratropium bromide potentiates bronchoconstriction induced by vagal nerve stimulation in the guinea-pig. *Eur J Pharmacol* 139: 187–191

55 D'Agostino G, Chiari MC, Grana E, Subisse A, Kilbinger H (1990) Muscarinic inhibition of acetylcholine release from a novel *in vitro* preparation of the guinea-pig trachea. *Naunyn-Schmiedebergs Arch Pharmacol* 342: 141–145

56 Kilbinger H, Schneider R, Siefken H, Wolf D, D'Agostino G (1991) Characterization of

prejunctional muscarinic autoreceptors in the guinea-pig trachea. *Br J Pharmacol* 103: 1757–1763

57 Ten Berge REJ, Santing RE, Hamstra JJ, Roffel AF, Zaagsma J (1995) Dysfunction of muscarinic M_2-receptors after the early allergic reaction: possible contribution to bronchial hyperresponsiveness in allergic guinea-pigs. *Br J Pharmacol* 114: 881–887

58 Fryer AD, Adamko DJ, Yost BL, Jacoby DB (1999) Effects of inflammatory cells on neuronal M_2 muscarinic receptor function in the lung. *Life Sci* 64: 449–455

59 Patel HJ, Barnes PJ, Takahashi T, Tadjkarimi S, Magdi HY, Belvisi MG (1995) Evidence for prejunctional muscarinic autoreceptors in human and guinea pig trachea. *Am J Respir Crit Care Med* 152: 872–878

60 Takahashi T, Belvisi MG, Patel H, Ward JK, Tadjkarimi S, Yacoub MH, Barnes PJ (1994) Effect of Ba 679 BR, a novel long-acting anticholinergic agent, on cholinergic neurotransmission in guinea-pig and human airways. *Am J Respir Crit Care Med* 150: 1640–1645

61 Roffel AF, Meurs H, Zaagsma J (1997) Muscarinic receptors in COPD and asthma. In: Barnes PJ and Buist AS (eds): *The role of anticholinergics in chronic obstructive pulmonary disease and chronic asthma*. Gardiner-Caldwell Communications Lim, Macclesfield, Cheshire, UK, 92–125

62 Watson N, Magnussen H, Rabe KF (1995) Antagonism of β-adrenoceptor-mediated relaxations of human bronchial smooth muscle by carbachol. *Eur J Pharmacol* 275: 307–310

63 Groeben H, Brown RH (1996) Ipratropium decreases airway size in dogs by preferential M_2 muscarinic receptor blockade *in vivo*. *Anaesthesiology* 85: 867–873

64 Yarbrough J, Mansfield LE, Ting S (1983) Immediate bronchoconstriction response to metered dose albuterol. *Ann Allergy* 50: 363

65 Yarbrough J, Mansfield LE, Ting S (1985) Metered-dose inhaler induced bronchospasm in asthmatic patients. *Ann Allergy* 55: 25–27

66 Cocchetto DM, Sykes S, Spector S (1991) Paradoxical bronchospasm after use of inhalation aerosol: a review of the literature. *J Asthma* 28: 49–53

67 Beasley CRW, Rafferty P, Holgate ST (1987) Bronchoconstrictor properties of preservatives in ipratropium bromide (Atrovent) nebuliser solution. *Br Med J* 294: 1197–1198

68 Bryant DH and Rogers P (1992) Effects of ipratropium bromide nebulizer solution with and without preservatives in the treatment of acute and stable asthma. *Chest* 102: 742–747

69 O'Callaghan C, Milner AD, Swarbrick A (1989) Paradoxical bronchoconstriction in wheezing infants after nebulised preservative free iso-osmolar ipratropium bromide. *Br Med J* 299: 1433–1434

70 O'Callaghan C, Milner AD, Swarbrick A (1989) Safer device with face mask attachment for giving bronchodilators to infants with asthma. *Br Med J* 298: 160–161

71 Szelenyi I, Boleski P (1989) Future trends in asthma therapy. *Drug News Perspect* 2: 270–277

72 Wellstein A, Pitschner HF (1988) Complex dose-response curves of atropine in man

explained by different functions of M_1- and M_2-cholinoceptors. *Naunyn-Schmiedeberg's Arch Pharmacol* 338: 19–27

73 Sertl K, Meryn S, Graninger W, Laggner A, Schlick W, Rameis H (1986) Acute effects of pirenzepine on bronchospasm. Intern J Clin Pharmacol Ther Toxicol 24: 655–657

74 Lammers JJ, Minette P, McCusker M, Barnes PJ (1989) The role of pirenzepine-sensitive (M_1) muscarinic receptors in vagally mediated bronchoconstriction in humans. *Am Rev Respir Dis* 139: 446–449

75 Mezzetti M, Colombo L, Marini MG, Crusi V, Pierfederici P, Mussini E (1990) A pharmacokinetik study on pulmonary tropism of ambroxol in patients under thoracic surgery. *J Em Surg* 13: 179–185

76 Kaiser C, Audia VH, Carter JP, McPherson DW, Waid PP, Lowe VC, Noronha-Blob L (1993) Synthesis and antimuscarinic activity of some 1-cycloalkyl-1-hydroxy-1-phenyl-3-(4-substituted piperazinyl)-2-propanones and related compounds. *J Med Chem* 36: 610–616

77 Miyachi H, Kiyota H, Segawa M (1998) Novel imidazole derivatives with subtype-selective antimuscarinic activity (2). *Bioorg Med Chem Letters* 8: 2163–2168

78 Alabaster VA (1997) Discovery and development of selective M_3 antagonists for clinical use. *Life Sci* 60: 1053–1060

79 Wallis RM (1995) Pre-clinical and clinical pharmacology of selective muscarinic M_3-receptor antagonists. *Life Sci* 56: 861–868

80 Maesen F, Smeets J, Smeets P, Gorter de Vries I, Hodges M (1996) A double blind, single dose, 3 way cross-over study comparing the efficacy of inhaled U.K.-112,166 with ipratropium bromide and placebo in patients with COPD. *Eur Respir J* 9 (Suppl 23): 29S

81 Beaumont KC, Cussans NJ, Nichols DJ, Smith DA (1998) Pharmacokinetics and metabolism of darifenacin in the mouse, rat, dog and man. *Xenobiotica* 28: 63–75

82 Cazzola M, Russo S, De Santis D, Principe P, Marmo E (1987) Respiratory responses to pirenzepine in healthy subjects. *Intern J Clin Pharmacol Ther Toxicol* 25: 105–109

83 Cazzola M, Matera MG, D'Amato G, De Santis D, Maione S, Lisa M, Cenicola ML, Marmo E (1989) Evidence of muscarinic receptor subtypes in airway smooth muscle of normal volunteers and of chronic obstructive pulmonary disease patients. *Intern J Clin Pharm Res* IX (1): 65–70

84 Ceyhan B, Çelikel T, Simsir S, Kandemir B (1993) Comparison of the bronchodilator efficacy of nebulized pirenzepine and ipratropium bromide in patients with airway obstructive lung disease. *Intern J Clin Pharmacol Ther Toxicol* 31: 510–513

85 Cazzola M, D'Amato G, Guidetti E, Staudinger H, Steinijans VW, Kilian U (1990) An M_1-selective muscarinic receptor antagonist telenzepine improves lung function in patients with chronic obstructive bronchitis. *Pulm Pharmacol* 3: 185–189

86 Ukena D, Wehinger C, Engelstätter R, Steinijans V, Sybrecht GW (1993) The muscarinic M_1-receptor-selective antagonist, telenzepine, had no bronchodilatory effects in COPD patients. *Eur Respir J* 6: 378–382

87 Howell RE, Laemont KD, Kovalsky MP, Lowe VC, Waid PP, Kinnier WJ, Noronha-Blob

L (1994) Pulmonary pharmacology of a novel, smooth muscle-selective muscarinic antagonist *in vivo. J Pharmacol Exp Ther* 270: 546–553

88   Miyachi H, Kiyota H, Segawa M (1998) Novel imidazole derivatives with subtype-selective antimuscarinic activity (1). *Bioorg Med Chem Letters* 8: 1807–1812

89   Muravchick S, Owens WD, Felts JA (1979) Glycopyrrolate and cardiac dysrhythmias in geriatric patients after reversal of neuromuscular blockade. *Can Anaesth Soc J* 26: 22–25

90   Gilman JG, Meyer L, Carter J, Slovis C (1990) Comparison of aerosolized glycopyrrolate and metaproterenol in acute asthma. *Chest* 98: 1095–98

91   Cydulka RK, Emerman CL (1995) Effects of combined treatment with glycopyrrolate and albuterol in acute exacerbation of chronic obstructive pulmonary disease. *Ann Emerg Med* 25: 470–473

92   Tzelepis G, Komanapolli S, Tyler D, Vega D, Fulambarker A (1996) Comparison of nebulized glycopyrrolate and metaproterenol in chronic obstructive pulmonary disease. *Eur Respir J* 9:100–103

93   Kubo S, Morikawa K, Yamazaki M, Matsubara I, Kato H (1981) Antispasmodic activities of 3-(di-2-thienylmethylene)-5-methyl-trans-quinolizium bromide (HSR-902) on smooth muscle organs and its organ selectivity. *Folia Pharmacol Jpn* 77: 87–98

94   Shioya C, Kagaya M, Sano M, Itaba M, Shindo T, Miura M (1996) Antimuscarinic effect of tiquizium bromide *in vitro* and *in vivo*. *Eur J Clin Pharmacol* 50: 375–380

95   Frith PA, Jenner B, Dangerfield R, Atkinson J, Drennan C (1986) Oxitropium bromide, dose-response and time-response study of a new anticholinergic bronchodilator drug. *Chest* 86: 249–253

96   Matera MG, Cazzola M, Vinciguerra A, Di Perna F, Calderaro F, Caputi M, Rossi F (1995) A comparison of the bronchodilating effects of salmeterol, salbutamol and ipratropium bromide in patients with chronic obstructive pulmonary disease. *Pulm Pharmacol* 8: 267–271

97   Mahler DA, Donohue JF, Barbee RA, Goldman MD, Gross NJ, Wisniewski ME, Yancey SW, Zakes BA, Rickard KA, Anderson WH (1999) Efficacy of salmeterol xinafoate in the treatment of COPD. *Chest* 115: 957–965

98   Littner MR, Ilowite JS, Tashkin DP, Friedman M, Serby CW, Menjoge SS, Witek TJ (2000) Long-acting bronchodilation with once daily dosing of tiotropium (Spiriva™) in stable COPD. *Am J Respir Crit Care Med* 161: 1136–1142

99   Casaburi R, Serby CW, Menjoge SS, Witek TJ for the American study group (1999) The spirometric efficacy of once daily dosing with tiotropium in stable COPD. *Am J Respir Crit Care Med* 159, A 524

100   Maesen FPV, Smeets JJ, Costongs MAL, Wald FDM, Cornelissen PJD (1993) BA 679 BR, a new long-acting antimuscarinic bronchodilator: a pilot dose-escalation study. *Eur Respir J* 6: 1031–1036

101   Disse B, Rominger K, Serby CW, Souhrada JF, Witek TJ (1999b) The pharmacokinetic (PK) profile of tiotropium during long term treatment in stable COPD. *Am J Respir Crit Care Med* 159: A524

102 O'Connor BJ, Towse LJ, Barnes PJ (1996) Prolonged effect of tiotropium bromide on methacholine-induced bronchoconstriction in asthma. *Am J Respir Crit Care Med* 154: 876–880

103 Wilson NM, Green S, Coe C, Barnes PJ (1987) Duration of protection by oxitropium bromide against cholinergic challenge. *Eur J Respir Dis* 71: 455–458

104 Maesen FPV, Smeets JJ, Sledsens TJH, Wald FDM, Cornelissen PJG (1995) Tiotropium bromide, a new long-acting antimuscarinic bronchodilator: a pharmacodynamic study in patients with chronic obstructive pulmonary disease (COPD). *Eur Respir J* 8: 1506–1513

105 American Thoracic Society (1987) Standards for the diagnosis and care of patients with chronic obstructive pulmonary disease (COPD) and asthma. *Am Rev Respir Dis* 136: 225–243

106 Van Noord JA, Smeets JJ, Maesen FP, Korducki L, Cornelissen PJ (1998) The onset of spirometric response following once daily inhalation of tiotropium in patients with COPD. *Eur Respir J* 12 (Suppl 28): 1S: A0124

107 Noord JA van, Bantje TA, Eland ME, Korducki L, Cornelissen PJG (2000) A randomised controlled comparison of tiotropium and ipratropium in the treatment of chronic obstructive pulmonary disease. *Thorax* 55: 289–294

108 American Thoracic Society (1991) Lung function testing: selection of reference values and interpretative strategies. *Am Rev Respir Dis* 144: 1202–1218

109 Tashkin DP (1995) Measurement and significance of the bronchodilator response: bronchodilation and inhibition of bronchoprovocation. In: Spector SHL (ed): *Provocation testing in clinical practice*. Marcel Dekker Inc, New York, 512–573

110 Myers GM (1996) Suggested guidelines for determining the trough-to-peak ratio of antihypertensive drugs. *Am J Hypertens* 9: 76S–82S

111 Hammer R, Giachetti A (1982) Muscarinic receptor subtypes: M_1 and M_2. Biochemical and functional characterisation. *Life Sci* 31: 2991–2994

112 Rabe KF, Lindén A (1997) Mechanism of duration of action of inhaled long-acting beta$_2$-adrenoceptor agonists. In: Pauwels R, O'Byrne PM (eds): *Beta$_2$-agonists in asthma treatment*. Marcel Dekker Inc, New York, 131–156

113 Schroeckenstein DC, Bush RK, Chervinsky P, Busse WW (1988) Twelve-hour bronchodilation in asthma with a single aerosol dose of the anticholinergic compound glycopyrrolate. *J Allergy Clin Immunol* 82: 115–119

114 Mitsuya M, Mase T, Tsuchiya Y, Kawakami K, Hattori H, Kobayashi K, Ogino Y, Fujikawa T, Satoh A, Kimura T et al (1999) J-104129, a novel muscarinic M_3 receptor antagonist with high selectivity for M_3 over M_2 receptors. *Bioorg Med Chem* 7: 2555–2567

Index

The PIR-Series
Progress in Inflammation Research

Homepage: http://www.birkhauser.ch

Up-to-date information on the latest developments in the pathology, mechanisms and therapy of inflammatory disease are provided in this monograph series. Areas covered include vascular responses, skin inflammation, pain, neuroinflammation, arthritis cartilage and bone, airways inflammation and asthma, allergy, cytokines and inflammatory mediators, cell signalling, and recent advances in drug therapy. Each volume is edited by acknowledged experts providing succinct overviews on specific topics intended to inform and explain. The series is of interest to academic and industrial biomedical researchers, drug development personnel and rheumatologists, allergists, pathologists, dermatologists and other clinicians requiring regular scientific updates.

Available volumes:
T Cells in Arthritis, P. Miossec, W. van den Berg, G. Firestein (Editors), 1998
Chemokines and Skin, E. Kownatzki, J. Norgauer (Editors), 1998
Medicinal Fatty Acids, J. Kremer (Editor), 1998
Inducible Enzymes in the Inflammatory Response, D.A. Willoughby, A. Tomlinson (Editors), 1999
Cytokines in Severe Sepsis and Septic Shock, H. Redl, G. Schlag (Editors), 1999
Fatty Acids and Inflammatory Skin Diseases, J.-M. Schröder (Editor), 1999
Immunomodulatory Agents from Plants, H. Wagner (Editor), 1999
Cytokines and Pain, L. Watkins, S. Maier (Editors), 1999
In Vivo *Models of Inflammation*, D. Morgan, L. Marshall (Editors), 1999
Pain and Neurogenic Inflammation, S.D. Brain, P. Moore (Editors), 1999
Anti-Inflammatory Drugs in Asthma, A.P. Sampson, M.K. Church (Editors), 1999
Novel Inhibitors of Leukotrienes, G. Folco, B. Samuelsson, R.C. Murphy (Editors), 1999
Vascular Adhesion Molecules and Inflammation, J.D. Pearson (Editor), 1999
Metalloproteinases as Targets for Anti-Inflammatory Drugs, K.M.K. Bottomley, D. Bradshaw, J.S. Nixon (Editors), 1999
Free Radicals and Inflammation, P.G. Winyard, D.R. Blake, C.H. Evans (Editors), 1999
Gene Therapy in Inflammatory Diseases, C.H. Evans, P. Robbins (Editors), 2000
New Cytokines as Potential Drugs, S. K. Narula, R. Coffmann (Editors), 2000
High Throughput Screening for Novel Anti-inflammatories, M. Kahn (Editor), 2000
Immunology and Drug Therapy of Atopic Skin Diseases, C.A.F. Bruijnzeel-Komen, E.F. Knol (Editors), 2000
Novel Cytokine Inhibitors, G.A. Higgs, B. Henderson (Editors), 2000
Inflammatory Processes. Molecular Mechanisms and Therapeutic Opportunities, L.G. Letts, D.W. Morgan (Editors), 2000
Cellular Mechanisms in Airways Inflammation, C. Page, K. Banner, D. Spina (Editors), 2000
Inflammatory and Infectious Basis of Atherosclerosis, J.L. Mehta (Editor), 2001